Bio-Based Materials: Contribution to Advancing Circular Economy

Bio-Based Materials: Contribution to Advancing Circular Economy

Editors

Maya Jacob John
Sabu Thomas

MDPI • Basel • Beijing • Wuhan • Barcelona • Belgrade • Manchester • Tokyo • Cluj • Tianjin

Editors
Maya Jacob John
Council for Scientific and Industrial Research
South Africa

Sabu Thomas
Mahatma Gandhi University
India

Editorial Office
MDPI
St. Alban-Anlage 66
4052 Basel, Switzerland

This is a reprint of articles from the Special Issue published online in the open access journal *Polymers* (ISSN 2073-4360) (available at: https://www.mdpi.com/journal/polymers/special_issues/Bio_Based_Mater).

For citation purposes, cite each article independently as indicated on the article page online and as indicated below:

LastName, A.A.; LastName, B.B.; LastName, C.C. Article Title. *Journal Name* **Year**, *Volume Number*, Page Range.

ISBN 978-3-0365-6047-2 (Hbk)
ISBN 978-3-0365-6048-9 (PDF)

Contents

About the Editors

Maya Jacob John

Dr. Maya Jacob John is Principal Researcher and leader of the Bioplastics and Biocomposites Platform at the Council for Scientific and Industrial Research (CSIR) in South Africa. Dr. John's research tackles the crucial problems of waste plastic pollution and focuses on creating a bio-based feedstock to replace materials derived from petroleum reserves. Dr. Maya has published more than 90 peer-reviewed research articles, was granted two patents and presented at a number of international conferences which includes several plenary and key-note lectures. The impact of her research is evident in her overall citation record with a total of more than 10,000 citations and current h-index of 37. Dr. Maya Jacob John has been ranked among the World's Top 100,000 scientists in a recent list prepared by Stanford University and is globally ranked at position 973 within the Polymer field. She has also received several awards which include a CSIR award for research excellence (Young Researcher category) and a Best Patent award from CSIR in the year 2011. She was the 1st runner-up in the Physical and Engineering Sciences category at DST's Women in Science Awards (Young Woman Scientist) and her work on bio-based composite panels for transport sector received Special Award under Import Replacement category at the National Advanced Manufacturing Innovation Awards in 2018. In 2019, Dr. John was a finalist in the materials for inclusive economic and sustainable development category at the National Science and Technology Foundation (NSTF) awards.

Sabu Thomas

Prof. Sabu Thomas is currently the Vice-chancellor of Mahatma Gandhi University, Kottayam, Kerala, India, Director of Energy Materials, former Professor of Polymer Science and Engineering at the School of Chemical Sciences, and the former Director of the International and Inter-university Centre for Nanoscience and Nanotechnology, Mahatma Gandhi University, India. In 1987, he received his PhD in Polymer Engineering from the Indian Institute of Technology (IIT), Kharagpur, India. He is a fellow of the Royal Society of Chemistry, London and a fellow of the European Academy of Sciences. Prof. Thomas has been ranked No.2 in India by Stanford University, USA and ranked number 114 globally in the area of polymer science. He has published over 1200 peer-reviewed international journal articles and book chapters, holds 15 patents and has edited more than 150 books. Prof. Thomas has a h-index of 111, total citations of 58,000 and has supervised more than 115 PhD students and 30 post docs.

polymers

MDPI

Editorial

Special Issue—"Bio-Based Materials: Contribution to Advancing Circular Economy"

Maya Jacob John [1,2,*] and Sabu Thomas [3,4,*]

1 Centre for Nanostructures and Advanced Materials, Council for Scientific and Industrial Research (CSIR), Pretoria 0001, South Africa
2 Department of Chemistry, Nelson Mandela University, Port Elizabeth 6001, South Africa
3 International and Interuniversity Centre for Nano Science and Nano Technology, Mahatma Gandhi University, Kottayam 686560, India
4 Basic Department of Biotechnology, School of Fundamental Biology and Biotechnology, Siberian Federal University, 79 Svobodnyi Av., 660041 Krasnoyarsk, Russia
* Correspondence: mjohn@csir.co.za (M.J.J.); sabuthomas@mgu.ac.in (S.T.)

Bio-based materials have a significant role to play in the implementation of a functional circular economy. The circular economy stresses the reduced use of raw materials, the reuse of products, waste streams (where waste is converted to a valuable resource), and the recycling of products, the combined effect of which will ensure that materials are retained in the loop as opposed to a linear economy which is based on a manufacture, use, and disposal model. In a circular society, bio-based materials are reused, repaired, recycled, and remanufactured. A circular approach for plastics will address the issue of plastic waste pollution on land and in oceans and the adverse health effects that microplastics have on marine and human life. This can be achieved by methods to improve recycling and developing bio-based materials as an alternative to petroleum-based feedstock. The concepts of the bio-based economy and the circular economy are similar in that they reduce the demand for fossil carbon and enhance the use of waste and side streams. Hence, it is quite clear that bio-based materials (bioplastics and biocomposites) can make a critical contribution to the implementation of the circular economy.

In this Special Issue, leading researchers from academia and industry were invited to submit reviews or their latest research on topics aligned to the development of sustainable materials from renewable resources. Sustainable materials include waste-derived, recyclable, and biodegradable materials. Studies dealing with recycling, waste conversion to bio-based products, the development of bio-based composites, and surface treatments on cellulose fibres have been included in this issue. The manuscripts were subjected to a rigorous review process, after which there was a compilation of thirteen research articles, reflecting the latest trends in bio-based and compostable materials. This issue consists of three review articles and ten research articles. The review articles include an account on the plastic circular economy and applications of chitosan-based materials.

As guest editors of this Special Issue, we acknowledge all the authors and reviewers who have contributed to its publication. We would also like to thank the technical support team at MDPI for their assistance in preparing this Special Issue.

Author Contributions: Conceptualization, S.T. and M.J.J.; writing—original draft preparation, M.J.J.; writing—review and editing, S.T. All authors have read and agreed to the published version of the manuscript.

Conflicts of Interest: The authors declare no conflict of interest.

Citation: John, M.J.; Thomas, S. Special Issue—"Bio-Based Materials: Contribution to Advancing Circular Economy". *Polymers* **2022**, *14*, 3887. https://doi.org/10.3390/polym14183887

Received: 13 September 2022
Accepted: 14 September 2022
Published: 17 September 2022

Publisher's Note: MDPI stays neutral with regard to jurisdictional claims in published maps and institutional affiliations.

polymers

MDPI

Article

3D Printing Parameter Optimization Using Taguchi Approach to Examine Acrylonitrile Styrene Acrylate (ASA) Mechanical Properties

Abdul Zubar Hameed [1,*], Sakthivel Aravind Raj [2,*], Jayakrishna Kandasamy [2], Muhammad Atif Shahzad [1] and Majed Abubakr Baghdadi [1]

1 Department of Industrial Engineering, Faculty of Engineering, King Abdulaziz University, Jeddha 21589, Saudi Arabia
2 Department of Manufacturing Engineering, School of Mechanical Engineering, Vellore Institute of Technology, Vellore 632014, Tamil Nadu, India
* Correspondence: aahameed@kau.edu.sa (A.Z.H.); aravindsakthivel@hotmail.com (S.A.R.)

Abstract: Polymer composites with different reinforcements have many applications. By adjusting process settings and adding fibers and fillers, composite properties can be improved. Additive manufacturing is popular in the polymer industry because it can manufacture intricately designed parts with fewer defects and greater strength with less material consumption. Composites use thermoplastics and thermosetting polymers. Thermoset plastics cannot be reused or recycled; therefore, they are disposed in landfills, creating pollution and environmental harm. In this work, thermoplastic ASA (Acrylonitrile Styrene Acrylate) polymer filament is used for FDM 3D printing. The specimens are made by varying five process parameters that affected the materials' mechanical properties. The tensile, flexural and impact specimens are made using MINITAB software and ASTM requirements. The L18 orthogonal array experimental design, specimens and results were optimized. Infill density and layer height were most influential. Maximum tensile strength of 51.86 MPa, flexural strength of 82.56 MPa and impact strength of 0.180 J/mm^2 were obtained by following the software-suggested input factors and compared with the predicted values. Final error percentage was obtained between the predicted and the experimental results and it was found to be under 3%, which is acceptable.

Keywords: ASA polymer; fused deposition modelling; infill density; L18 orthogonal array; 3D printing

Citation: Hameed, A.Z.; Aravind Raj, S.; Kandasamy, J.; Shahzad, M.A.; Baghdadi, M.A. 3D Printing Parameter Optimization Using Taguchi Approach to Examine Acrylonitrile Styrene Acrylate (ASA) Mechanical Properties. *Polymers* **2022**, *14*, 3256. https://doi.org/10.3390/polym14163256

Academic Editors: Maya Jacob John and Sabu Thomas

Received: 16 July 2022
Accepted: 4 August 2022
Published: 10 August 2022

1. Introduction

As early as the Paleolithic period, humans were able to produce things for their everyday needs; they had a material sense and used it to create tools for their own gain. However, the manufacturing process for the tools was based on subtractive techniques, in which the original components are turned into useful goods by removing surplus bulk material until the final object is achieved. The subtractive method of production has progressed substantially and new techniques continue to emerge. In the early 1980s, a paradigm known as additive manufacturing was established. It has proven to be an excellent technique for creating any geometry [1].

In the FDM process, different thermoplastic materials are utilized to print different complex shaped objects. Different materials used in the FDM process include Polylactic acid, PLA-based graphene, Acrylonitrile butadiene styrene, Polycarbonate, NinjaFlex 85 A, Polycarbonate–ABS, Polyurethane elastomer, thermoplastic Polyurethane and Carbon-Silicone rubber [2]. PLA is derived from renewable resources and is compostable. It will not produce toxic fumes during printing or when incinerated. It is basically derived from crops. ABS is usually used for processing foods and has improved resistance against impact, abrasion and chemicals. It has better machinability, it can be thermoformed and it is flexible enough to print complex shaped jobs. It has better electrical properties and dimensional stability also. Polycarbonate has better thermal insulation when compared to other thermoplastic polymers and is light in weight. It has UV protection properties. NinjaFlex 85A is used in making sporting goods, in healthcare and in industrial manufacturing. It features good flexibility, excellent elongation properties, low friction and abrasion resistance. Thermoplastic Polyurethane has good load bearing capability, resistance to oil and ozone, and has good abrasion resistance also. Very few materials are used in making 3D-printed membranes, and for printing the membranes only a few additive manufacturing processes are used. Processes like electrospinning, sintering, stretching, track etching and phase inversion are used in printing membranes [3].

3D printing has introduced a new approach for layer-by-layer creation of items from the bottom up. By adding successive layers of material in response to computer-controlled instructions, additive manufacturing makes parts. People have grasped the significance and benefits of additive manufacturing as its popularity has increased relative to conventional techniques. Compared to conventional or subtractive manufacturing techniques, additive manufacturing has progressed and produced more promising results. It can be employed in small-scale production for customization and personalization, whereas conventional methods are limited to mass production.

To improve ASA's thermal insulation and weather resistance, chemicals have been added to ASA resin used as a construction material. The modified ASA has been subjected to an infrared radiation test, which revealed that the surface temperature of the specimen decreased by 6.4 °C. In addition, pigments with NIR reflectance enhanced the weather resistance of ASA. The setup for the radiation test to measure weather resistance with the help of a near infrared lamp is shown in Figure 1a, and the corresponding heat curves for pure and modified ASA are shown in Figure 1b [4].

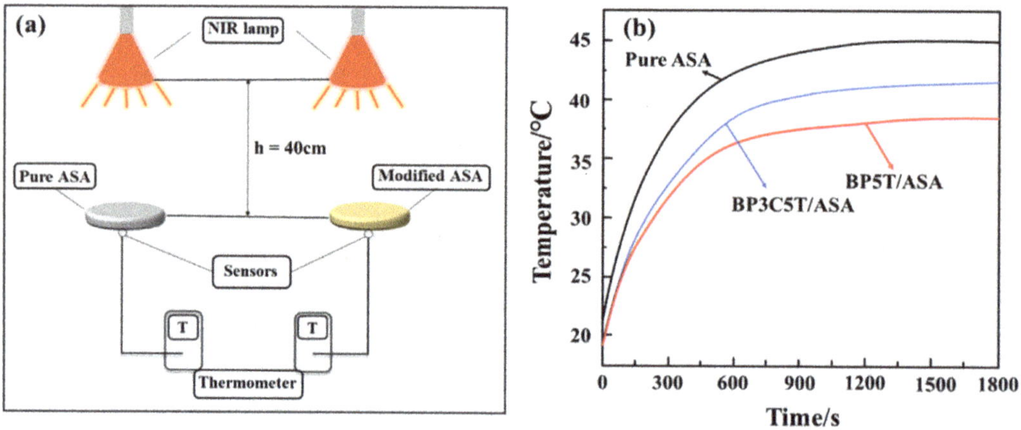

Figure 1. (**a**). NIR lamp radiation test setup, (**b**). Heating curves of pure ASA and modified ASA [4] Copyright 2021 with permission from Elsevier.

The difference in interfacial tension between polymer pairs allows for the incorporation of two different elastomers, CPE and BR, into the core of ASA. Higher impact strengths were demonstrated by the mixes as a result of their testing [5]. After reinforcing ASA with carbon fiber and measuring its mechanical and thermal properties, the materials had superior tensile and flexural strength, and the thermal conductivity increased by 500 percent. Figure 2a depicts SEM images captured after a tensile test conducted at liquid nitrogen temperatures, whereas Figure 2b depicts the interaction between fiber and resin following flexural tests. The presence of a gap between the fiber and resin shows a lack of compatibility and interface strength, which results in fiber pullouts and degraded mechanical qualities [6].

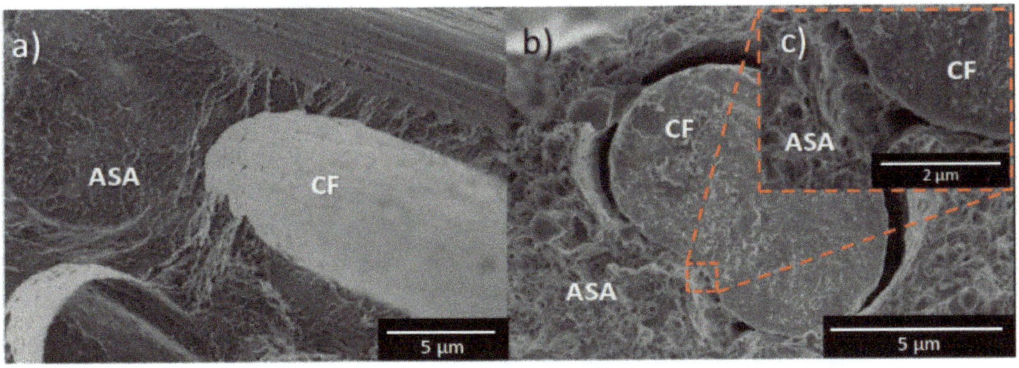

Figure 2. (**a**). SEM images showing flexural fracture at lower temperatures, (**b**). Tensile tested specimen, (**c**). Details of interface between carbon fiber and ASA [4] Copyright 2020 with permission from Elsevier.

ASA is combined with barium titanate and evaluated as a roof-cooling and solar-reflecting material. The introduction of barium titanate increased the NIR reflectance by two to three times, while the interior temperature dropped by 10 °C. Figure 3 depicts the increase in temperature of the specimen's inner surface, which is used to measure solar reflectance. ASA with barium titanate (1068 nm) exhibited superior solar reflectance compared to other combinations, including glass [7].

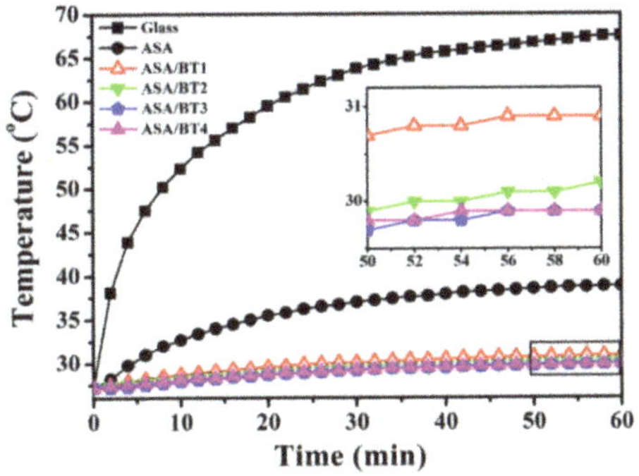

Figure 3. Inner temperature difference with respect to time [7] Copyright 2018 with permission from Elsevier.

The addition of antimony trioxide to ASA resin enhanced the cooling effect. Antimony trioxide exhibits solar-reflective qualities. When antimony trioxide is added to ASA, the water contact angle increases, indicating that the material can be used for roof cooling and composited as cool materials [8]. The use of 3D printing in the production of biological components is rising fast due to demand. PLA was discovered to be a biodegradable material suitable for use in the FDM production of bio-implants. The biodegradation test findings reveal good results throughout 30 to 60 days, and the weight loss was found to be satisfactory [9]. Stamp sand deposited as mine waste is combined with recycled ASA to create composites whose mechanical properties have been evaluated. It was noted that the tensile strength was reduced by half compared to recycled ASA in its pure form, and that stamp sand in conjunction with ABS produced positive results [10]. Metamaterials such as chiral, Re-entrant, and ASA are printed using FDM, and mechanical and ex situ tests revealed that structural alterations improved the materials' properties, although inter-laminar fracture was seen during a compression test [11]. To determine the relationship between the morphologies of ASA composites and their mechanical properties, RAFT emulsion polymerization is used to generate block copolymers from ASA. By increasing the spherical particle size, the inter-particle distance can be decreased and the tensile characteristics can be improved. Additionally, the impact resistance was enhanced in comparison to triblock copolymer materials [12]. The technique of melt blending is used to create composites composed of polycarbonate, ASA and acrylic resin. It was noticed that the addition of acrylic resin boosted the impact strength and abrasion resistance. Additionally, it was discovered that the addition of acrylic resin reduced the tensile, hardness and melt flow rates marginally. The crack path in the composite was tracked with SEM and LSCM images as shown in Figure 4a,b [13].

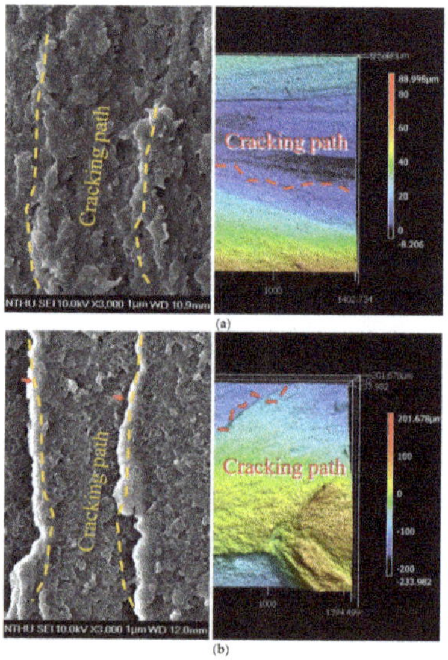

Figure 4. (**a**). Laser scanning confocal microscopy (Right). (**b**). SEM images of impact tested specimen showing the crack path [13] Reproduced with permission from MDPI.

Analyses of the effect of ASA addition to eucalyptus/PVC composites revealed that adding less than 15 percent ASA improves impact, tensile and flexural strengths. Adding 15 to 20 percent ASA to composites increased their thermal stability in the early stages but weakened them later due to ASA's thermal degradation [14]. Thus, in this study, the ASA polymer is 3D printed utilizing the fused deposition modelling technique by altering five process parameters at three distinct levels while keeping other parameters constant. In earlier 3D printing research work, a maximum of three process parameters were taken into account and varied. In this work, however, a total of five process parameters were taken into consideration and their influence on mechanical properties was studied. The curiosity to understand the combined influence of the input factors motivated this work and makes it novel when compared with previous work carried out in this field. By evaluating the mechanical characteristics of the manufactured specimens, the influence of the parameters is examined and the results are optimized based on the analysis.

2. Materials and Manufacturing Process

2.1. Materials

River Polymer Industries of Gujarat, India, provided ASA polymer (acrylonitrile-co-styrene-co-acrylate). According to temperature tests, ASA's strong solar reflectance prevents heat conduction and offers a more effective cooling effect in sunshine than glass. ASA features a high service temperature, low thermal conductivity and resistance to the elements. ASA has a density of 1.06–1.1 g/cm^3 at 20 °C and a softening point of 85 to 100 °C, making it suitable for outdoor applications because it ignites at 400 °C. Low melting temperatures, which are 200–230 °C, 200–280 °C, and 200–250 °C, make it suitable for pipe extrusion, injection molding and thermoforming. Solvents including chloroform, dichlorobenzene, diethyl ether, ethyl benzoate, ethyl chloride, mesityl oxide, methyl chloride, methyl propyl ketone and xylene, dissolve ASA well. ASA is suited for outdoor sheet furniture, exterior cable enclosures and wide screen displays. ASA is a scratch-resistant polymer that can be utilized in automobile interiors and electrical devices. ASA has superior solar reflectivity and can be used in roof cooling materials [15]. The mechanical and physical properties of ASA are listed in Table 1.

Table 1. Properties of ASA [6]. Copyright 2020 with permission from Elsevier.

Properties	Units	Neat ASA
Ultimate tensile strength	MPa	45.7 ± 0.9
Ultimate flexural strength	MPa	52.6 ± 1.9
Impact strength	kJ/m^2	140 ± 30
Material flow rate	g/10 min	47
Glass transition temperature	°C	103
Thermal conductivity	W/mK	0.17

2.2. Fabrication Method and Machinery Details

Using additive manufacturing, Acrylonitrile Styrene Acrylate (ASA) specimens for experimental work were printed on a "Creality Ender—3" FDM-based 3D printing system. The printer has a printing size of 220 × 220 × 250 mm and is based on the Cartesian system, with all three axes moving independently utilizing stepper motors. The ASA filament used in the experiment was a 1.75 mm-diameter wire that was kept on its mounting spool. All three samples were modelled in the CAD software Autodesk Fusion 360 and produced in the stl file format [16]. Then, the file was sliced in Ultimaker Cura slicing software according to the parameters, its G-code (machine code for the printer) was generated, and this was inserted into the 3D printer machine. Figure 5 depicts the 3D printer with which the specimens were printed.

Figure 5. "Creality Ender—3" printer.

2.3. Selected Process Parameters and Their Levels

The mechanical properties of components are greatly affected by printing parameters. Printing temperature, infill density, layer height, raster angle and printing orientation were chosen as experimental design factors to determine their effect on the specimen's mechanical properties. The remaining printing parameters were adjusted to the same value for each sample. The printing temperatures for the setup were determined to be 245 °C, 255 °C and 265 °C. The infill density ranged from 30%, 70% and 100%. The height of the print layer ranged between 0.08 mm, 0.12 mm and 0.16 mm. The raster angle values for the experiment were 15°, 30° and 45°, and two print orientations, X-90 and Y-90, were obtained [17–20]. An experiment was designed by systematically altering each of these values, and samples were subjected to mechanical testing to determine the optimal printing conditions for achieving maximum strength. All other printing parameters, including printing speed of 60 mm/s, number of perimeters at 4, print bed temperature of 90 °C, 40 percent print cooling and the retraction settings, were held constant for all samples. The selected process parameters and their respective levels are listed in Table 2.

Table 2. Process variables and their various stages.

S. No.	Variables	Units	Stages		
			1	2	3
1	Part orientation	axis	X	Y	-
2	Printing temperature	°C	245	255	265
3	Layer height	mm	0.08	0.12	0.16
4	Infill density	%	30	70	100
5	Raster angle	°	15	30	45

Using MINITAB, the experimental design for five components and three levels was prepared according to Taguchi's experimental design. To avoid confusion and elicit precise results from a minimum number of trials, experimental designs were developed. By utilizing good experimental design, time, energy and money can be saved as fewer experiments will be required to attain the desired findings. The extra benefit of MINITAB software is that manual programming is not required, as the programs are already inbuilt into the software. A mixed-level design with two levels for the first factor (part orientation), and three levels for the remaining four factors, was selected in the software and an experimental design was created. The experimental design for the L18 orthogonal array is illustrated in Table 3. In accordance with the experimental design, the parameters were selected and the specimens were 3D-printed.

Table 3. L18 orthogonal array experimental design.

Trial No.	Process Parameters				
	Part Orientation (Axis)	Printing Temperature (°C)	Layer Height (mm)	Infill Density (%)	Raster Angle (°)
1	X	245	0.08	30	15
2	X	245	0.12	70	30
3	X	245	0.16	100	45
4	X	255	0.08	30	30
5	X	255	0.12	70	45
6	X	255	0.16	100	15
7	X	265	0.08	70	15
8	X	265	0.12	100	30
9	X	265	0.16	30	45
10	Y	245	0.08	100	45
11	Y	245	0.12	30	15
12	Y	245	0.16	70	30
13	Y	255	0.08	70	45
14	Y	255	0.12	100	15
15	Y	255	0.16	30	30
16	Y	265	0.08	100	30
17	Y	265	0.12	30	45
18	Y	265	0.16	70	15

2.4. Mechanical Properties Evaluation Methods

For evaluating the mechanical properties of 3D-printed ASA, the components for the tensile test, flexural test and impact test were selected in accordance with ASTM standards. For tensile testing, an ASTM D638 sample of 165 mm in length, 19 mm in width and 3 mm in thickness was selected. Figure 6 depicts the dumbbell-shaped specimen's dimensions, whereas Figure 7 depicts the printed tensile test specimens. Figure 8 depicts an Instron 8801 UTM, the device utilized for tensile and flexural testing. The specimen was created in accordance with the experimental design by altering the process parameters. The printed samples were tidily packaged to prevent environmental contact and moisture contamination. The tensile test was conducted using an Instron 8801 model UTM machine with a 100 kN load capacity, two columns equipped with M30 load cells and actuators and a load capacity of 100 kN. The specimen was grasped in the grippers and testing commenced. During the test, the cross-head velocity and strain rate were both maintained at 5 mm/min. The tensile strength of the printed specimen can be found using Equation (1) [21].

$$\text{Tensile strength (MPa)} = \text{Maximum load during breaking (Newton)}/\text{Cross-sectional area of the specimen (mm}^2) \quad (1)$$

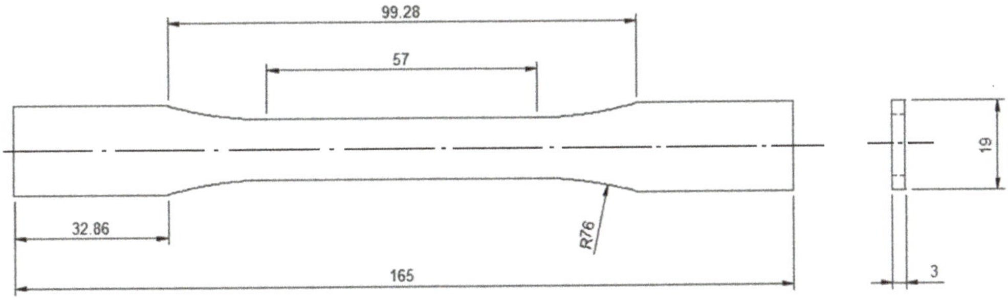

Figure 6. ASTM D638 specimen for tensile testing (all dimensions are in mm) [22].

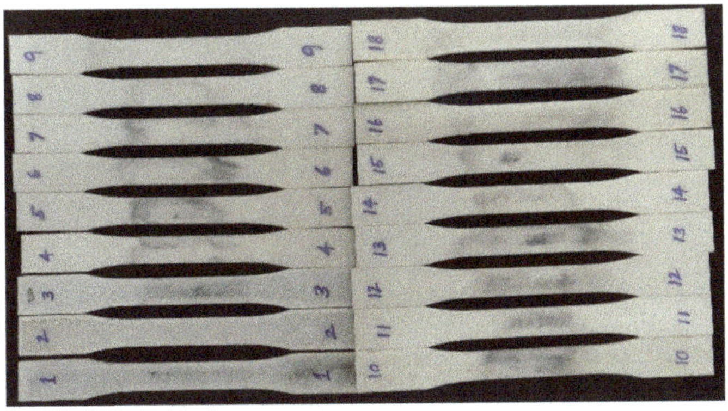

Figure 7. Printed specimens for tensile test as per the experimental design.

For impact testing, an ASTM D6110 sample with dimensions of 110 mm in length, 10 mm in width and thickness, and a 45° notch angle at the specimen's midpoint was selected. The impact test was conducted using FIT-300-D model pendulum-type impact testing equipment with a capacity of 300 J and a minimum count of 0.1 J. Chennai's Herenba Instruments and Engineers supplied and calibrated the equipment. The test was conducted at 37 °C and normal atmospheric pressure. Initially, the striker was released without the location of the specimen and its error was evaluated. Figure 9 depicts the specimen dimensions according to ASTM standards, and Figure 10a,b depict the printed specimens prior to the impact test. Figure 11 depicts the machine configuration, and the specimen was placed in the horizontal work holding fixture with the notch facing away from the striker. The striker was unleashed and caused to strike the specimen in its center. After breaking, the energy absorbed by the specimen was recorded and the specimen's impact strength was calculated using Equation (2).

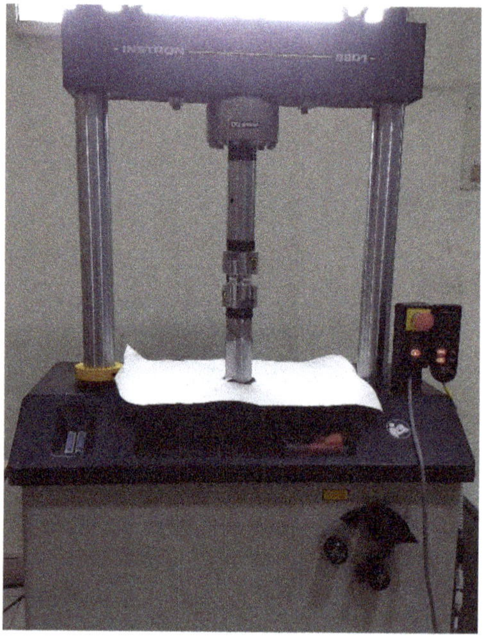

Figure 8. Instron 8801 UTM for tensile and bending test.

$$\text{Impact strength (Charpy test) (J/mm}^2\text{)} = \text{Energy absorbed (Joules)/Cross-sectional area (mm}^2\text{)} \quad (2)$$

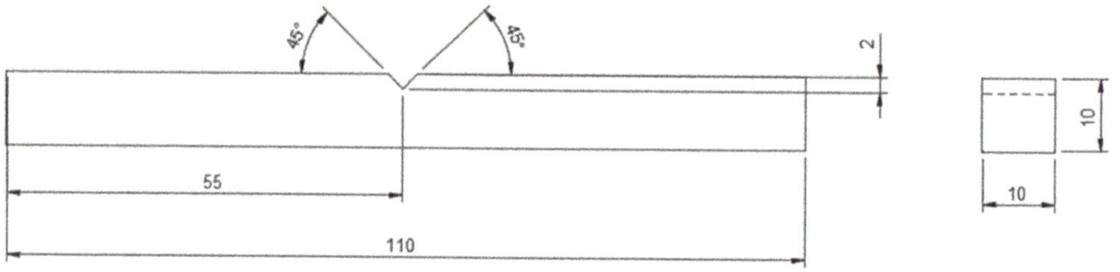

Figure 9. ASTM D6110 specimen for impact testing.

For flexural testing, an ASTM D790 sample of 130 mm in length, 12.7 mm in width and 3 mm in thickness was chosen. The flexural test was conducted on the same Instron 8801 model UTM equipped with a unique attachment for three-point bending testing. Figure 12 depicts the dimensions of the specimen, whereas Figure 13a,b depict the printed specimens for the three-point bending test according to the experimental design. Using Equation (3), the specimen's flexural strength was determined.

$$\text{Flexural strength (MPa)} = 3 \times \text{Load (N)} \times \text{Span of support (mm)}/2 \times \text{Specimen width (mm)} \times (\text{specimen height})^2\ \text{(mm)} \quad (3)$$

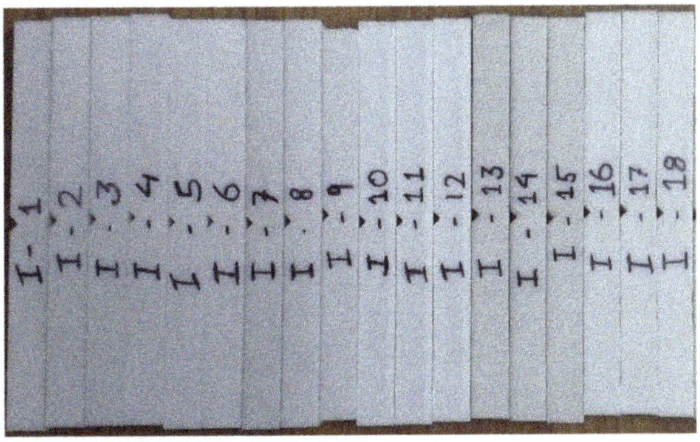

(**a**)

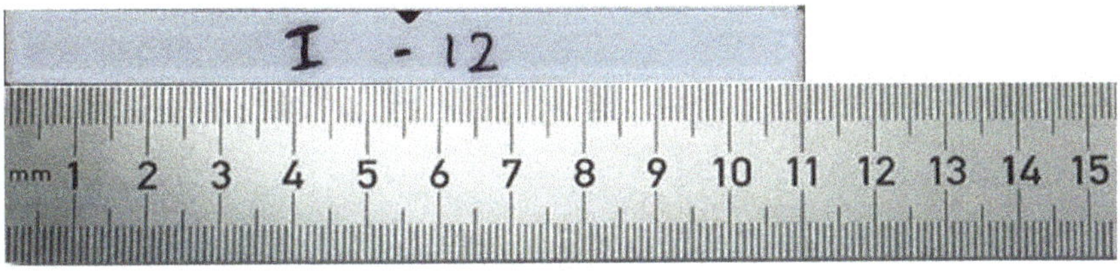

(**b**)

Figure 10. (**a**) Printed specimens for impact test as per the experimental design. (**b**). Impact specimen with scale.

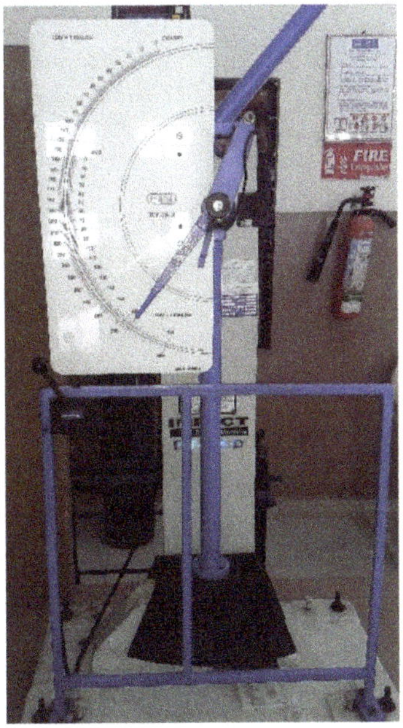

Figure 11. FIT-300-D model impact tester.

Figure 12. ASTM D790 specimen for flexural testing [23].

(**a**)

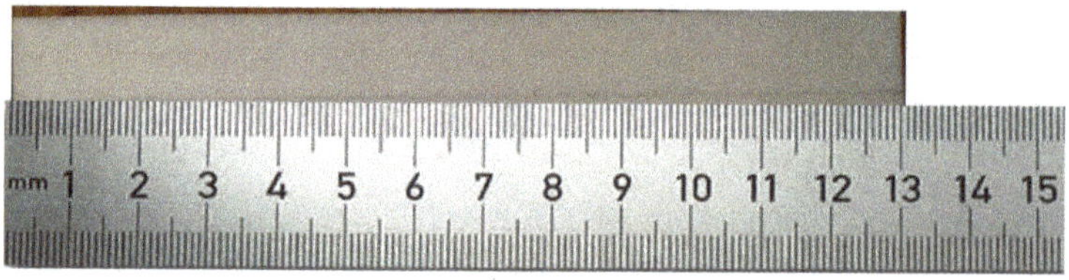

(**b**)

Figure 13. (**a**). Printed specimens for flexural test as per the experimental design. (**b**). Printed flexural specimen with scale.

3. Results and Discussion

Table 4 contains the experimental design that varies five process parameters and the results of the tests that were conducted using that design. When compared to others, specimen 7 demonstrated greater values for all the test findings and had the greatest tensile strength rating, which was 50.57 MPa. It demonstrated the second greatest output in the flexural test, which shows that this particular design parametric combination produces superior mechanical qualities.

Table 4. Experimental design with output responses.

Trial No.	Process Parameters					Responses		
	Part Orientation (Axis) (A)	Printing Temperature (°C) (B)	Layer Height (mm) (C)	Infill Density (%) (D)	Raster Angle (°) (E)	Tensile Strength (MPa)	Flexural Strength (MPa)	Impact Strength (J/mm^2)
1	X	245	0.08	30	15	46.84	77.45	0.14
2	X	245	0.12	70	30	43.20	73.11	0.11
3	X	245	0.16	100	45	43.54	74.35	0.10
4	X	255	0.08	30	30	45.50	74.74	0.12
5	X	255	0.12	70	45	43.41	72.28	0.10

Table 4. *Cont.*

Trial No.	Process Parameters					Responses		
	Part Orientation (Axis) (A)	Printing Temperature (°C) (B)	Layer Height (mm) (C)	Infill Density (%) (D)	Raster Angle (°) (E)	Tensile Strength (MPa)	Flexural Strength (MPa)	Impact Strength (J/mm^2)
6	X	255	0.16	100	15	48.19	79.34	0.13
7	X	265	0.08	70	15	50.57	81.43	0.17
8	X	265	0.12	100	30	48.95	80.90	0.15
9	X	265	0.16	30	45	42.19	71.23	0.10
10	Y	245	0.08	100	45	44.72	77.32	0.14
11	Y	245	0.12	30	15	42.76	71.60	0.11
12	Y	245	0.16	70	30	39.97	72.53	0.09
13	Y	255	0.08	70	45	43.46	75.12	0.13
14	Y	255	0.12	100	15	46.72	78.32	0.15
15	Y	255	0.16	30	30	39.67	69.82	0.09
16	Y	265	0.08	100	30	49.88	82.87	0.17
17	Y	265	0.12	30	45	43.26	70.36	0.11
18	Y	265	0.16	70	15	45.32	75.21	0.13

3.1. Tensile Strength

The experimental data are provided as input to the MINITAB software for regression analysis in order to identify the most influential parameter which serves as the determining factor in producing output results. The tensile strength was determined to range between 39.97 MPa and 50.57 MPa. The majority of specimens subjected to tensile loads exhibited superior extension properties, as shown by the lengthening of the specimens prior to failure [24]. The majority of specimens exhibited plastic deformation prior to failure or cracking. The failure of the specimens happened approximately at the specimen's center, indicating that the specimens are defect free and that the load acts at the proper place. It also demonstrated that the tests yield precise findings. The experimental values were then inputted into the software for regression analysis in order to determine the most influential parameter affecting the results. The coefficient of regression is 93.82%, which is closer to the acceptable value. The most influential criteria for tensile strength were found to be layer height and infill density. The specimens subjected to tensile testing demonstrated an extended fracture failure mechanism rather than a brittle fracture. Figure 14 depicts the deformed portion following a tensile test, which has an extended section due to plastic deformation. Due to enhanced tensile characteristics, the specimen did not fail completely while undergoing deformation. Table 5 represents the elongation of the specimen at break and yield point values. The ANOVA findings are presented in Table 6, where the Fisher value of each selected parameter reflects the extent of its influence.

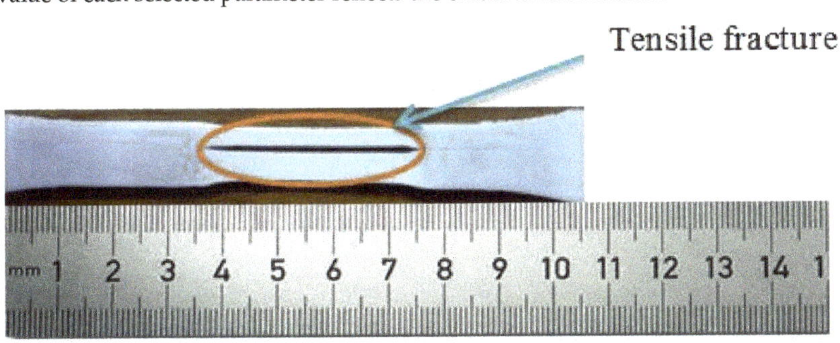

Figure 14. Tensile tested specimen.

Table 5. Elongation at break and yield point values for tensile test results.

Trial No.	Responses		
	Tensile Strength (MPa)	**Elongation at Break (mm)**	**Yield Point (MPa)**
1	46.84	25	52.32
2	43.20	12	48.87
3	43.54	9	45.35
4	45.50	14	51.67
5	43.41	7	47.75
6	48.19	24	53.83
7	50.57	34	59.13
8	48.95	21	56.82
9	42.19	15	47.26
10	44.72	18	50.1
11	42.76	26	47.98
12	39.97	10	45.73
13	43.46	16	50.9
14	46.72	28	53.42
15	39.67	7	46.82
16	49.88	29	57.12
17	43.26	15	50.13
18	45.32	20	52.47

Table 6. Tensile test response ANOVA.

Source	DF	Adj SS	Adj MS	F-Value	*p*-Value
Regression	5	156.8	31.3595	36.44	0
Printing temperature	1	30.53	30.5283	35.47	0
Layer height	1	40.66	40.664	47.25	0
Infill density	1	37.51	37.505	43.58	0
Raster angle	1	32.74	32.736	38.04	0
Part orientation	1	15.36	15.3643	17.85	0.001
Error	12	10.33	0.8606		
Total	17	167.12			
S	R-sq		R-sq (adj)	R-sq (pred)	
0.92769	93.82%		91.25%	85.64%	

Figure 15a–d depict the main effects, interaction, contour and contribution plots for tensile tested results. The larger is better concept was followed when plotting the S/N ratio's main effects. It was projected that 'X' part orientation, printing temperature of 265 C, layer height of 0.08 mm, 100 percent infill density and a raster angle of 15 (A1, B3, C1, D3, E1) would result in the highest tensile strength. The interaction plot reveals that there is no interaction between printing temperature and part orientation, other than the fact that all components are dependent on one another. The contour plot was drawn between the layer height and the infill density, as these were the most influential parameters among the others. In order to achieve higher tensile strengths, the contour plot indicates that the layer height and infill density must be 0.08 mm and 100 percent, respectively.

3.2. Flexural Strength

The specimen with the highest tensile strength also possessed superior flexural properties. Specimen number 16, having a 'Y' axis part orientation printed at 265 °C, with a layer height of 0.08 mm, a 100 percent infill density and a raster angle of 30 degrees, exhibited a maximum flexural strength of 82.87 MPa. Figure 16 depicts the specimen following the flexural test, which demonstrates that even after failure the cracks are not visible to the naked eye. Due to the formation of microscopic fractures that do not propagate further, the specimen did not fail completely or shatter into two pieces. It demonstrates the flexural stiffness of the specimen and the behavior of ASA prior to failure. The flexural strength ANOVA findings are presented in Table 7, and according to the Fisher value, infill density was the most influential parameter, followed by layer height. In addition, the regression coefficient was determined to be greater than 90 percent, demonstrating the significance of the model.

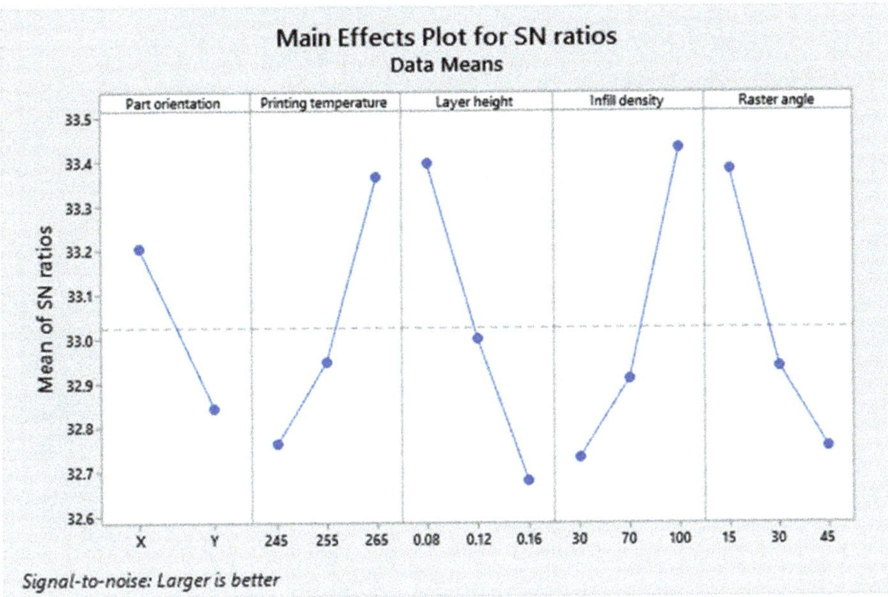

(a)

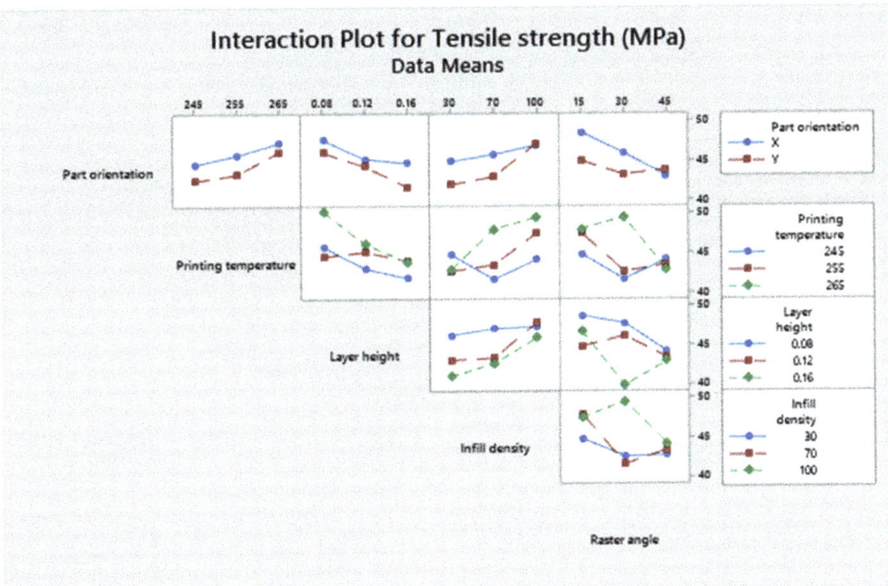

(b)

Figure 15. *Cont.*

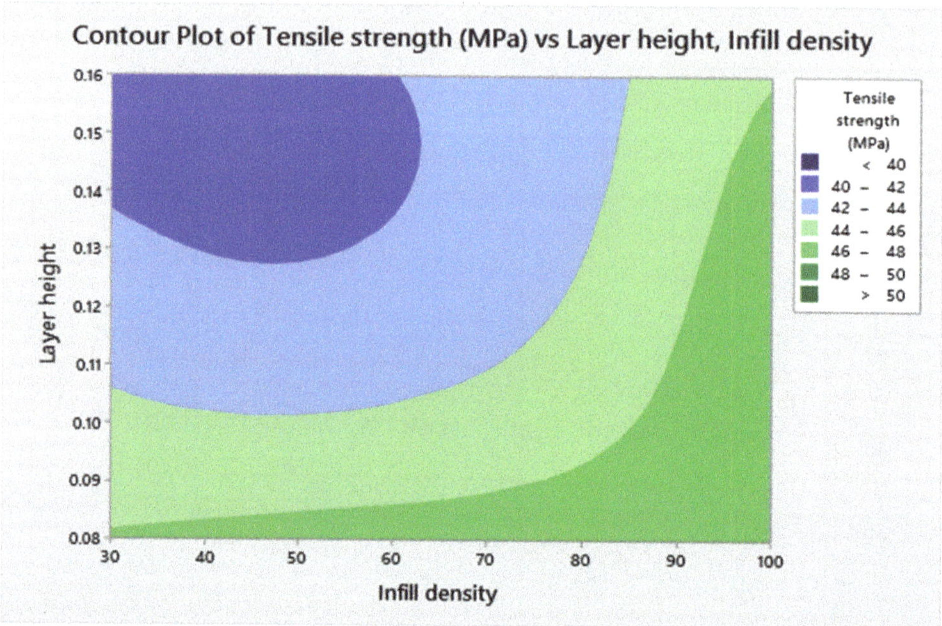

(c)

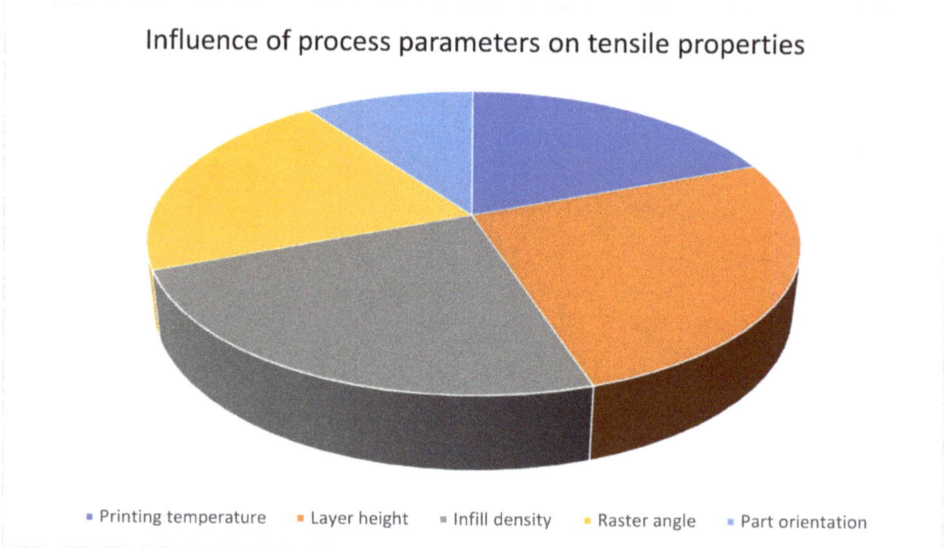

(d)

Figure 15. (**a**). Main effects plot for tensile test responses. (**b**). Interaction plot for tensile test responses. (**c**). Contour plot for maximum tensile strength. (**d**). Contribution plot of parameters for tensile strength.

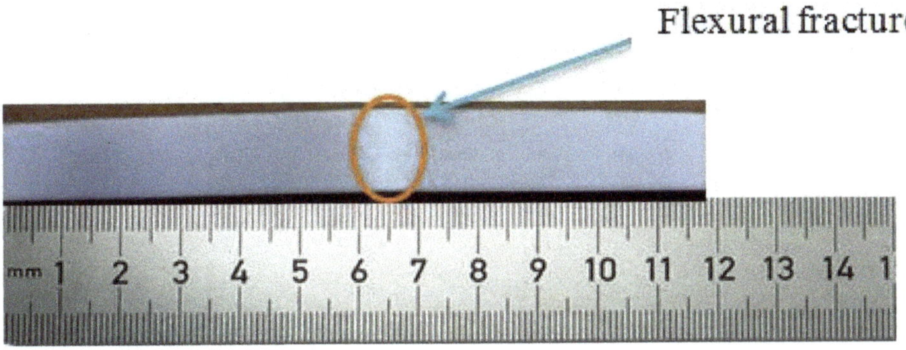

Figure 16. Deformed specimen after flexural test.

Table 7. Flexural test response ANOVA.

Source	DF	Adj SS	Adj MS	F-Value	*p*-Value
Regression	5	245.4	49.08	26.59	0
Printing temperature	1	20.384	20.384	11.04	0.006
Layer height	1	58.314	58.314	31.6	0
Infill density	1	116.216	116.216	62.97	0
Raster angle	1	42.903	42.903	23.25	0
Part orientation	1	7.583	7.583	4.11	0.065
Error	12	22.147	1.846		
Total	17	267.547			
S	R-sq		R-sq (adj)	R-sq (pred)	
1.35853	91.72%		88.27%	81.70%	

A1, B4, C1, D3 and E1 were the anticipated optimal values for maximal flexural strength, based on the main effects figure. The same predictions were made regarding the tensile property. It illustrates the relationship between tensile and flexural properties, which are directly proportional. The interaction plot provides the same information as tensile responses. The orientation of the part and the printing temperature are unrelated. Figure 17d depicts the effect of each parameter on the flexural properties as a percentage. Figure 17a–c depict the main effects, interaction and contour plots for the flexural strength results.

3.3. Impact Strength

The specimens with the highest impact resistance to sudden loads were numbers 7 and 16. The maximum impact strength observed was 0.17 J/mm^2, whereas the minimum was 0.09 J/mm^2. Figure 18 depicts the front view of a specimen that was fractured due to impact stresses with the effect of layer height and infill density evident. The crack began at the location of the notch and the layers impeded its progression. There are irregularities in the fracture propagation as the specimen was not shattered in a straight line. The specimen remained intact during the impact test, demonstrating its resistance to impact loads. The layer height appears to be the most influential factor in impact test analysis, as shown in Table 8 of the ANOVA results. With 93.01 percent, the regression coefficient is adequate.

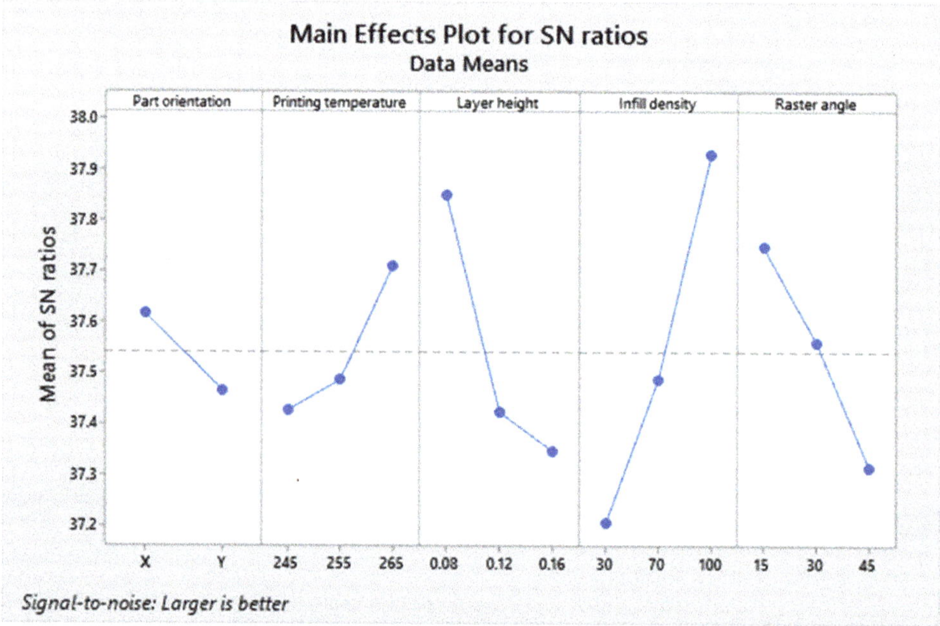

(a)

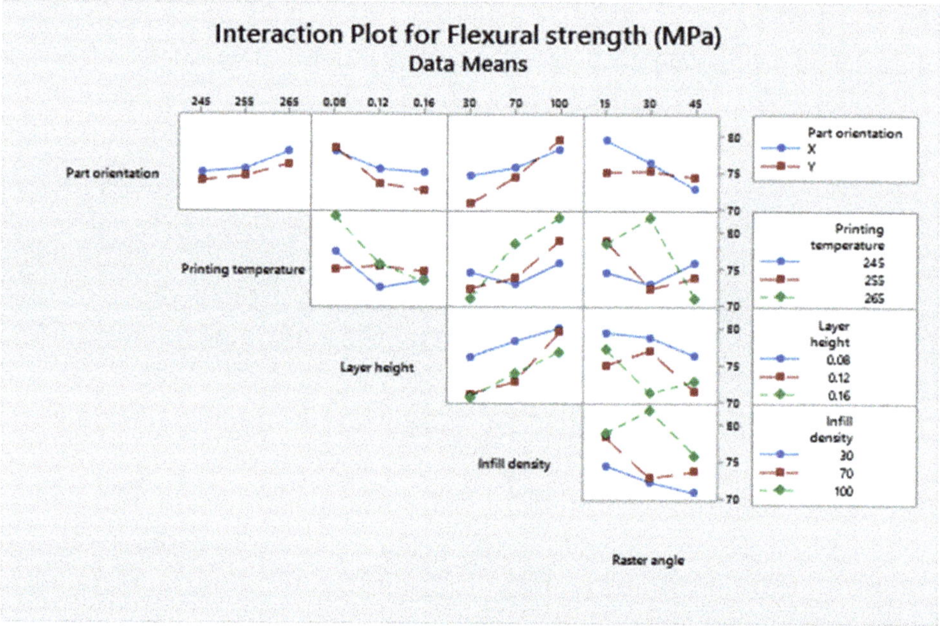

(b)

Figure 17. *Cont.*

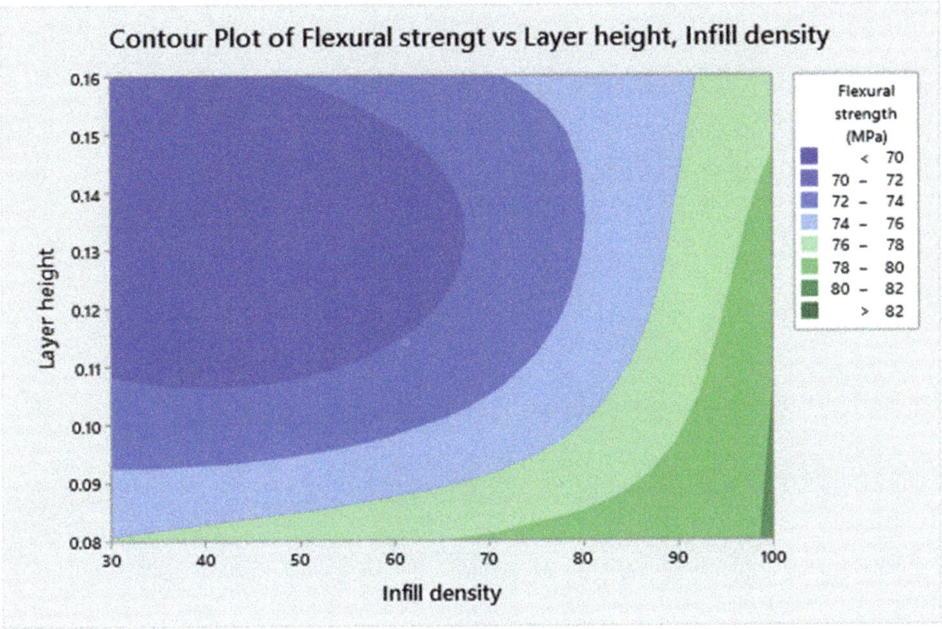

(**c**)

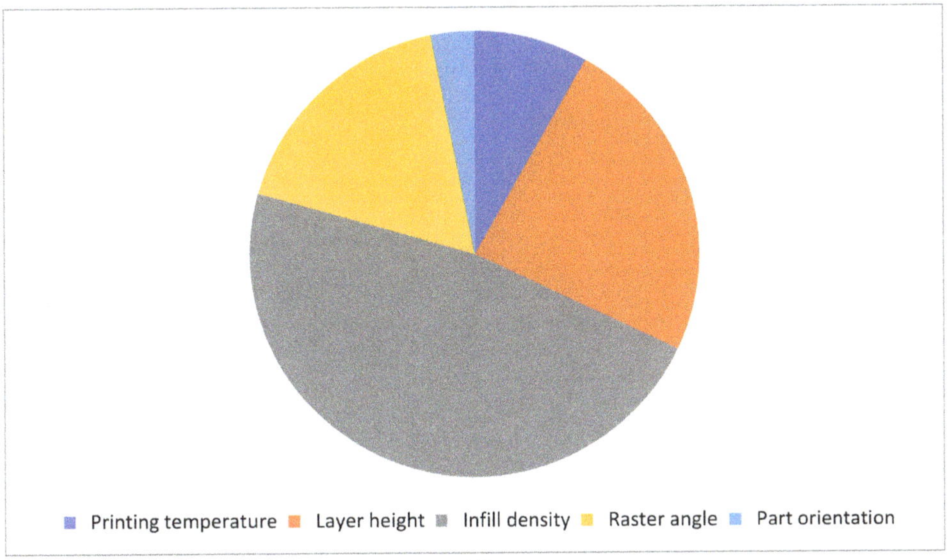

(**d**)

Figure 17. (**a**). Main effects plot for bending test responses. (**b**). Interaction plot for bending test responses. (**c**). Contour plot for maximum flexural strength. (**d**). Contribution of each factor for maximum flexural strength.

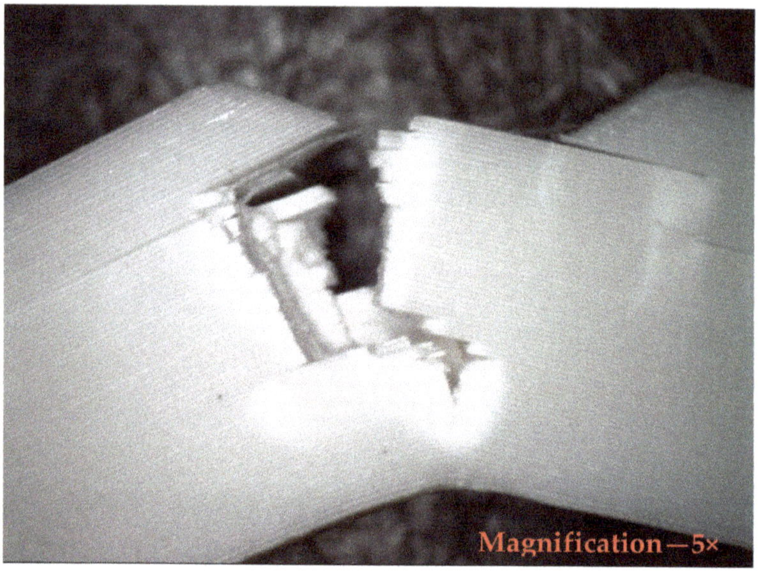

Figure 18. Optical microscopic view of the deformed specimen after impact test, side view.

Table 8. Impact test response ANOVA.

Scheme	DF	Adj SS	Adj MS	F-Value	*p*-Value
Regression	5	0.010085	0.002017	31.94	0
Printing temperature	1	0.001633	0.001633	25.86	0
Layer height	1	0.004332	0.004332	68.59	0
Infill density	1	0.002032	0.002032	32.17	0
Raster angle	1	0.00208	0.00208	32.94	0
Part orientation	1	0.000008	0.000008	0.13	0.728
Error	12	0.000758	0.000063		
Total	17	0.010843			
S	R-sq		R-sq (adj)	R-sq (pred)	
0.007947	93.01%		90.10%	82.68%	

Figure 19a–d represent the main effects plot, the interaction plot and the contour plots, respectively. According to the forecast of the main effects plot, A1, B3, C1, D3 and E1 are the optimal values that result in greater impact strengths. It is apparent from the interaction plot that all the components are interrelated. Contour plots were created for infill density and layer height, as well as for layer height and raster angle, since these are the most influential elements in terms of impact strength [25]. The contribution and effect of each factor can be understood from Figure 19e.

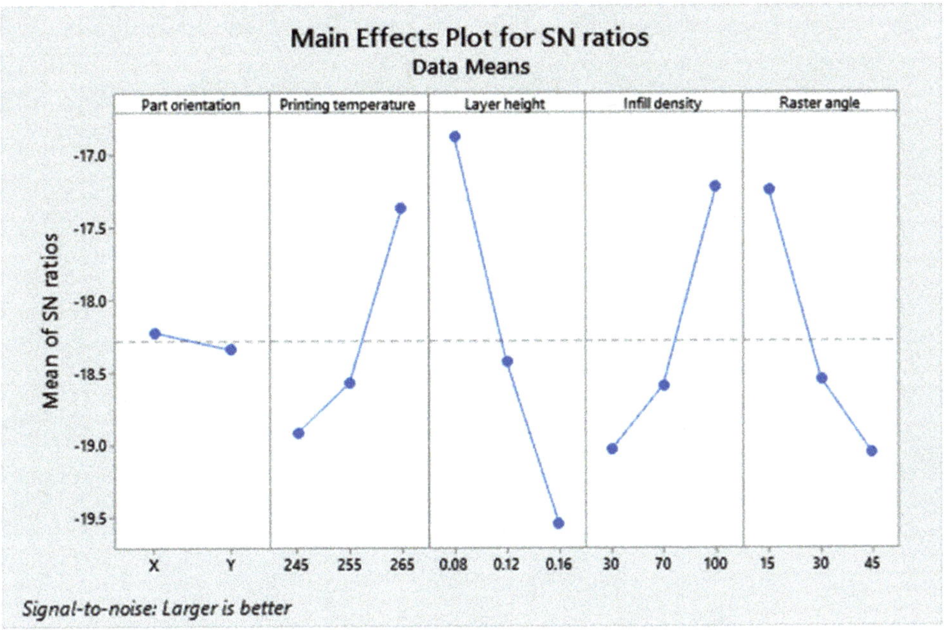

(a)

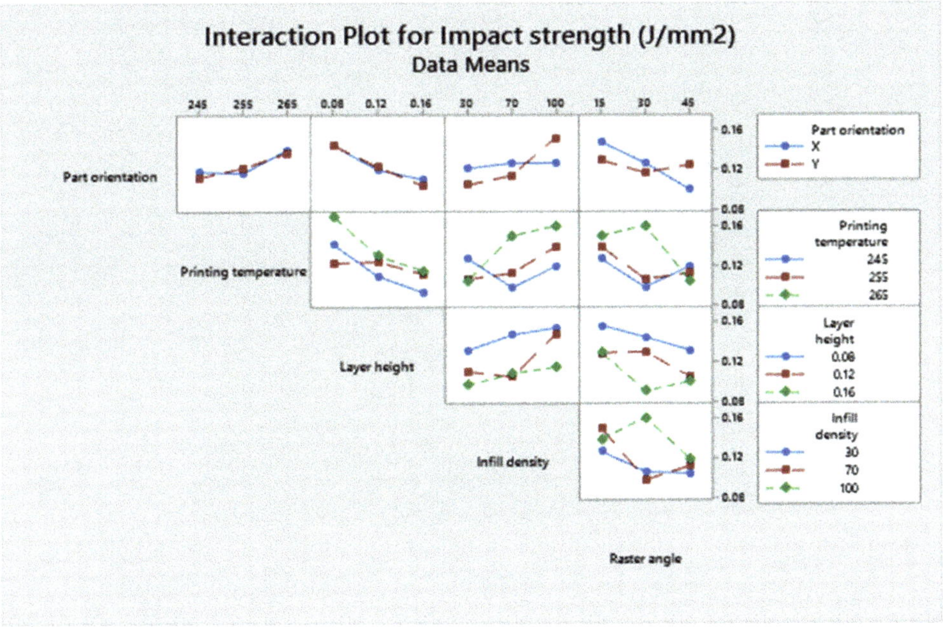

(b)

Figure 19. *Cont.*

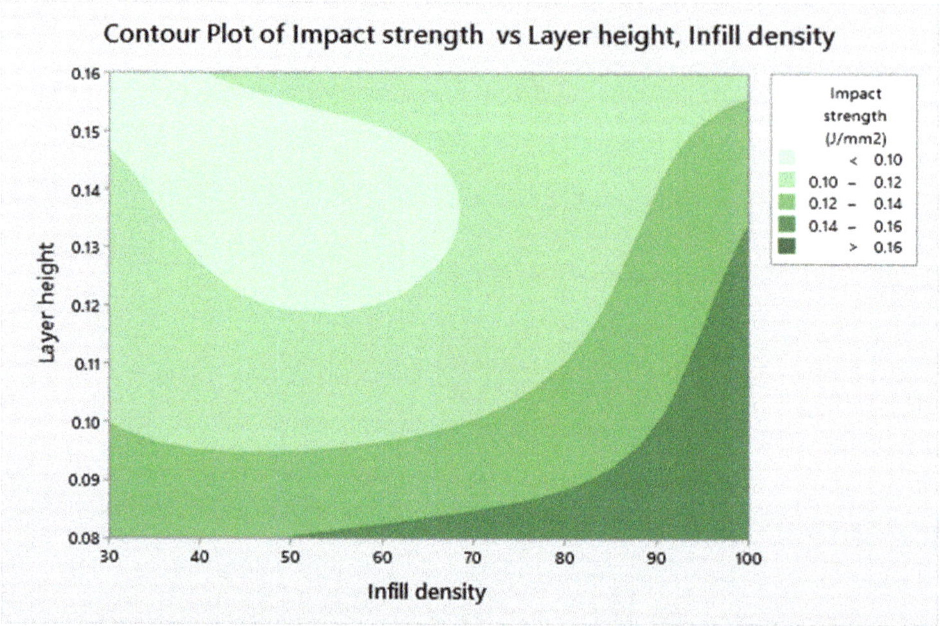

(c)

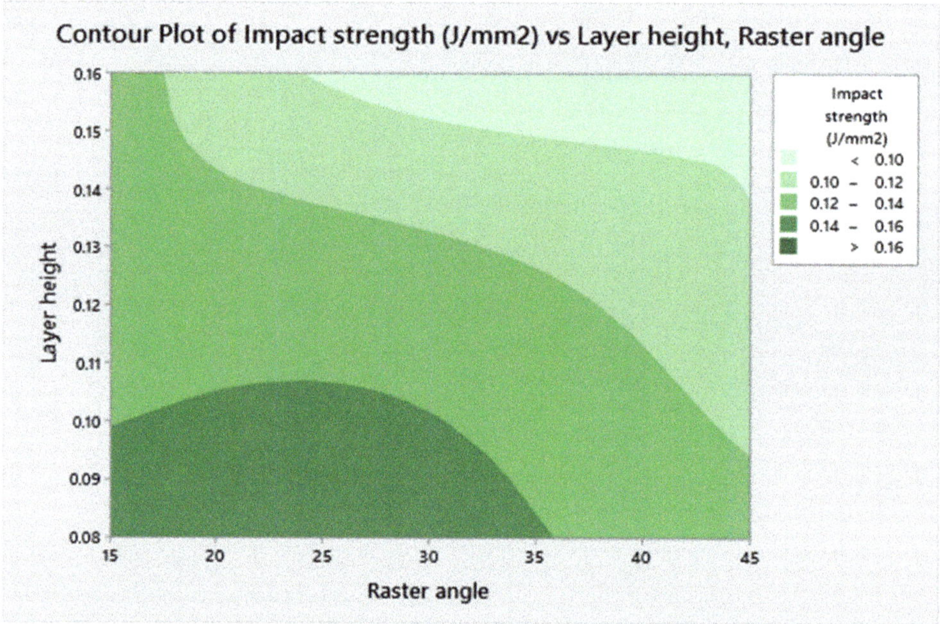

(d)

Figure 19. *Cont.*

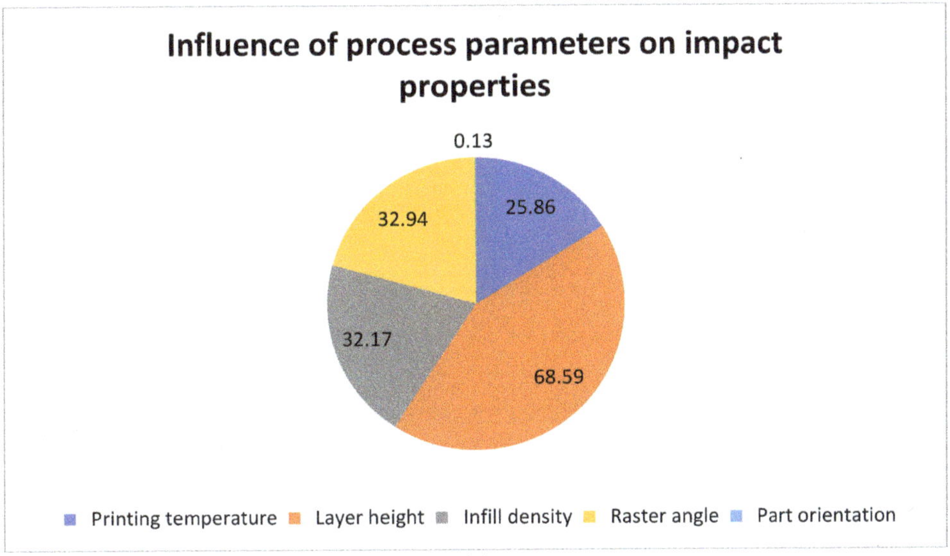

(e)

Figure 19. (**a**). Main effects plot for impact test responses. (**b**). Interaction plot for impact test responses. (**c**). Contour plot between layer height and infill density for maximum impact strength. (**d**). Contour plot between layer height and raster angle for maximum impact strength. (**e**). Contribution plot of various factors for maximum impact strength.

4. Validation Test Details

From the experimental data that were fed into the software, optimization outcomes and the factors that influence outputs were determined. The software predicted a result for each test and provided the optimal process parameters for the predicted result. Figure 20a–c depict the optimized parameters and the expected output for the optimal values. The comparison of anticipated and experimental values, together with the percentage of error, is displayed in Table 9. Specimens were prepared for each test using the optimum process parameters. For each test, three specimens were manufactured with identical process settings and their mean values were used to determine the final output. Validation was conducted by comparing the outcomes of each test, and the percentage of error was determined by computing the difference between the expected and experimental outcomes [26]. It was noticed that the error rate was less than 3 percent, which is within acceptable parameters.

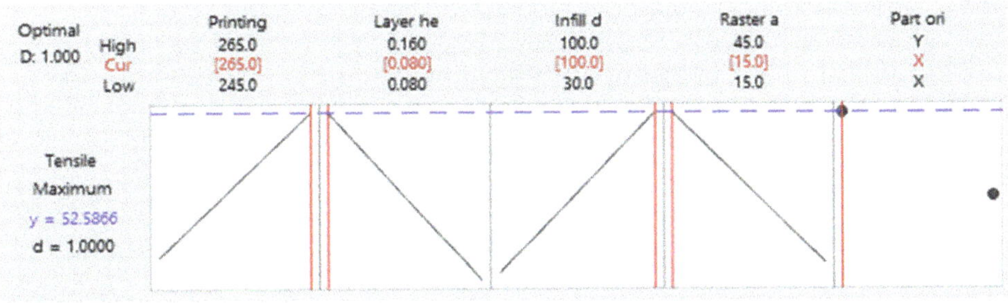

(a)

Figure 20. *Cont.*

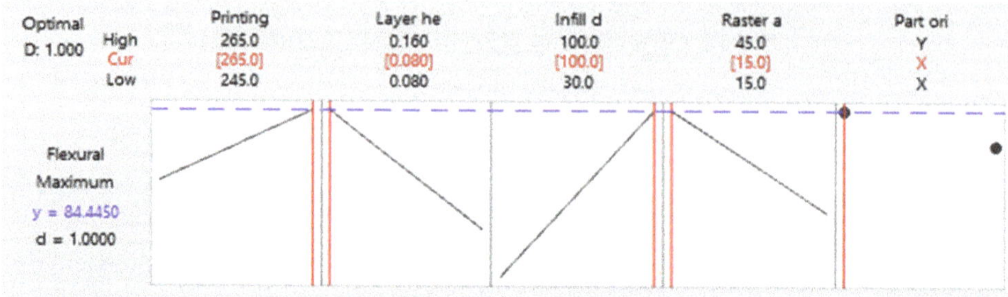

(**b**)

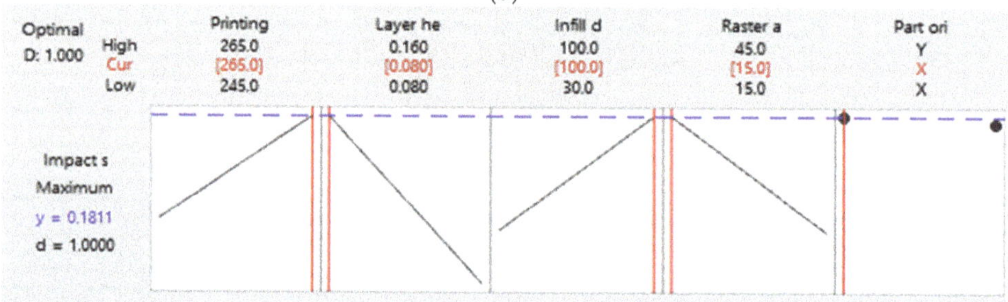

(**c**)

Figure 20. (**a**–**c**). Predicted results and optimum process parameters.

Table 9. Predicted and experimental results with error percentage.

Output Factors	Optimal Control Parameters		Error (%)
	Predicted	Experimental	
Tensile strength in MPa	A1B3C1D3E1	A1B3C1D3E1	1.31
	52.58	51.89	
Flexural strength in MPa	A1B3C1D3E1	A1B3C1D3E1	2.22
	84.44	82.56	
Impact strength in J/mm^2	A1B3C1D3E1	A1B3C1D3E1	0.55
	0.181	0.180	

5. Conclusions

- The L18 orthogonal array was constructed using the software MINITAB and specimens were produced in accordance with the experimental design.
- Specimens created with a 3D printer were tested to establish their mechanical properties in compliance with the standards and the results were then optimized.
- The optimized results enabled the identification of the process factors that had the largest impact on the output of each test, as well as the process parameters that produced the best results overall.
- The enhanced process parameters were then applied to the production of three further specimens, which were subsequently evaluated using the same criteria.
- The outcomes of the experiment were validated by comparing them to the predicted values based on the experiment's results.
- When the specimen was built in accordance with the predicted input factors, tensile strength of 51.86 MPa, flexural strength of 82.56 MPa and impact strength of 0.180 J/mm^2 were obtained.

- The proportion of inaccuracy discovered between the predicted and experimental results fell within the permitted range.
- A1B3C1D3E1 were the software's projected input factors, and the software's proposal was followed to make the specimens and validate the results.
- It was determined that the infill density and the layer height have the most effect on the outcomes when it comes to the fabrication of 3D-printed objects.

6. Future Scopes of Inquiry

Already, elements and things that have been created using a 3D printer are being used in every industry. The advantages of 3D printing include the ease with which difficult profiles can be manufactured, less material waste, shortened production times and lower overall energy requirements. In the not-too-distant future, virtually all items and parts will be produced using 3D printing. A pie chart has been used in Figure 21 to show the many different applications for 3D printing technology. The quality of the items produced by additive manufacturing is also being improved through the utilization of a great number of recently developed methods and procedures. To further extend the scope of this work, the mechanical properties of the printed specimens could be improved by applying ceramic or metal coatings to them. The ASA polymer can be mixed with several fillers, each of which has the potential to contribute to an improvement in the polymer's fundamental properties. Wear, water absorption, abrasion resistance, conductivity and flammability are some additional tests that can be carried out in addition to the ones already mentioned. It is possible to reinforce ASA with high performance fibers in order to determine the extent of the improvement in all qualities necessary for specific applications.

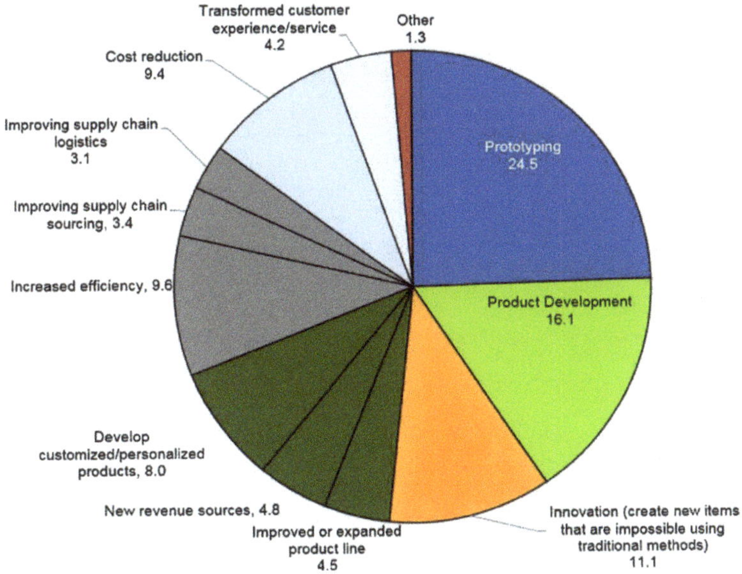

Figure 21. Contribution of 3D printing in various applications [27] Copyright 2018 with permission from Elsevier.

Author Contributions: Conceptualization, A.Z.H., M.A.S. and M.A.B.; Funding acquisition, A.Z.H., M.A.S. and M.A.B.; Methodology, S.A.R. and J.K.; Validation, S.A.R. and J.K.; Writing—original draft, S.A.R. All authors have read and agreed to the published version of the manuscript.

Funding: The authors extend their appreciation to the Deanship for Research and Innovation, Ministry of Education in Saudi Arabia, for funding this research work through the project number IFPRC-022-135-2020 and King Abdulaziz University, DSR, Jeddah, Saudi Arabia.

Informed Consent Statement: Not applicable.

Data Availability Statement: Not applicable.

Conflicts of Interest: The authors declare no conflict of interest.

Nomenclature

ASA	Acrylonitrile styrene acrylate
NIR	Near infrared radiation
CPE	Chlorinated Polyethylene
BR	Polybutadiene rubber
3D	Three dimensional
FDM	Fused deposition modelling
ASTM	American Society for Testing and Materials
SEM	Scanning electron microscopy
RAFT	Reversible addition–fragmentation chain-transfer
LSCM	Laser scanning confocal microscopy
PVC	Polyvinyl chloride
ANOVA	Analysis of variance
PLA	Polylactic acid
ABS	Acrylonitrile butadiene styrene

References

1. Jakus, A.E. An Introduction to 3D Printing-Past, Present, and Future Promise. *3D Print. Orthop. Surg.* **2018**, 1–15. Available online: https://www.sciencedirect.com/science/article/pii/B9780323581189000014 (accessed on 16 July 2022). [CrossRef]
2. Delda, R.N.M.; Basuel, R.B.; Hacla, R.P.; Martinez, D.W.C.; Cabibihan, J.-J.; Dizon, J.R.C. 3D Printing Polymeric Materials for Robots with Embedded Systems. *Technologies* **2021**, *9*, 82. [CrossRef]
3. Thiam, B.G.; El Magri, A.; Vanaei, H.R.; Vaudreuil, S. 3D Printed and Conventional Membranes—A Review. *Polymers* **2022**, *14*, 1023. [CrossRef]
4. Tian, M.; Chen, C.; Han, A.; Ye, M.; Chen, X. Estimating Thermal Insulation Performance and Weather Resistance of Acrylonitrile-Styrene-Acrylate Modified with High Solar Reflective Pigments: Pr^{3+}/Cr^{3+} Doped $BaTiO_3$. *Sol. Energy* **2021**, *225*, 934–941. [CrossRef]
5. Mao, Z.; Zhang, J. Largely Improved the Low Temperature Toughness of Acrylonitrile-Styrene-Acrylate (ASA) Resin: Fabricated a Core-Shell Structure of Two Elastomers through the Differences of Interfacial Tensions. *Appl. Surf. Sci.* **2018**, *444*, 345–354. [CrossRef]
6. Sánchez, D.M.; de la Mata, M.; Delgado, F.J.; Casal, V.; Molina, S.I. Development of Carbon Fiber Acrylonitrile Styrene Acrylate Composite for Large Format Additive Manufacturing. *Mater. Des.* **2020**, *191*, 108577. [CrossRef]
7. Xiang, B.; Zhang, J. A New Member of Solar Heat-Reflective Pigments: BaTiO3 and Its Effect on the Cooling Properties of ASA (Acrylonitrile-Styrene-Acrylate Copolymer). *Sol. Energy Mater. Sol. Cells* **2018**, *180*, 67–75. [CrossRef]
8. Qi, Y.; Zhang, J. Chemically Modified Sb_2O_3, a New Member of High Solar-Reflective Material Family, Incorporating with ASA (Acrylonitrile-Styrene-Acrylate Copolymer) for Fabrication of Cooling Composite with Lower Wetting Behavior. *Compos. Part B Eng.* **2019**, *162*, 112–121. [CrossRef]
9. Raj, S.A.; Muthukumaran, E.; Jayakrishna, K. A Case Study of 3D Printed PLA and Its Mechanical Properties. *Mater. Today Proc.* **2018**, *5*, 11219–11226. [CrossRef]
10. Meyer, T.K.; Tanikella, N.G.; Reich, M.J.; Pearce, J.M. Potential of Distributed Recycling from Hybrid Manufacturing of 3-D Printing and Injection Molding of Stamp Sand and Acrylonitrile Styrene Acrylate Waste Composite. *Sustain. Mater. Technol.* **2020**, *25*, e00169. [CrossRef]
11. Magadum, S.; Gilorkar, A.; Amol Deepak, M.; Rakshith, B.S. Design, Simulation and Experimental Investigation of 3D Printed Mechanical Metamaterials. In Proceedings of the 2021 International Solid Freeform Fabrication Symposium 2021, Austin, TX, USA, 2–4 August 2021; University of Texas at Austin: Austin, TX, USA, 2021; pp. 2–12. [CrossRef]
12. Huang, J.; Luo, Y.; Gao, X. Morphology and Mechanical Properties of Acrylonitrile-Styrene-Acrylate Toughened Plastics with Block Copolymer Chain Structure. *Polym. Eng. Sci.* **2019**, *59*, 389–395. [CrossRef]
13. Huang, J.; Kuo, C.; Tsai, H. Stiffness Enhancement, Anti-Aging, and Self-Forming Holes in Polycarbonate/Acrylonitrile-Styrene-Acrylic by the Core-Shell Structure of Acrylic ResinStructure of Acrylic Resin. *Polymers* **2022**, *14*, 782. [CrossRef] [PubMed]
14. Zhang, K.; Hamza Bichi, A.; Yang, J. Effect of Acrylonitrile Styrene Acrylate on Mechanical, Thermal and Three-Body Abrasion Behaviors of Eucalyptus Fiber Reinforced Polyvinyl Chloride Composite. *Mater. Res. Express* **2021**, *8*, 025305. [CrossRef]
15. Xiang, B.; Zhang, J. Effects of Content and Surface Hydrophobic Modification of BaTiO 3 on the Cooling Properties of ASA (Acrylonitrile-Styrene-Acrylate Copolymer). *Appl. Surf. Sci.* **2018**, *427*, 654–661. [CrossRef]

16. Veselý, P. Nozzle Temperature Effect on 3D Printed Structure Properties. In Proceedings of the ELEKTROTECHNOLÓGIA 2019, Zuberec, Slovakia, 21–23 May 2019.
17. Dey, A.; Yodo, N. A Systematic Survey of FDM Process Parameter Optimization and Their Influence on Part Characteristics. *J. Manuf. Mater. Process.* **2019**, *3*, 64. [CrossRef]
18. Ramezani Dana, H.; Barbe, F.; Delbreilh, L.; Ben Azzouna, M.; Guillet, A.; Breteau, T. Polymer Additive Manufacturing of ABS Structure: Influence of Printing Direction on Mechanical Properties. *J. Manuf. Process.* **2019**, *44*, 288–298. [CrossRef]
19. Popescu, D.; Zapciu, A.; Amza, C.; Baciu, F.; Marinescu, R. FDM Process Parameters Influence over the Mechanical Properties of Polymer Specimens: A Review. *Polym. Test.* **2018**, *69*, 157–166. [CrossRef]
20. Raam Kumar, S.; Sridhar, S.; Venkatraman, R.; Venkatesan, M. Polymer Additive Manufacturing of ASA Structure: Influence of Printing Parameters on Mechanical Properties. *Mater. Today Proc.* **2020**, *39*, 1316–1319. [CrossRef]
21. Ganapathy, S.B.; Sakthivel, A.R.; Sultan, M.T.H.; Shahar, F.S.; Shah, A.U.M.; Khan, T.; Sebaey, T.A. Effect of Prosopis Juliflora Thorns on Mechanical Properties of Plastic Waste Reinforced Epoxy Composites. *Polymers* **2022**, *14*, 1278. [CrossRef]
22. *ASTM D638-14;* Standard Test Method for Tensile Properties of Plastics. ASTM International: West Conshohocken, PA, USA, 2014. [CrossRef]
23. *ASTM D790-17;* Standard Test Methods for Flexural Properties of Unreinforced and Reinforced Plastics and Electrical Insulating Materials. ASTM International: West Conshohocken, PA, USA, 2017. [CrossRef]
24. Guessasma, S.; Belhabib, S.; Nouri, H. Microstructure, Thermal and Mechanical Behavior of 3D Printed Acrylonitrile Styrene Acrylate. *Macromol. Mater. Eng.* **2019**, *304*, 1800793. [CrossRef]
25. Ezeh, O.H.; Susmel, L. Fatigue Strength of Additively Manufactured Polylactide (PLA): Effect of Raster Angle and Non-Zero Mean Stresses. *Int. J. Fatigue* **2019**, *126*, 319–326. [CrossRef]
26. Sakthi Balan, G.; Nandha Gopan, S.; Santhosh Kumar, V.; Ravichandran, M. Effect of Chemical Treatment on Mechanical Properties of Prawn Antenna Reinforced Waste Plastic Particulates Filled Polymer Composites. *Mater. Today Proc.* **2020**, *33*, 3668–3675. [CrossRef]
27. Dizon, J.R.C.; Espera, A.H.; Chen, Q.; Advincula, R.C. Mechanical Characterization of 3D-Printed Polymers. *Addit. Manuf.* **2018**, *20*, 44–67. [CrossRef]

polymers

MDPI

Review

Towards a Circular Economy of Plastics: An Evaluation of the Systematic Transition to a New Generation of Bioplastics

Elda M. Melchor-Martínez [1], Rodrigo Macías-Garbett [1], Lynette Alvarado-Ramírez [1], Rafael G. Araújo [1], Juan Eduardo Sosa-Hernández [1], Diana Ramírez-Gamboa [1], Lizeth Parra-Arroyo [1], Abraham Garza Alvarez [2], Rosina Paola Benavides Monteverde [2], Karen Aleida Salazar Cazares [2], Adriana Reyes-Mayer [3], Mauricio Yáñez Lino [4], Hafiz M. N. Iqbal [1,*] and Roberto Parra-Saldívar [1,*]

1. Tecnologico de Monterrey, School of Engineering and Sciences, Monterrey 64849, Nuevo Leon, Mexico; elda.melchor@tec.mx (E.M.M.-M.); rodrigo.macias@tec.mx (R.M.-G.); a00814259@tec.mx (L.A.-R.); rafael.araujo@tec.mx (R.G.A.); eduardo.sosa@tec.mx (J.E.S.-H.); diana.ramirez.gamboa@tec.mx (D.R.-G.); a01036078@tec.mx (L.P.-A.)
2. Cadena Comercial OXXO S.A de C.V., Monterrey 64480, Nuevo Leon, Mexico; abraham.garza@oxxo.com (A.G.A.); rosina.benavides@oxxo.com (R.P.B.M.); karen.salazar@oxxo.com (K.A.S.C.)
3. Centro de Caracterización e Investigación en Materiales S.A. de C.V., Jiutepec 62578, Morelos, Mexico; areyes@cecim.com.mx
4. Polymer Solutions & Innovation S.A. de C.V., Jiutepec 62578, Morelos, Mexico; myanezl@polysol.com.mx

* Correspondence: hafiz.iqbal@tec.mx (H.M.N.I.); r.parra@tec.mx (R.P.-S.)

Citation: Melchor-Martínez, E.M.; Macías-Garbett, R.; Alvarado-Ramírez, L.; Araújo, R.G.; Sosa-Hernández, J.E.; Ramírez-Gamboa, D.; Parra-Arroyo, L.; Alvarez, A.G.; Monteverde, R.P.B.; Cazares, K.A.S.; et al. Towards a Circular Economy of Plastics: An Evaluation of the Systematic Transition to a New Generation of Bioplastics. *Polymers* **2022**, *14*, 1203. https://doi.org/10.3390/polym14061203

Academic Editor: Maya Jacob John

Received: 5 February 2022
Accepted: 10 March 2022
Published: 17 March 2022

Abstract: Plastics have become an essential part of the modern world thanks to their appealing physical and chemical properties as well as their low production cost. The most common type of polymers used for plastic account for 90% of the total production and are made from petroleum-based nonrenewable resources. Concerns over the sustainability of the current production model and the environmental implications of traditional plastics have fueled the demand for greener formulations and alternatives. In the last decade, new plastics manufactured from renewable sources and biological processes have emerged from research and have been established as a commercially viable solution with less adverse effects. Nevertheless, economic and legislative challenges for biobased plastics hinder their widespread implementation. This review summarizes the history of plastics over the last century, including the most relevant bioplastics and production methods, the environmental impact and mitigation of the adverse effects of conventional and emerging plastics, and the regulatory landscape that renewable and recyclable bioplastics face to reach a sustainable future.

Keywords: bioplastics; biobased plastics; plastic degradation; bioplastic sources; plastic business case; bioplastic legislation

1. Introduction

Since the creation of the first synthetic polymers in the 20th century, plastics have become an essential material in human activities due to their valuable characteristics, such as their low cost of production and broad potential applications. It is estimated that 370 billion tons of plastic materials are produced each year, a trend that is not expected to diminish in upcoming years [1]. Plastics are highly praised due to the ease of their production and diversity of applications thanks to their properties of durability, sturdiness, and lightweightedness, which are unmatched by other material categories and the current proposed alternatives [2].

Although there are an estimated 60,000 different plastic formulations, only six polymers account for 90% of the total plastic production: polypropylene (PP), high-density polyethylene (HDPE), low-density polyethylene (LDPE), poly(ethylene terephthalate) (PET), polyvinyl chloride (PVC), polystyrene (PS), and polyurethane (PU) [3]. The most common synthetic polymers are manufactured from petroleum-derived compounds, a nonrenewable resource that has raised concerns over its long-term sustainability, as its fabrication model uses declining resources at increasing prices [4]. Moreover, these resources are not being used efficiently. Single-use plastics consume 1.6 billion liters of oil and are disposed of immediately after use, with a limited possibility of recycling [5].

The lack of management strategies for plastics once they reach the end of their lives, coupled with the inherent durability of polymeric materials, has given rise to the improper disposal of plastic products and the accumulation of plastic-related particles. This issue has been exacerbated by single-use plastics and an increase in the production of personal protection equipment brought on by the COVID-19 pandemic [6]. Plastic is the predominant type of human waste found in freshwater reservoirs, accounting for 57% of the total debris by weight [7]. Discarded plastics subjected to natural degradation turn into secondary microplastics, an emerging contaminant category of concern that often appears in modern research due to its potential effects on the environment and wildlife, as well as its status as a vector for adjunct pollutive particles such as heavy metals and hormone disruptors [8].

The critical impact of plastics on the environment as well as on human health, coupled with the need for sustainable methods to produce synthetic plastic analogues to alleviate the global demand, has been considered by policymakers and scientists in recent years. In the UN Global Sustainable Development goals, Goal #14 specifically mentions the need to mitigate the impact of plastic (Indicator 14.1.1 (b)—extraordinary efforts are required to reduce microplastic discharge into marine and freshwater ecosystems [9]), and another 12 goals relate to the alleviation of the problems created by the plastic industry in the human health and environmental spheres [10]. Nevertheless, the plastic alternatives that are currently presented lack widespread adoption. This is mainly due to their perceived inferiority, the lack of large-scale production processes to sustain their demand, and the higher costs associated with their production [11]. In recent years, bioplastics have been proposed as a promising alternative to conventional plastics because of their similar properties. The fabrication of this branch of polymeric materials has certain advantages over petroleum-based resins, such as their renewable nature and lower costs of industrial-grade production [12]. Bioplastics have recently been adopted for a few applications, such as single-use plastics and 3D-printing materials; however, their sustainability and ecocompatibility have remained a topic of discussion. Their widespread usage has been hindered by technical limitations in the production processes, as they offer lower yields than their petroleum-based parallels [13].

This article summarizes the historical relevance of plastic materials, their adoption and evolution, their current shortcomings, and the emerging trends in biobased plastic manufacturing and implementation. The environmental effects of conventional and emerging plastics are approached, as well as their mitigation strategies and the main applications of recent alternative materials. The challenges and opportunities of these biological plastics are discussed, as well as the relevant market in the context of the modern demands for plastic materials. The relevant legislation that concerns plastic production, certification, and the regulation of its disposal is also briefly discussed.

2. Historical Development of Plastic Materials: A Brief Timeline

The field of plastics is currently extremely prevalent in research and innovation, particularly with relation to new polymeric products, with a focus on their application and performance. Nevertheless, some concerns remain regarding their production, commercialization, and final disposal. The synergy between biotechnology, bioeconomy, and the chemical industry provides possible solutions to these problems [14].

Records of the use of plastics date from as early as a few centuries ago. Ancient civilizations around the world used materials such as resins, insoluble oils, and amber similarly to how we use plastics today. The first reference to rubber concerned cultures native to Central America who congealed latex to create waterproof shoes. In 1839, Charles Goodyear, an American inventor, discovered the elasticity and resistance of rubber heated with sulfur (Figure 1). This process was termed vulcanization and patented in 1844. Subsequently, ebonite was created and became relevant because it was thermosetting and was prepared from a natural material, rubber, though with larger quantities of sulfur. In the 1850s, the conditions for controlling the nitration of cellulose were optimized in Europe. A solid residue was produced from solvent evaporation, which demonstrated elastic and waterproofing properties. In 1862, Parkesine was prepared by the dissolution of cellulose nitrate in minimal solvent [15]. In 1863, two materials that had similar characteristics to Parkesine were developed: xylonite and ivoride. In 1869, billiard balls were being made out of cloth, ivory dust, and shellac in the US. In 1872, the term celluloid was first used to describe a material obtained from cellulose nitrate and camphor. The first protein-based member of the natural polymer family was developed in Germany in 1897 out of casein that was reacted with formaldehyde. Casein was separated from milk by coagulation. The formalized casein was used in buttons, dress ornaments, necklaces, manicure sets, pens, and other decorations. In 1899, ebonite was patented in the UK. It was made by reacting phenol and aldehyde resins and was used as an electrical insulation. In 1927, a nonflammable replacement for celluloid was created, called cellulose acetate, which was extensively used for artificial fibers [16].

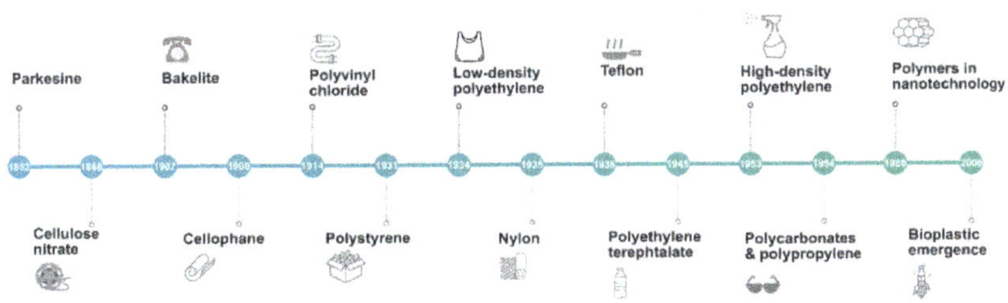

Figure 1. Timeline of plastics from 19th to 21st century. Designed online at Freepik.com [Last accessed 23 February 2022].

Between 1930 and 1940, the current major industrial thermoplastics derived from ethylene were created: polystyrene (PS); poly(vinyl chloride) (PVC); and polyolefins, such as polyethylene (PE), polypropylene (PP), and polymethylmethacrylate (PMMA). During World War II (WWII), PMMA, a rigid, transparent plastic, was used for aircraft glazing and, to a lesser extent, in denture manufacture. The first polyurethane was synthesized by Otto Bayer in Leverkusen, Germany, in 1937 [17]. The diversity of raw materials that could act as sources for polyurethane production, as well as its wide range of uses, enabled the wide application of plastics in settings such as buildings, automobiles, coatings, and sealants [18].

More materials were designed before World War II. One was nylon, a sticky, bendable material, first formulated in 1933. Afterwards, during WWII, polyamide 66 and Teflon were discovered in 1941. During the mid-1950s, high-density polyethylene (PE) was produced, followed by polypropylene (PP). In 1956, polycarbonates were developed in the United States and Germany simultaneously. A variety of copolymers and blends were produced during the period 1960–2000. In the 1960s, there was a growth in the variety of synthetic fibers. Two important fibers to note were Nomex®® and Kevlar®®, the first meta-aramid and para-aramid fibers created by DuPont™. Nomex®® had a higher melting temperature [19], while Kevlar®®· had a structure that allowed it to create composites [20], replacing steel fibers in racing tires and later being used for consumer products and human armor [21].

Polybutylene terephthalate was introduced in 1969, and later, polycyclohexylenedimet hylene terephthalate, which was a plastic polyester with a higher melting temperature. High-performance thermoplastics were introduced during the 1970s and 1980s and could withstand temperatures of above 200 °C. In 1977, polyetherether ketone (PEEK) was invented, followed by polyether sulfone (PES) in 1983. In 1990, polyhydroxybutyrate was commercially produced under name "Biopol". In the late 1990s, ethylene, propylene, and styrene were first introduced [22]. In the 2000s, the field of plastics was focused on materials produced from vegetable sources. Considerable research was carried out in bioplastics such as polyhydroxyalkanoate and poly-lactic acid (PLA) extracted from sugarcane, corn, and rice [23]. They were considered alternatives to petroleum-based plastics and could be biodegraded. Biobased polymers such as bio-polyethylene (bio-PE) and bio-poly(ethyleneterephthalate) (bio-PET) have been produced for its functionality and considerable capacity in packaging application. In 2019, biobased polypropylene was produced at a commercial scale, and its production capacity is expected to quadruple by 2025 [12,24].

3. Environmental Impact and Health Effects of Synthetic Plastics and Bioplastics

3.1. Plastic Degradation and Insertion into the Environment: An Overview

The affordability and desirability of plastic have enabled its ubiquity in all aspects of human development. Plastic production reached 368 million metric tons in 2019 and is projected to reach a total of 1.1 billion metric tons by 2050 [1,25]. Despite recent efforts, local and global policies for plastic disposal and handling after products have reached the end of their lifetimes have lagged behind production and consumption patterns, promoting the unregulated discharge of plastic materials into the ecosystem [26]. The stability conferred by the polymeric nature of plastics hinders their degradation in the environment by natural means. This results in the accumulation of plastics in the ecosystem as well as the macroscopic fragmentation of plastic particles through mechanical erosion, ultraviolet weathering, and biological assimilation, meaning that extended periods are needed for their complete breakdown beyond just their chemical degradation [27].

Once a plastic product is disposed of, its final fate can vary according to its geographical location; the available waste management infrastructure; the economy; and the intrinsic properties of the discarded product, such as its morphology and composition. Plastic waste can either be primed for reuse or recycling, managed through landfilling and incineration, or directly disposed of into the environment [1]. Unlawful means of plastic disposal such as littering and unregulated landfills also represent a significant source of plastic entry into ecosystem elements such as rivers and soil [28]. The plastics' morphology and chemical nature also determine the polluting potential of discarded products. The diversity and complexity of the interactions between plastic sources, compositions, morphologies, entry pathways, and degradative mechanisms hamper the development of a holistic understanding of the global plastic pollution issue, preventing practical and integral actions to mitigate its environmental impact (Figure 2) [29].

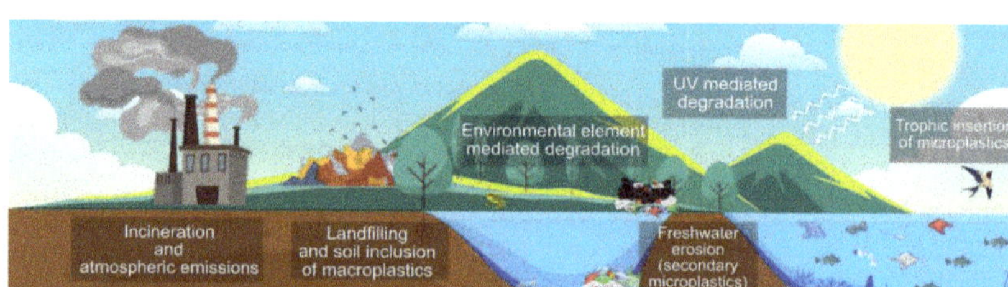

Figure 2. Insertion pathways of plastic-related pollutants into the environment. Designed online at flaticon.com [Last accessed: 23 February 2022].

As mentioned previously, the chemical nature of plastic hinders its quick breakdown in the environment. It tends to accumulate in soil, freshwater reservoirs, and oceans. Plastic permanence is estimated to last hundreds or even thousands of years, depending on factors that facilitate passive ageing and degradation. The onset of polymer decay is triggered by ultraviolet degradation and thermo-oxidative reactions promoted by the sun and environmental components such as beaches and pavements [30]. Plastics may also degrade mechanically through erosion caused by continuous contact with rocks, wind, and water, which shear and tear plastic particles [31]. Finally, plastics may also degrade through biological means, either by the chewing and digestion of macroscopic animals or by the microscopic biodegradation mediated by bacteria, fungi, and other biologic actors (Figure 3) [32].

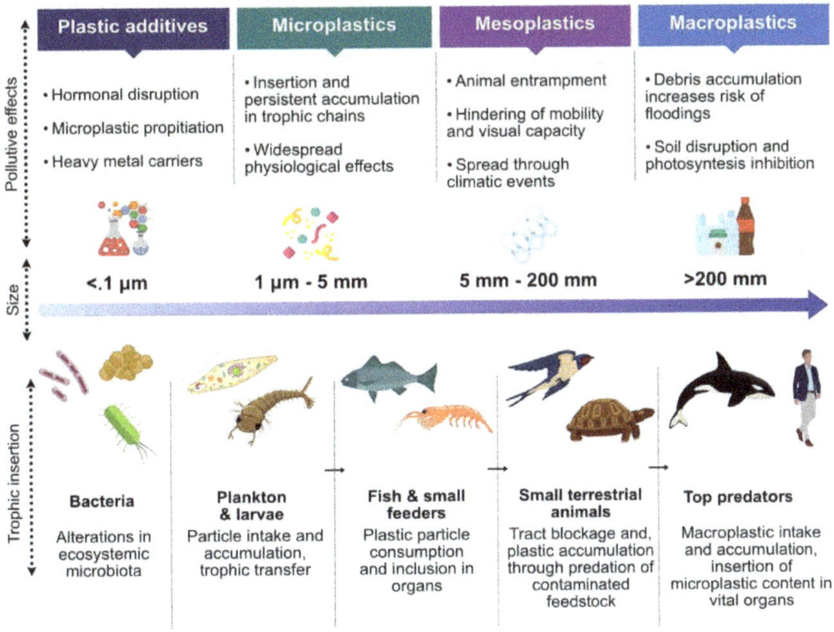

Figure 3. Size, features, and related effects of plastic-derived pollutants in the trophic chain. Designed online at flaticon.com [Last accessed: 23 February 2022].

The natural routes of watersheds transfer plastic waste from soil-based landfills into water bodies. Although pollution models have traditionally considered oceans as the final destination of plastic waste, freshwater networks have recently been recognized as both an active transport vector and a retention platform for polymeric waste [33]. The size and composition distribution of plastic pollutants in rivers is heterogeneous and characterized by retention in riverbeds, sediments, and vegetation reported throughout the world in landmarks such as the Great Lakes [34], the Danube River [35], the Ganges River [36], and the Thames River [37], to cite some examples. Polymer particles may then be ingested or absorbed by organisms, inserting plastics into the ecosystem's trophic chain [38]. The presence of plastic particles in freshwater also poses a threat to human health, as it is the primary source of drinking water worldwide, and the presence of microplastics has been reported in local water-supply networks [39,40]. Oceans represent the major worldwide sink for discarded plastic through natural transport pathways, and this is exacerbated by the indiscriminate product disposal caused by human activities such as fishing, commerce, industry, and tourism. Oceans are thus cited as the ecosystem most affected by human plastic pollution, which impacts the marine biosphere in all degrees and interactions [41]. Marine plastic debris floats and is diffused by global currents, acting as a carrier and a passive modifier of local ecosystems, introducing foreign particles and organisms, and potentially disrupting local environment interactions. Marine plastic is also deposited on beaches and shores. The persistence of marine plastic also favors material inflow as opposed to outflow mechanisms such as degradation, leading to product accumulation, with the Great Pacific Patch being the most well-known marine plastic reservoir in the world [42]. The uptake of plastic particles by living organisms has also been studied and reported at macroscopic as well as microscopic trophic levels; nonetheless, the fate of this uptaken plastic, whether retained in tissues or excreted, is poorly understood [43].

Plastic fragments in soil are accumulated and distributed according to their particle size and the local environment dynamics, becoming a feature that alters soil properties and health. Microplastics can hinder soil density and bulk volume, affecting water retention and promoting the diffusion of other pollutants by creating surfaces with high adsorption coefficients [44]. The presence of plastic particles in terrestrial ecosystems negatively impacts biological developments at macroscopic and microscopic levels. Plastic debris may be ingested by soil-based organisms such as earthworms that act as microplastic concentrators, accumulating in their tissues [45]. Microplastics can thus be found in the terrestrial food supply chain through the ingestion of such organisms by poultry, later leading to negative implications for human health and posing a severe risk of microplastic exposure and ingestion [46]. An investigation conducted by Sun et al. [47] demostrated that polymer particles can also impact soil bacterial consortia, changing the composition of bacterial communities directly related to microplastic concentration and morphology.

The pathways through which plastic and related contaminants enter the atmosphere have been proposed, yet there still remains a knowledge gap. The incineration of plastic products is a common waste-management strategy for energy recovery, along with recycling and landfilling, as these are proposed as low-cost solutions that require minimal space [48]. Nevertheless, plastic incineration leads to the emission of hazardous contaminants and poses a risk to human and animal health while disrupting the environment through the release of carcinogenic and mutagenic compounds [49]. Plastic debris may also enter the atmosphere by mechanisms of natural erosion of predominantly synthetic textiles in the form of microfibres [50]. The fate of airborne plastic particles may vary according to the local ecosystem conditions, and they can be deposited into water ecosystems and soil, where they can be become airborne again in a dynamic cycle [51]. Plastic particles can also be inhaled and deposited in animal respiratory tracts, providing another pathway for trophic accumulation and transport [52].

3.2. Macroplastics

Macroplastics constitute a significant component of human litter and are commonly defined as plastic pieces over 25 mm in size [53]. This class of plastic debris impacts landscapes and ecosystem biology and serves as a source of secondary microplastics [54]. It is estimated that macroplastics kill 1 million marine animals through ingestion and entanglement [55], and the recent onset of the COVID-19 pandemic has exacerbated the disposal of single-use plastic products in the form of personal protection equipment, packaging, disposable cutlery, and containers [56,57].

Macroplastics have a negative impact on the environment due to their several hazardous effects. Depending on their nature, discarded plastic products have the potential to entrap organisms, hindering their mobility and visual capacity [58]. Macroplastics also act as vectors for the conveyance of undesirable ecosystem elements and pollutants through agglomeration and passive transport through water and air currents, exacerbated through climatic events such as rain and hurricanes [59], and may introduce nonindigenous species to foreign biospheres, such as mollusk stocks and algae dispersal through a mechanism known as rafting [60].

Animals may actively ingest macroscopic products which can be regurgitated, excreted, or retained for extensive periods [61]. In the latter case, plastic pieces large enough may cause gut blockage, starvation, or reduced nutrient absorption, leading to animal mortality [62]. Over time, plastic ingestion may act as an evolutionary trap, affecting the viability of animal populations and overall species through the persistent risk of prey resemblance and ever-increasing animal–plastic encounters [63].

Sizeable plastic debris is also likely to accumulate at drainage basin constrictions, causing blockages and leading to increased risks of flooding and raised water levels [64]. In urban settings, the blockage of drainage and sewer systems causes undesirable effects such as bad smells in streets and propitiates the spread of diseases such as cholera and typhoid through contaminated drinking water [65]. In natural ecosystems, water obstructions promoted by plastic accumulation may promote changes in the hydrodynamics of rivers and lakes, altering the immediate landscape by the accidental creation of ponds during rainfall, leading to biofouling [66]. Floods caused by plastic blockages also provide a breeding opportunity for vector arthropods, increasing the risk of infectious viral and parasitic diseases in prolonged inundated zones [67].

Macroplastics can have adverse effects on soil, as they disrupt its physical and chemical properties. Plastic particles on land can alter the floor salinization and the root insertion of neighboring plants, with potential effects on the emergence of crops of agricultural interest [68]. Macroscopic plastic debris has also been evidenced to directly block photosynthesis and entangle plant seedlings, having a considerable impact in plant survivability in ecosystems deeply affected by littering [69]. Soil can also preserve plastic through earth-borne biota, such as earthworms, which are reported to burrow plastic particles that hinder solar-dependent degradation through UV radiation and lead to slow plastic decomposition [70].

3.3. Microplastics

The term microplastic has been commonly used in literature to describe a broad assortment of plastic-derived particles classified according to their size. Although the term is widespread and standard, the complexity and variety of microplastic compositions and morphologies are often overlooked in categorizing these contaminants [71,72]. This simplification has led to the inadequacy of studies and protocols that assess the environmental effect of microplastics, the irreproducibility of strategies for microplastic detection, and the ineffectiveness of pollutant-mitigation governance strategies [73].

Microplastic particles have become ubiquitous in all environments, and their presence has been reported in air, water, and soil ecosystems, as well as in food [74]. The primary sources of entry into the environment are either the release of synthetically designed plastic microparticles included in functional products such as cosmetics, drugs, and pellets (primary microplastics) or by the natural degradation of larger plastic debris through the previously discussed processes (secondary microplastics) [75]. The composition of the plastic in addition to its physical properties, environmental effects, and exposure time define a microplastic particle [30,76]. Its interaction with environmental elements, accumulation, and association with other plastic and pollutive particles, given by its physical properties, define its effects on the ecosystem [77]. Microplastic particles have a variety of shapes, including spheres, fibers, fragments, and films, each related to their source and composition with a varying pollutive potential through their macroscopic and microscopic biological effects [78].

The emission of plastic residues into aquatic environments is estimated to be between 4.8 and 12.7 million metric tons annually [79]. A fraction of such debris is degraded to create microparticles, along with the direct discarding of primary microplastics through water drainages. Once in oceans and freshwater reservoirs, microplastic can be ingested by fish and plankton and be inserted into food chains through trophic exchange into larger predators and humans, ultimately affecting these species through the consumption of contaminated prey [52]. Microplastics can float and drift in the water surface of oceans and rivers, interfering with surface microbiota and inhibiting the photosynthetic efficiency of algal plants, diminishing available nutrients in the marine ecosystem [80]. Degraded microplastic particles can also be deposited in the marine environment. Once there, they act as pollutant liberators, slowly releasing adjunct chemical pollutants such as endocrinal disruptors (BPA), while their further degradation is limited due to the inhibition of ultraviolet degradation that converts river and ocean beds to central microplastic repositories [81,82].

Microplastics possess a large surface area and adsorptive potential, which allow them to interact with other organic molecules, heavy metals, and other microplastics, making them a pollutant vector [83]. In the case of heavy metals, PVC and PS microparticles have been reported to have a solid interaction with copper and zinc, common elements found in chemical additives such as paint [84,85]. Bioactive molecules such as pesticides and antibiotics have also been reported to have sorption activity through microplastic molecules, making them a bigger hazard in estuaries compared to oceans and freshwater due to a higher concentration of contaminant particles [86].

Human life is threatened by microplastics through several means. The ingestion of plastic particles through contaminated food is considered the main route of human exposure and can have molecular effects in cancer, obesity, and other oxidative-stress-related conditions. Additives found in microplastics, such as bisphenol A, are also endocrine disruptors that hinder proper development and have adverse reproductive effects [87]. Microplastic particles can also be inhaled, accumulating in lung tissue and causing cytotoxic effects in pulmonary cells, resulting in bronchial inflammation, fibrosis, allergic reactions, and interalveolar lesions [88]. Human exposure to airborne microplastics is estimated at 272 particles per day and is influenced by the material's nature, the performed activity, the quality of ventilation, and the particular season [89]. It is estimated that the primary microplastic morphology in the atmosphere is microfibers composed of nylon, polyester, and acrylic [90].

3.4. Other Plastic-Derived Pollutants

Besides the polymeric backbone, other chemicals are frequently added to synthetic plastics to heighten the qualities of the final product, such as the color, mechanical features, and stability. The classification of additives is as diverse as that of plastic polymers and includes plasticizers, antioxidants, dyes, lubricants, and fillers [87]. The release of chemical pollutants in conjunction with plastic debris particles may occur at all steps of the plastic lifecycle. The release of such compounds is mediated by the composition of the plastic and additive blend, the dynamic of the plastic particle with the additive, and the interaction of the compound with the environmental features (e.g., leachate solubility) [91].

Plasticizers are frequently employed to improve polymeric mechanical features, such as the flexibility and processability of polymer resins [84]. Over 90% of produced plasticizers are used to enhance PVC properties, expected to reach 59 million tons in 2020 [92,93]. Phthalic acid esters are the most used plasticizers globally, di(2-ethylhexyl) phthalate (DEHP) being the most pervasive plasticizer added to plastic [94]. Phthalates are not covalently associated with the polymeric matrix, and the mechanical wearing of the plastic product allows its migration to food and the environment [93]. Phthalate esters are recognized as a predominant category with contaminant effects, and their occurrence in all aspects of the environment has been comprehensively reported. This category of plastic additives has been recognized as potentially hazardous to the environment and human health through endocrine disruption [95]. DEHP also has a low toxicity threshold (LC_{50} of 0.50 ppm), causing embryonic mortality and necrosis in potentially exposed animals such as zebrafish [96,97].

The deleterious effect of phthalates on human health is documented and includes the alteration of estrogenic levels, infertility, asthma, diabetes, hyperglycemia, and endometriosis [98]. Phthalate exposure has also been associated with an increased BMI and the diagnosis of obesity in children [99]. The effects of DEHP exposure may also increase multigenerational hepatic fibrosis risk through the alteration of regular DNA methylation patterns [100]. Microplastics are considered a source of phthalate emission in the environment, and the high hydrophobicity and surface area of plastic particles enable the adsorption and transport of phthalates into water, soil, and sediment deposition [101]. The most frequent human entry pathway for such a family of compounds is ingesting contaminated foods that have come into contact with phthalate-supplemented packaging [102], followed by microplastic consumption through the drinking of contaminated water sources [103].

Different kinds of technologies can also promote the oxidative degradation of synthetic plastic based on fossil resources. This approach is frequently proposed as a pathway to minimize the environmental impact through the enhanced breakup of plastic resins without compromising their functionality [104]. Pro-oxidative additives are based on the salts of transition metal ions such as Co^{2+}, Fe^{2+}, or Ni^{2+}, which stimulate the degradation of polymeric chains in the presence of oxygen and heat or radiation [105]. The most widely used commercial oxodegradable additives are d2w and TDPAs (totally degradable plastic additives); both are based on a combination of manganese, iron, cobalt, and nickel salts [106]. Oxodegradable plastics have enjoyed widespread adoption recently due to the technical feasibility of their implementation in already established manufacturing processes [107].

Concern has been raised over using transition metal compounds as additives designed to degrade environmental elements. Oxoplastics may then contribute and accelerate the production and accumulation of microplastics worldwide, and legislative regulations have recently been imposed due to the ambiguous nature of the conditions needed to degrade the modified resins completely [108]. The incomplete fragmentation of oxodegradable plastics also challenges their proenvironmental viability, as studies conflict with the true degradative capabilities of discarded oxoplastics. The usage of pro-oxidative additives does not ensure the complete degradation of plastics and may require special conditions nonreplicable in the environment [109]. However, transition metal salts have been regarded as safe due to their design as fatty acid salts of potential nutrient microelements [110].

Research has argued that the degradation pathway of oxoplastics may be further assisted by biological mechanisms. The term oxo-biodegradable plastics refers to the two-stage degradation of additive-enhanced polymers. The first stage of degradation occurs by physical phenomena such as ultraviolet light and erosion, and the resulting plastic particles are then degraded by microorganisms through enzymatic pathways [111]. The addition of an oxodegradable additive has been found to enhance complete biodegradation with no toxicological effects on the obtained compost [107]. The resulting oligomers from photo-oxidized plastics have also been used as a carbon source for bacterial β-oxidation. Such a phenonmenon was observed when Rhodoccocus rhodochrous was observed to assimilate polyethylene film oligomers, leading to their biodegradation [112]. Both degradation mechanisms are considered biologically compatible, and the degradation of oxoenabled waste can be accelerated by metabolic action through compatible enzymes [113]. The findings of such reports show that the use of oxidant additives to promote plastic degradation may be justified from a combined physical and biological mechanism standpoint.

The Parliament of the European Union has carried out a ban on oxodegradable plastics under the "lack of evidence that oxo-degradable plastics are fully degraded in a reasonable time, are not appropriate for recycling or composting, and there is a risk that small plastic pieces will not completely biodegrade" [104]. Bioplastics are a better alternative in terms of complete product degradation versus oxo-bioplastics [114].

4. New Generations of Plastics

Bioplastics are polymers that can be biodegradable or not. Additionally, they can be synthesized from renewable biological sources such as bacterial, plant, and algal sources. However, their biodegradability depends on the chemical structure, but not on the sources used [115,116]. Nowadays, bioplastics represent only around one percent of the total plastic produced annually. However, it is expected that the demand for bioplastics will increase [117]. Figure 4 shows the current stages of bioplastic production, from the research and development stages to the commercial stage.

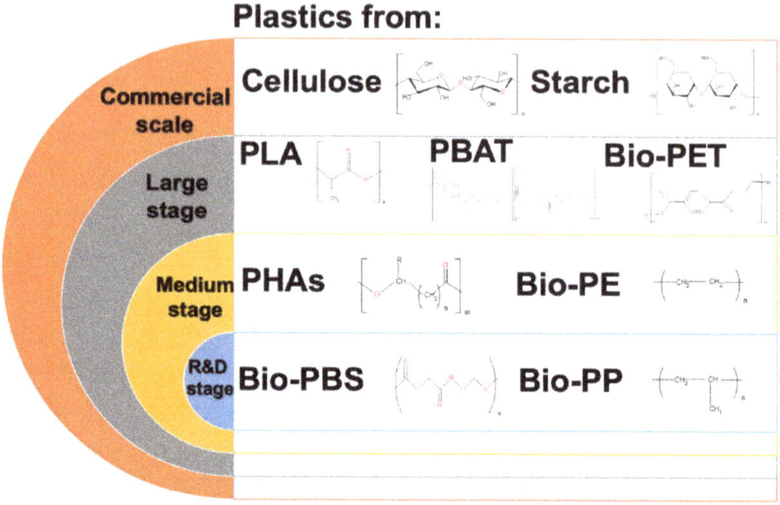

Figure 4. Stages of bioplastic production. PLA—polylactic acid; PBAT—poly(1,4-butylene adipate-co-1,4-butylene terephthalate); Bio-PET—bio-poly(ethylene terephthalate); PHAs—polyhydroxyalkanoates); Bio-PE—bio-polyethylene; Bio-PBS—bio-polybutylene succinate; Bio-PP—bio-polypropylene.

Bioplastics are generally reported in three categories (Figure 5): (i) fossil-resource-derived but biodegradable, (ii) partially biobased or biobased and non-biodegradable, and (iii) biobased and biodegradable [118,119].

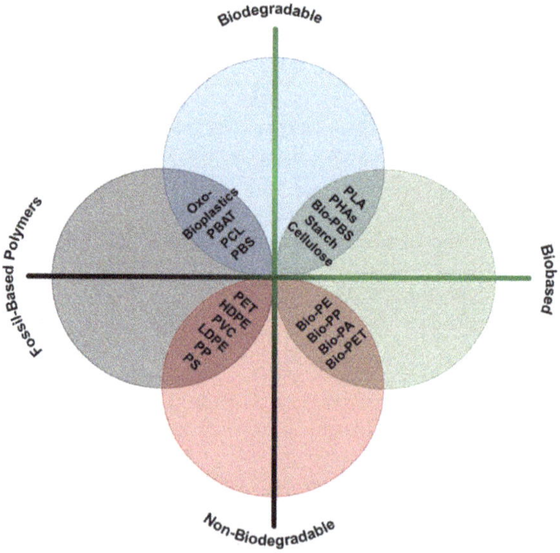

Figure 5. Scheme of bioplastic classification. PLA—polylactic acid; PHAs—polyhydroxyalkanoates; Bio-PBS—bio-polybutylene succinate; Bio-PE—bio-polyethylene; Bio-PP—bio-polypropylene; Bio-PA—bio-polyamide; Bio-PET—bio-poly(ethylene terephthalate); PET—poly(ethylene terephthalate); HDPE—high-density polyethylene; PVC—polyvinyl chloride; LDPE—low-density polyethylene; PP—polypropylene; PS—polystyrene; PBAT—poly(1,4-butylene adipate-co-1,4-butylene terephthalate); PCL—polycaprolactone; PBS—polybutylene succinate.

4.1. Biodegradable Plastics from Fossil Resources

Generally, synthetic polymers are obtained from crude oil or natural gas. Most of these polymers are not biodegradable. Nevertheless, degradability can be achieved by integrating unstable bonds (e.g., amide, ester, or ether) [120]. Examples of biodegradable fossil-based plastics are poly(1,4-butylene adipate-co-1,4-butylene terephthalate) (PBAT) and polycaprolactone (PCL). PBAT is a polyester that is synthesized by polycondensation from the combination of dicarboxylic acids and diols. PBAT, due to its aliphatic unit in its molecule chain, has good biodegradability and excellent mechanical properties [121]. PCL is a hydrophobic and biodegradable polymer with sufficient mechanical strength and flexibility [122]. It has excellent biocompatible properties that enable its use for medical applications. To enhance their properties, PCL and PBAT can be used in blends with biobased and biodegradable materials such as polylactic acid (PLA) and polyhydroxyalkanoates (PHAs), and polybutylene succinate (PBS) [120].

Oxo-biodegradable plastics are composed of petroleum-based polymers such as polyethylene (PE) with pro-oxidant additives that promote the degradation process [14]. The materials used as additives are generally transition metals such as iron, manganese, cobalt, and nickel. The additive's function is to break down the large molecular polymer chain into smaller fragments that microorganisms can process and convert into biomass and carbon dioxide (CO_2) [104,123]. The degradation of these plastics can be initiated by UV light, moisture, heat, and microorganisms. Usually, the degradation can be evaluated by measuring the changes in physical properties such as the loss of molecular weight, the amount of CO_2 evolved, and the microbial growth on the polymer surface. The degradation of oxo-biodegradable plastics generally takes months to years; however, this length of time is occasionally unpredictable, because it depends on climate factors such as temperature and the intensity of solar radiation. The most commercially successful additives used as pro-oxidants are totally degradable plastic additives (TDPAs), Renatura, AddiFlex, d2W, and Reverte. AddiFlex is employed to produce single-use plastic bags. Reverte is used for bottle production, and different companies use d2W for their consumables, such as Pizza Hut, Walmart, and KFC [104].

4.2. Biobased Non-Biodegradable Plastics

Biobased plastics that are not degradable represent a group of biopolymers such as bio-polyethylene (bio-PE), bio-poly(ethylene terephthalate) (bio-PET), bio-polyamides (bio-PAs), and bio-polypropylene (bio-PP) [115]. They can be synthesized from biobased sources and offer a nearly equal chemical structure and properties to their fossil equivalents. Among them, bio-PET is the most widely produced bioplastic [124]. Since 2010, the precursors of PET, ethylene glycol and terephthalic acid (TPA), have been obtained from biological sources. Nowadays, the Toyota Tsusho Corporation, Japan and Futura Polyesters, the Coca-Cola–Gevo Venture, and the PepsiCo–Virent Venture are the principal producers of bio-PET [115]. Bio-PE is obtained from biological resources by the dehydration of bioethanol, obtained from glucose. Different natural feedstocks can be used to obtain glucose, such as maize, wheat, sugar cane, and sugar beet. The biopolymer is identical in its chemical, mechanical, and physical properties to fossil-based PE [115]. According to European Bioplastics [117], biobased but non-biodegradable plastics represent almost 40% of the global bioplastic production capacity; nevertheless, it is predicted that in the following years their participation will be lower.

4.3. Biobased and Biodegradable Plastics

4.3.1. Starch

Starch is an essential polysaccharide that plants synthesize and store in their structure as an energy reserve. It is considered one of the best biopolymers with extraordinary potential, because it is biodegradable, renewable, and available in huge quantities at a low cost. Starch could be extracted from different sources such as corn, wheat, potato, rice, tapioca, tam, and barley. However, most starch is produced from maize [104,125]. Starch polymers are divided into two principal forms, amylose and amylopectin; the composition of these components can affect the properties of starch-based films [126]. Native starches have limits in their mechanical properties, thermal stability, and brittleness; that is why plasticizers (sorbitol, glycol, and glycerol) are necessary to improve the bioplastic functionality, transforming starch into thermoplastic starch [104]. Additionally, to enhance starch's characteristics, attempts have been made to mix it with synthetic polymers, lignocellulosic biomass, or agricultural waste (Table 1) [125]. Different studies demonstrate the applicability of starch bioplastics in shopping bags, food packaging, agriculture, and medicine.

Table 1. Biopolymer production from plants.

Biopolymer	Source	Reinforcement	Plasticizer	Reference
Starch	Corn and cassava	*Cola cordifolia*	Glycerol	[127]
	Rice and corn	Ethanol, rice, and olive oil	Sorbitol	[128]
	Tapioca	Sugarcane bagasse fiber	Glycerol	[129]
	Banana peel		Glycerol	[130]
	Corn, potato, and cassava	Recycled newspaper pulp fiber	Glycerol	[131]
	Cassava	Microcrystalline cellulose	Sorbitol	[132]
	Tapioca		Acetyl Tributyl Citrate	[133]
	Corn	Microalgae *Nannochloropsis*	Glycerol	[134]
	Microalgae *Spirulina* sp.	Poly(vinyl alcohol)		[135]
	Microalgae *Chlamydomonas reinhardtii* 11-32A		Glycerol	[136]
Cellulose acetate	Cotton linters		Polyethylene glycol 600	[137]
	Flax fibers		Polyethylene glycol 600	[137]
	Parthenium hysterophorus weed		Polyethylene glycol 600	[138]

4.3.2. Cellulose

Cellulose is a polymer which is renewable and is present in large quantities. It is widely used because it is inexpensive and biodegradable. Cellulose is a polysaccharide chain that contains D-glucopyranose units joined by β-1,4-glycoside linkages. Cellulose from plants can be found from wood and non-wood plant lignocellulosic biomasses. It is the main constituent of plant fiber [139]. Cellulose is a potential film material with a highly crystalline structure; however, it is not soluble in water or common organic solvents. For this reason, it needs to be modified to transform it into water-soluble materials [139]. Cellulose acetate is a derivative of cellulose obtained from a chemical modification of cellulose. It is extensively used in the production of membranes, cigarettes, and food packaging [138].

Although cellulose is a plant material, some bacteria can produce cellulose, which is known as bacterial cellulose (BC) [139]. This is an interesting material due to its unique properties, making it an ideal candidate for industrial-scale production [140]. It has excellent mechanical strength and degradability. Additionally, it is purer than natural cellulose, and it has a higher water-holding capacity. These characteristics make it a promising natural polymer with multiple applications in medicine, food, electronics, and other fields [141]. Few companies produce this polymer on a large scale, such as Bowil Biotech (Poland) [140]. However, the high-cost production has limited its applications. Bacterial cellulose can be synthesized by multiple bacteria species, such as *Acetobacter* sp. and *Gluconacetobacter* sp. [141]. Using by-products and wastewater as raw materials for the synthesis of biopolymer is a strategy for large-scale production at a low cost (Table 2) [142]. The production of BC from agricultural and industrial waste uses food wastes [141,143,144], wastewater, and crude effluents [142,145–147]. An important advantage of using these residues is the reduction in both the cost of production and human waste [146].

4.3.3. Polybutylene Succinate

PBS is an aliphatic polyester that is synthesized by the condensation of succinic acid and 1,4-butanediol. Traditionally, the monomers were synthesized from fossil resources; however, butanediol can be prepared from renewable sources such as renewable biomass resources and succinic acid from sugar fermentation (Table 2) [148]. Furthermore, PBS has good mechanical properties comparable to PE and PP, resulting in multiple applications for food packaging, shopping bags, agriculture mulch film, and hygiene products [120]. Nevertheless, the production cost is high, so future developments are needed to make it more economically viable [51].

4.3.4. Polylactic Acid

Polylactic acid (PLA) is an aliphatic polyester produced by chemical or biological methods from cellulosic biomass [120]. It is biodegradable, and its mechanical properties are equivalent to those of petroleum-based plastics such as PET and PS, in addition to properties such as biocompatibility, innocuity, and compostability. The Food and Drug Administration (FDA) allowed the direct contact of the material with food. PLA is produced from lactic acid through the fermentation of renewable resources such as rice, corn, and potatoes. The fermentation is performed principally by lactic acid bacteria; however, different microorganisms have also been used (Table 2) [104]. PLA can be composed of D- or L-lactic acid isoforms or by both. It is necessary to obtain both isomers in separate processes [149], as the crystallinity and thermal stability of the final product depends on the isomer proportion.

PLA is probably the most well-known bioplastic and can be synthesized from renewable and natural resources. Its high manufacturing cost must be reduced in order to produce more bioplastics. Several works in the literature have reported the use of lignocellulosic wastes and food-derived wastes, such as spent coffee grounds [150], fruit wastes [149,151,152], alfalfa silages [153], and corn cobs [154] (Table 2). Despite PLA being extensively adopted, it still has drawbacks such as its cost and its structural breakability and fragility. Different approaches have been studied to improve its properties, such as the incorporation of nanoparticles [155].

4.3.5. Polyhydroxyalkanoates

PHAs are natural aliphatic polyesters that consist of R-3-hydroxyalkanoate groups. There are synthesized by fungal and bacterial strains [156] from different substrates such as industrial by-products, oils, fats, lignocellulosic raw materials, agroindustrial waste materials, sugars, and wastewater (Table 2) [157]. PHAs are less porous and therefore trap less O_2, CO_2, and H_2O, making them suitable for the production of packing materials such as films, coatings, bottles, and bags [156]. Additionally, PHA has good thermal-mechanical characteristics similar to synthetic polymers; therefore, it can substitute for PE and PP [118]. PHBs are one of the most studied polyhydroxybutyrates [158]. PHAs are commercialized by different companies, for example, Metaboli (Woburn, MA, USA); Procter & Gamble Co., Ltd. (Cincinnati, OH, USA); Tianjin Green Bioscience Co., Ltd. (Tianjin, China); Bio-on (Emilia Romagna, Italy); Biocycle PHB Industrial S.A. (Serrano, SP, Brazil); and Goodfellow Cambridge, Ltd. (Huntingdon, UK). The commercialization of this biopolymer has been increased due to the progress in purification technologies. However, in 2020, PHA biopolymers represented only 1.70% of the worldwide production of bioplastics [117]; the high production costs represent the principal reason for the limited production. To make the production of PHA competitive, it is necessary to use low-value waste materials as a substrate [159]. Different waste streams have been evaluated (Table 2), such as waste vegetable oil [160,161] and fruit waste [159,162,163].

Table 2. Biopolymer production by microorganisms using agroindustrial residues as a substrate.

Biopolymer	Microorganism	Production Scale	Employed Substrate	Productivity	Reference
Bacterial cellulose	*Gluconacetobacter xylinum BC-11*		Wastewater	1.77 g/L	[147]
	Gluconacetobacter xylinus		Wastewater	0.659 g/L	[145]
	Komagataeibacter saccharivorans	Static production in flasks	Crude distillery effluent	1.24 g/L	[142]
	Gluconacetobacter oboediens	1 L	Crude distillery effluent	0.85 g/100 mL	[164]
	Gluconacetobacter sucrofermentans B-11267	Flask	Whey	5.45 g/L	[143]
	Gluconaceter xylinus BNKC19		Pineapple peel	12.3 g/L	[141]
	Gluconacetobacter xylinum CGMCC No.2955		Wastewater of candied jujube-processing industry	2.25 g/L	[146]
	Bacillus cabrialesii		Grass straw, grass husk, wheat husk, and corn cobs		[144]
PHA [1]	*Pseudomonas putida KT2440*	[4] BB (3 L)	Waste vegetable oil	1.91 g/L	[160]
	Pseudomonas chlororaphis 555	Pulse-fed batch fermentation (5 L)	Waste cooking oil	13.87 g/L	[161]
	Pseudomonas resinovorans	[4] BB (15 L)	Grease-trap waste	0.41 g/g maximum mcl-PHA[2] 61.8%	[165]
	Pseudomonas chlororaphis subsp. *Aurantiaca*	[4] BB (2 L)	Diluted fruit pulp waste	0.15 g/g maximum mcl-PHA[2] 49%	[166]
	Halomonas campisalis MCM B-1027	[5] SF 250 mL	Banana and orange peel	0.329 g/L (banana) 0.11 g/L (orange)	[162]
PHB [3]	*Bacillus cereus*	[5] SF 250 mL	Grape peel	0.53 g/L	[159]
	Bacillus subtilis	[5] SF	Papaya and orange peels	11.65 g/L (papaya) 9.68 g/L (orange)	[167]
	Klebsiella pneumoniae	[5] SF 125 mL	Watermelon, papaya, orange, and banana peels	22.61 g/L 23.72 g/L 23.38 g/L 25.11 g/L	[163]
PHB [3] and mcl-PHA [2]	*Cupriavidus necator, Pseudomonas citronellolis*	[4] BB (10 L)	Apple pulp waste	3.03 g/L	[168]
L-lactic	*Bacillus coagulans*		Sugarcane bagasse	1.7 g/L·h	[152]
	Bacillus coagulans		Corn cob residue	79 g/L	[169]
	Enteroccus mundtii	350 mL flask	Spent sulfite liquor	56.3 g/L	[170]
Lactic acid	*Bacillus subtillis and Lactobacillus buchneri*		Alfalfa silage	44.2 g/L	[153]
	Lactobacillus bulgaricus, Strepto- coccus thermophilus, Lactobacillus acidophilus, Lactobacillus plantarum, and Lactobacillus casei.	1000 mL bottles	Swine manure with apple waste	28 g/L	[151]
	Lactobacillus rhamnosus B103		Dairy industry waste	143.7 g/L	[171]
D-lactic acid	*Lactobacillus delbrueckii* ssp. *delbrueckii CECT286*	[4] BB (1 L)	Orange peel wastes	6.72 g/L·h	[149]
	Saccharomyces cerevisiae		Spent coffee grounds	13.4 g/L	[150]

[1] PHA—polyhydroxyalkanoates; [2] bmcl-PHA—medium-chain-length polyhydroxyalkanoate; [3] PHB—polyhydro xybutyrates; [4] BB—batch bioreactor; [5] SF—shake flask.

4.3.6. Other Natural Sources

Seaweed is an excellent alternative for bioplastic production because seaweeds can grow fast and are easy to harvest and cheap. To improve their properties, they can be mixed with other species or materials [172]. Additionally, the use of seaweeds for bioplastics can reduce the impact on the food chain [123]. Standard methods include washing, milling, drying, alkalinization, acidification, neutralization, filtration, and precipitation to produce seaweed bioplastics. Nevertheless, the traditional methods are expensive and have a low yield. For these reasons, green production methods have gained attention due to their high potential and viability [172]. Seaweed polysaccharides can be used for food industry applications, packaging materials, and coatings [123]. Microalgae also represent a promising feedstock for bioplastic production due to their fast growth rates. Numerous researchers have implemented the use of whole microalgae cells such as *Spirulina* sp. [135], *Chlamydomonas* sp. [135], and *Scenedesmus* sp. [173].

Similarly, fungal mycelia from seed peels and corn stalks were used to produce biodegradable packaging and tilling; the mycelia were composed of polysaccharides, chitin, proteins, and lipids. A New York company named Evocative and the Swedish company IKEA have used this bioplastic [123,174]. Chitin and cellulose from crab shells and fiber trees, respectively, have been used to contain liquids and foods [175].

5. Green Industry of Plastics

5.1. Global Market, Business Cases, and Applications

Currently, plastic products are the material of choice for different industrial, medical, and personal applications, among others. Most plastic products are derived from petrochemicals and are single-use, representing losses of 95% of the material value, which is reflected in annual economic losses between USD 80 and 120 billion [117]. To ensure greater environmental sustainability, it is necessary to exchange non-biodegradable plastics of petrochemical origin for biodegradable and compostable biobased materials. Bioplastics are attractive alternatives due to their rapid biodegradation in the environment into CO_2 and H_2O, causing fewer negative effects on the environment [176]. Environmental awareness about the impact of the use of plastics of petrochemical origin has increased both at the business and societal level; however, for economic reasons, bioplastics only represent about 1% of the 360 million tons that are produced annually. The advancement of knowledge and the development or improvement of biopolymers allows the reduction in production costs, and in turn the increased demand for bioplastics is generating an annual growth of 20–30% [177].

Currently, the production of bioplastics uses mainly carbohydrates from plants such as sugar cane or corn as a raw material. Nevertheless, the use of these crops, called primary crops, for bioplastics is in competition with the safety and economic stability of the food industry. Another strategy to produce bioplastics is the use of non-food raw materials, known as second-generation, such as cellulose from crop residues such as corn and wheat stubble, and third-generation raw materials such as micro- and macroalgae. The current production of bioplastics, 2.11 million tons, translates into approximately 0.7 million hectares of arable land, which represents 0.02% of the world's agricultural area. However, the growing demand for bioplastics will require a greater cultivation area, which will generate greater food competitiveness and is why it is important to implement the sustainable production of bioplastics through effective biotechnological processes from food waste, non-food crops, and cellulosic or algal biomass [117].

The production of bioplastics began due to the need to replace the non-biodegradable plastic materials of single-use products, such as bags, plates, bottles, plastic films, packaging, containers, and cutlery. However, only in recent years has this transition increased due to the decrease in the production costs of biodegradable biopolymers derived from new technological advances and new renewable sources for the production of biopolymers, in parallel with public policies and social conscientization, which have made it possible to produce economically viable bioplastics. Table 3 summarizes the properties of biopolymers in a relation to their applications. Biopolymers on their own do not provide the properties demanded by the market. The costs per kilogram were estimated according to the literature [176].

Table 3. Commercial applications of biobased polymers and their properties.

Biopolymer	Applications	Properties	Cost USD/kg	Reference
Starch	Translucent film, net packaging, bags, containers, egg boxes, sandwich bags, capsules, carrier bags, drinking straws, drug-release films	Sealable, durable, fine finishing, barrier for water	0.5–2.0	[118]
Cellulose	Packaging films, films, transparent films, barrier films, cups for cold drinks, plates and dishes, cups for hot drinks, labels	Sealable, barrier for water, transparent, approved for direct food contact	1.8–4.0	[118]
PLA [1]	Bottles, cups, transparent films, containers, dishes, fruit nets, top-covering films, trays, tea bags, ice cream cups, carrier bags	Approved for direct contact, transparent, sealable, durable, barrier for water and oxygen	4.0–6.0	[118]
PHA [2]	Disposable cups, plates, and cutlery; Tetra Pak covers; tubes to produce vegetable seedlings; agrochemical packaging; textile fibers; electronic equipment components	Physical properties like conventional plastics; insoluble in water, nontoxic, and biocompatible; present piezoelectric properties; some PHA films exhibit gas-barrier properties	2.4–5.5	[157]
Bio-PE [3]	Food packaging, cosmetics, personal care, automotive and toy applications	Equal in its chemical, physical, and mechanical properties to fossil-based PE	2.3	[115]
PBS [4]	Biopackaging, tissue-engineering, and medical materials; agriculture mulch film; plant pots; hygiene products	High processability, good mechanical properties, thermal properties	4.0–10.0	[120,178]
PLC [5]	Drug delivery systems and tissue-engineering scaffolds	High toughness and flexibility, biocompatibility, and slow degradation in in vivo conditions	4.5–10.0	[120]
PBAT [6]	Compostable organic waste bags, agricultural mulch films, packaging (wrapping) films, disposable tableware	Excellent toughness, improved wear and fracture resistance, good chemical resistance to water and oils, high strain at break	3.8–5.8	[120]

[1] PLA—polylactic acid; [2] PHA—polyhydroxyalkanoates [3] Bio-PE—biobased polyethylene; [4] PBS—polybutylene succinate; [5] PCL—polycaprolactone; [6] PBAT—polybutylene adipate terephthalate.

Throughout the world, various companies have been established that develop and produce biodegradable biopolymers/bioplastics for different applications, which in turn has created greater competitiveness in the bioplastics market to reduce the costs of biodegradable biopolymers (Table 4). The most widely produced biopolymers are mainly PLA, PHA, and starch, which have been used in various applications from packaging to bioplastic materials for toys, parts for the automotive industry, construction, electronics, and agriculture, among others.

The Qmilk®® company has developed one of the most innovative biopolymers, managing to produce a 100% natural polymer from whey with antimicrobial, biodegradable, and compostable properties; it is also a flame-retardant with a low density and high hydrophobicity. This product is created by casein and forms an excellent barrier to gases such as oxygen and CO_2 and aromas. All these properties give it a great advantage against other biopolymers for food and grocery packaging and textile-fiber production with silklike characteristics [179].

Table 4. Commercial bioplastics and business cases.

Company	Bioplastic	Applications	Properties	Country
Plantic®®[1]	Starch	Food and goods packing, agricultural plastics	Biodegradable and compostable	Australia
Mater-Bi®®-Novamont[2]	Starch	Bags, toys, food, and cosmetic containers	Biodegradable and compostable	Italy
BIOPAR®®[3]	Starch	Bags and flexible packaging	Biodegradable	Portugal
Biofase®®[4]	Starch-based	Cutlery	Biodegradable	Mexico
Solany®®[5]	Starch-derived	Flowerpots, tomato clips, cultivation tubes, promotional items, toys, CD and DVD trays, protection covers for packaging, cup holders, plant stakes, golf tees	Biodegradable and compostable	Canada
Bionolle Starcla™-Showa Denko[6]	Starch- and PLA-based	Bioplastics	Biodegradable and compostable	Japan
BIOFRONT-Teijin[7]	Stereocomplex PLA[13]	Automotive, films and packaging, molded parts for civil engineering and construction, parts for electronic devices	Biodegradable	Japan
Ingeo™-Nature Works[8]	PLA[13]	Bottles, gift cards, durable goods, films, layers of paper, cups and containers for food, fabrics, clothing, disposables, and base material for many compounds	Biodegradable and compostable	USA
WeforYou[9]	PLA[13]	Reusable bags	Biodegradable and compostable	Austria
Total-Corbion[10]	PLA[13]	Biopolymer	Biodegradable and compostable	Netherlands/Thailand
Danimer Scientific[11]	PHA[14]	Straws, cups, lids, bottles, produce bags, shopping bags, cutlery, diaper linings, plates, wipes, toys, trash bags, seals, labels, glues, and much more	Biodegradable and compostable	USA[15]
Qmilk[12]	Milk protein	Textile fibers	Compostable	Germany

[1] Website: https://plantic.com.au/ (accessed on 3 February 2022); [2] Website: https://materbi.com/ (accessed on 3 February 2022); [3] Website: https://unitedbiopolymers.com/ (accessed on 3 February 2022); [4] Website: https://biofase.com.mx/ (accessed on 3 February 2022); [5] Website: https://solanylbiopolymers.com/ (accessed on 3 February 2022); [6] Website: https://www.sdk.co.jp/ (accessed on 3 February 2022); [7] Website: https://www.teijin.co.jp/ (accessed on 03 February 2022); [8] Website: https://www.natureworksllc.com/ (accessed on, 3 February 2022); [9] Website: https://weforyou.pro/ (accessed on 3 February 2022); [10] Website: https://www.total-corbion.com/ (accessed on, 3 February 2022); [11] Website: https://danimerscientific.com/ (accessed on 3 February 2022); [12] Website: https://www.qmilkfiber.eu/ (accessed on 3 February 2022). [13] PLA—Polylactic acid; [14] PHA—Polyhydroxyalkanoates; [15] USA—United States of America.

Biofase®® is a Mexican company that has also stood out for the development of biobased biopolymers with avocado seed starch. This company uses an agroindustrial waste widely produced in Mexico to manufacture single-use bioplastics for applications such as cutlery, straws, and food containers [180].

Ingeo®® is a biopolymer developed by NatureWorks®® based on environmental sustainability and a circular economy, using greenhouse gases such as CO_2 as a raw material to generate a polymer that is biodegradable and compostable. Ingeo®® polymers are currently obtained from sugars produced by plants such as cassava, sugar cane, corn, and beets, which function as CO_2-sequestering plants, making them sustainable. These polymers are produced by a chain of technological bioprocesses that convert sugars into PLA and are used to create different products such as coffee capsules and electronic components [181].

The development and production of different polymers from biological origins with biodegradable and compostable characteristics have been carried out worldwide, but with greater impact and production in developed countries such as the US and European and Asian nations. Mexico is an example of a developing country where there are still a few limitations to the use of petrochemical plastics. However, the change of some single-use products (bags and packaging) to biobased or oxo-biodegradable products that have a lesser impact on the environment is the way towards a sustainable use of plastics.

Advances in the production chains of biopolymers and the environmental and social commitments of companies have been the result of public policies that limit the use of fossil-derived plastics to encourage a switch to the new generation of plastics. However, with these great changes come new challenges and a need to enhance the production of biopolymers.

5.2. Strategies for Plastic Reinsertion and Environmental Impact Mitigation

The recycling and reuse of plastic waste creates new challenges for strategies of collecting, separating, and treating materials of different chemical natures in mixed-waste streams [182]. The diversity of discarded plastic materials and the loss of mechanical properties in recycled polymers have given raise to alternative treatment strategies, such as chemical recycling, which turns plastic waste into chemical products of high value [183]. Biological routes of plastic waste degradation allow its reinsertion into natural carbon cycles, mitigating its environmental impact [184]. Current trends in modern plastic material design, such as biobased plastics, integrate recycling and degradation pathways in the polymer composition [185]. Recent chemical and biological strategies for assimilation are discussed in Table 5.

Table 5. Chemical and biological strategies for degradation.

Plastic	Biodegradation Conditions (Chemical/Biological)	Biodegradation	Reference
Cassava-based bioplastic	Burial-soil pH measurement, 14 days (using microorganisms)		[127]
Starch (TPS)–PLA [1]	*Ulomoides dermestoides*, 5 days	TPS biodigestion—biodegradation (80%) and PLA biodisintegration (50%)	[186]
PHA [2]	Alluvial-type soil, 35% soil moisture, 60 days	35%	[187]
HDPE [3] LDPE [4] PP [5] PS [6]	Incubation with microbial consortium, 357 days	15% 4.96% 6.7% 5.29%	[188]
PHB [7] PHBV [8] PBS [9] PCL [10] PLA [1]	Incubation under standard test aerobic and anaerobic conditions	Aerobic conditions, 117 days PHB [7] 83% PBHV [8] 87.4% PCL [10] 77.6% Anaerobic, 77 days PHB [7] 83.9% PBHV [8] 81.2%	[189]
PET [11]	PET[7] hydrolase enzyme, 10 h	90%	[190]
PET [11]	Recombinant bacterial polyester hydrolase TfCut2, expressed in *Bacillus subtilis*, 70 °C, 96 h	50%	[191]
LDPE [2] and HDPE [3]	Thermal degradation (pyrolysis), 30 to 550 °C at 5 °C min^{-1}	1-oleofins and n-paraffins if C2–C6 were the major products	[192]
LDPE [2] and PS [6]	Pyrolysis, 300–500 °C, nitrogen pressure of 0.3 MPa	LDPE [2] was degraded to oil at 425 °C PS [6] was degraded at around 350 °C	[193]

Table 5. *Cont.*

Plastic	Biodegradation Conditions (Chemical/Biological)	Biodegradation	Reference
PS [6]	Pyrolysis, room temperature 800 °C under inert atmosphere	70%	[194]
Polyethylene (HDPE) [3] pellets	Thermal pyrolysis, 350 °C	81%; the oil consisted mainly of paraffinic hydrocarbons, most of which contained between 6 and 16 carbon atoms	[195]
Cellulose	Enzymatic degradation (endoglucanases, β-glucosidases, endoxylanases, β-xylosidases, mannosidases), 7 days	0.5% (*w/v*)	[196]
PCL [10]	Enzymatic degradation (external PCL [5] depolymerase), 10 days	>80%	[197]

[1] PLA—polylactic acid, [2] PHA—polyhydroxyalkanoate, [3] HDPE—high-density polyethylene; [4] LDPE—low-density polyethylene; [5] PP—polypropylene; [6] PS—polystyrene; [7] PHB—poly(3-hydroxybutyrate-co-3-hydroxyvalerate); [8] PHBV—poly(butylene succinate); [9] PBS—poly(butylene adipate-co-terephthalate); [10] PCL—poly(ε-caprolactone); [11] PET—polyethylene terephthalate.

6. Challenges and Opportunities for Biodegradable Plastics from Production to Degradation, and Further Perspectives under Circular Economy

The world demand for plastics has grown rapidly in recent years due to their economic advantages, properties, and applications, with the packaging sector representing about 40% of the demand. As a result of this demand, the production of plastics is growing, while recycling systems are limited [198]. For a long time, there was no need to recover used plastic, since there was not a large demand for these raw materials, the costs were very low, and the impact of the presence of plastics in the environment had not been determined or recognized [199].

In recent years, challenges have arisen regarding the scale of production, the cost of production, industrial implementation, recycling, and the effects of degradation and contaminants derived from plastics, as well as issues relating to their regulation. The scale of the production of biodegradable plastics is small compared to plastics from fossil sources. The processes for the production of plastics from agroindustrial wastes such as corn crops and sugarcane bagasse are limited due to issues relating to the logistics of collecting, drying, and storing the raw materials. For example, in PLA production, optimization requires the selection of a microorganism that maximizes the yield [200]. Additionally, the microbial fermentation of lactic acid is associated with high costs derived from the pretreatment of the agroindustrial wastes and their conversion into fermentable sugars for the downstream biotechnological process. The optimization of cost–performance sustainability for bioplastics produced from organic wastes is needed before there can feasibly be a switch from fossil-based plastics to bioplastics [201].

The issues in the industrial implementation of biodegradables plastics are very clear in the food sector. For food packaging, biodegradable plastics should be a barrier against the transfer of gases and water vapor, preferably be transparent for the visual attraction of the customers, and form films with proper tensile and mechanical strength. Currently, the majority of these plastics are opaque, with limited tensile properties and a susceptibility to UV-triggered oxidation [202].

Biodegradable plastics are more susceptible to hydrolysis than petroleum-based plastics, depending on their composition. In the process of degradation, they exhibit a reduction in their molecular weight, leading to a low recycling quality. Their physical and chemical characteristics limit the possibility of mechanically recycling conventional and biobased plastics. The technologies and equipment used and the specific plastic waste influences the recyclability [120].

Additionally, the contamination of plastic waste with organic materials leads to the formation of thin films that often limit the recyclability of petro- and biobased plastics [203]. Of all the plastic produced, only 7% is recycled, about 8% is incinerated, and the rest reaches landfills or the environment [123]. Recyclability is a key element of the circular economy and is of great importance in biodegradable materials such as bioplastics. In the case of plastics of petrochemical origin, the recycling process is already well-established, while for bioplastics, it is still under development, since a fully effective and viable process has not yet been achieved [204]. Most bioplastics of biological origin (PLA, PHA, starch, etc.) may require separation based on their chemical composition; otherwise, they may suffer a loss in their quality and physical integrity due to contamination with other polymers. Currently, the most effective process for the treatment of bioplastics of biological origin is biodegradation [205]. The recycling of bioplastics for food-grade applications is a process that requires a higher degree of processing for the removal of contaminants, which is sometimes not economically viable when compared to the production of enriched compost, which is a more profitable process [176]. Biobased plastics, such as bio-PE, can be recycled by the same processes as their fossil-fuel counterparts, as well as with other plastics of petrochemical origin that are structurally similar [206].

The main concerns for the use of petrochemical plastic materials are their low rate of decomposition in the environment, as well as the release of microplastics and various toxic organic compounds that are known to be harmful to the environment and human health. Bioplastics, on the other hand, have a low environmental impact, since their biodegradation does not lead to the release of highly toxic compounds [207]. Different biopolymers obtained from natural resources are easily converted into bioplastics with biodegradable and compostable properties; however, these new bioplastics need regulations that verify and certify these properties to guarantee their safety of use and the lack of harm they pose to the environment [208]. On the other hand, the oxidation of biopolymers during burning releases the same amount of CO_2 into the atmosphere as petroleum-based polymers. In this sense, strategies to control plastic use and its release are still needed to solve this environmental problem.

Despite the many efforts made in the design of standards to evaluate the compostability and biodegradability of bioplastics, at the industrial level, they have still not been extensively explored and remain unregulated [209].

From the perspective of a circular economy, which aims to properly use renewable natural resources without affecting the environment, biobased bioplastics are the most studied plastic materials, since they present the most advantages in their disposal after use [176]. Adopting a circular economy incorporating biobased resources hopes to reduce greenhouse-effect emissions while increasing the ecoefficiency of resources, valorizing waste and by-products, and thus reducing dependence on nonrenewable resources to reduce the impact of climate change and carbon footprints [210].

The European Commission (EC) implemented an action plan for the circular economy of plastics in 2015. The EC published two legal documents for the packaging industry and its waste management: Directive (EC) 2018/851 of the European Parliament and of the Council of May 30, 2018 which modifies Directive 2008/98/EC on waste, and Directive (EC) 2018/852 of the European Parliament and of the Council of 30 May 2018, which modifies Directive 94/62/EC on packaging and packaging waste. These regulations, based on a circular economy, focused on the reuse and recovery of plastic containers to allow a longer life for the products while minimizing waste and reducing the environmental impact; however, the full recycling potential of the European Union remains untapped, since only 30% of plastics are recycled, with hopes to reach 55% by 2025 [211]. The goal of the "Plastics 2030 Voluntary Commitment" is to recycle 60% of all plastic used and reduce the production of petroleum-based plastic by 100% by 2040 [212].

Biorefineries focus on the comprehensive use of different organic wastes such as food, agroindustrial waste, forestry, and algae biomass, which are of great interest for generating new precursors of bioplastics. Through different strategies, such as bioprocesses with microorganisms, the reduction in the use of nonrenewable products has been notable. Different natural resources can be used or explored to obtain biopolymers for bioplastic fabrication, such as PLA, PHA, and PBS. These approaches advance society towards a circular economy and produce different single-use bioplastic products that are transformed into natural compostable compounds without any negative effects on the environment [210].

Biodegradable plastics are not the only solution to the problem of garbage accumulation in landfills and the environment. However, the use of biodegradable bioplastics that are degraded by biological agents under certain conditions and within certain time-frames allows a reduction in their environmental impact, which will leave a significant ecological footprint in the future [114].

7. Legislation and Certifications for Biodegradable Plastics

Due to the evolution of bioplastics over the last few decades, their regulation has become imperative. Experts from different fields have been working with organisms across the world to provide standards to analyze the properties, composition, and composability of different bioplastics using specific techniques and conditions that can be replicated. The major organizations creating these standards are the International Organization for Standardization (ISO), the European Committee for Standardization (CEN), and the American Society for Testing and Materials (ASTM). Table 6 is a compilation of the most used and cited standards from different organizations regarding plastics. The United States of America and the European Union currently have many regulations in place, and these are regarded with significant concern, since they have a great impact at the international level. In comparison, in Mexico, there are few reports of standards for bioplastics; meanwhile, the key standardization bodies have a wide variety of standards covering most of the bioplastic family known to this date.

Table 6. Standards related to bioplastics from selected countries.

Country	Nomenclature of the Standard	Title of Standard
Mexico	NMX-E-273-NYCE-2019	Plastic Industry—Compostable plastics—Specifications and essay methods
	NMX-E-267-CNCP-2016	Plastic industry—Biobased plastics—Essay methods
USA	ASTM D5071-06(2013)	Standard practice for exposure of photodegradable plastics in a xenon arc apparatus
	ASTM D5208-14	Standard practice for fluorescent ultraviolet (UV) exposure of photodegradable plastics
	ASTM D5272-08(2013)	Standard practice for outdoor exposure testing of photodegradable plastics
	ASTM D5338-15	Standard test method for determining aerobic biodegradation of plastic materials under controlled composting conditions, incorporating thermophilic temperatures
	ASTM D5511-18	Standard test method for determining anaerobic biodegradation of plastic materials under high-solids anaerobic-digestion conditions
	ASTM D5526-18	Standard test method for determining anaerobic biodegradation of plastic materials under accelerated landfill conditions
	ASTM D5988-18	Standard test method for determining aerobic biodegradation of plastic materials in soil
	ASTM D6400-19	Standard specification for labeling of plastics designed to be aerobically composted in municipal or industrial facilities
	ASTM D6691-17	Standard test method for determining aerobic biodegradation of plastic materials in the marine environment by a defined microbial consortium or natural sea water inoculum
	ASTM D6866-21	Standard test methods for determining the biobased content of solid, liquid, and gaseous samples using radiocarbon analysis
	ASTM D6868-21	Standard specification for labeling of end items that incorporate plastics and polymers as coatings or additives with paper and other substrates designed to be aerobically composted in municipal or industrial facilities
	ASTM D6954-18	Standard guide for exposing and testing plastics that degrade in the environment by a combination of oxidation and biodegradation
	ASTM D7444-18a	Standard practice for heat and humidity aging of oxidatively degradable plastics
	ASTM D7475-20	Standard test method for determining the aerobic degradation and anaerobic biodegradation of plastic materials under accelerated bioreactor landfill conditions
	ASTM D7991-15	Standard test method for determining aerobic biodegradation of plastics buried in sandy marine sediment under controlled laboratory conditions

Table 6. *Cont.*

Country	Nomenclature of the Standard	Title of Standard
UK	BS 8472:2011	Methods for the assessment of the oxo-biodegradation of plastics and of the phyto-toxicity of the residues in controlled laboratory conditions
	BS ISO 16620-1:2015	Plastics. Biobased content General principles
	BS ISO 16620-2:2019	Plastics. Biobased content Determination of biobased carbon content
	PD CEN/TR 16721:2014	Biobased products. Overview of methods to determine the biobased content (British standard)
	BS ISO 16620-3:2015	Plastics. Biobased content Determination of biobased synthetic polymer content
	BS ISO 22526-3:2020	Plastics. Carbon and environmental footprint of biobased plastics Process carbon footprint, requirements, and guidelines for quantification
	BS ISO 23517:2021	Plastics. Soil biodegradable materials for mulch films for use in agriculture and horticulture. Requirements and test methods regarding biodegradation, ecotoxicity and control of constituents
	BS ISO 5412	Biodegradable plastic shopping bags for industrial composting
EU	CSN EN ISO 10210	Plastics—Methods for the preparation of samples for biodegradation testing of plastic materials
	DIN EN 13432	Requirements for packaging recoverable through composting and biodegradation—Test scheme and evaluation criteria for the final acceptance of packaging
	CSN EN ISO 14851	Determination of the ultimate aerobic biodegradability of plastic materials in an aqueous medium—Method by measuring the oxygen demand in a closed respirometer
	CSN EN ISO 14852	Determination of the ultimate aerobic biodegradability of plastic materials in an aqueous medium—Method by analysis of evolved CO_2
	CSN EN ISO 14853	Plastics—Determination of the ultimate anaerobic biodegradation of plastic materials in an aqueous system—Method by measurement of biogas production
	CSN EN ISO 14855-1	Determination of the ultimate aerobic biodegradability of plastic materials under controlled composting conditions—Method by analysis of evolved CO_2—Part 1: General method
	CSN EN ISO 14855-2	Determination of the ultimate aerobic biodegradability of plastic materials under controlled composting conditions—Method by analysis of evolved CO_2—Part 2: Gravimetric measurement of CO_2 evolved in a laboratory-scale test
	CSN EN 14995	Plastics—Evaluation of compostability—Test scheme and specifications
	CSN EN ISO 15985	Plastics—Determination of the ultimate anaerobic biodegradation under high-solids anaerobic-digestion conditions—Method by analysis of released biogas
	CSN EN 16640	Biobased products—Biobased carbon content—Determination of the biobased carbon content using the radiocarbon method
	CSN EN 16760	Biobased products—Life Cycle Assessment
	EN 16785-1	Biobased products—Biobased content—Part 1: Determination of the biobased content using the radiocarbon analysis and elemental analysis
	CSN EN 16785-2	Biobased products—Biobased content—Part 2: Determination of the biobased content using the material balance method
	CSN EN ISO 16929	Plastics—Determination of the degree of disintegration of plastic materials under defined composting conditions in a pilot-scale test
	CSN EN ISO 17556	Plastics—Determination of the ultimate aerobic biodegradability of plastic materials in soil by measuring the oxygen demand in a respirometer or the amount of CO_2 evolved
	CSN EN 17417	Determination of the ultimate biodegradation of plastics materials in an aqueous system under anoxic (denitrifying) conditions—Method by measurement of pressure increase
	CSN EN ISO 18830	Plastics—Determination of aerobic biodegradation of nonfloating plastic materials in a seawater/sandy sediment interface—Method by measuring the oxygen demand in closed respirometer
	CSN EN ISO 19679	Plastics—Determination of aerobic biodegradation of nonfloating plastic materials in a seawater/sediment interface—Method by analysis of evolved CO_2
International	ISO 14851	Determination of the ultimate aerobic biodegradability of plastic materials in an aqueous medium—Method by measuring the oxygen demand in a closed respirometer
	ISO 14852	Determination of the ultimate aerobic biodegradability of plastic materials in an aqueous medium—Method by analysis of evolved CO_2
	ISO 14853	Plastics—Determination of the ultimate anaerobic biodegradation of plastic materials in an aqueous system—Method by measurement of biogas production

Even when standards are in place, they are not mandatory but are a set of rules and tests to analyze the bioplastic. The portion of the industry sector that analyzes their bioplastic is motivated to obtain a certification for customer approval, particularly when their customers prefer brands with sustainable commitments [213]. The process to obtain the certification is to test the product by an independent certification laboratory that determines whether it complies with the standards and is deserving of a certification. Multiple organizations and associations offer these services and guide the business to obtain certifications that are recognized by governments and society in general. The Biodegradable Product Institute, Tüv Austria (formerly Vinçotte), DIN Certco, the Australasian Bioplastics Association, and the Japan BioPlastics Association are the most in-demand organizations that offer this service (Figure 6). For instance, the Tüv Austria group offers diverse certification logos, which include the Seedling logo, OK compost INDUSTRIAL, and OK compost HOME, all of which are related to the compostability of organic material; the first two logos are acquired when products are in compliance with the standard EN 13432, and the later was developed by the Tüv Austria group without referring to a specific standard, but it contains all the technical requirements that a product must meet to serve as a basis for the creation of other standards across Europe and Australia [214]. Tüv Austria also offers logos that describe the biodegradation of plastic materials depending on the environment in which this process occurs, the logos being OK biodegradable MARINE, OK biodegradable SOIL, and OK biodegradable WATER. All of these are based on the idea of stopping littering and giving information as to the right environment for the biodegradation of a particular material to occur [215]. OK biobased is another logo that rewards companies that manufacture their products with an alternative to fossil-based raw materials. This certification presents the biobased percentage of renewable raw materials used as a base in the product, and the product can be rated up to four stars [216]. Lastly, the logo NEN BIO-BASED CONTENT refers to the biomass content in materials and products in general; this certification is based on the European standard EN 16785-1 and was recently added to the list of the TÜV AUSTRIA group [217].

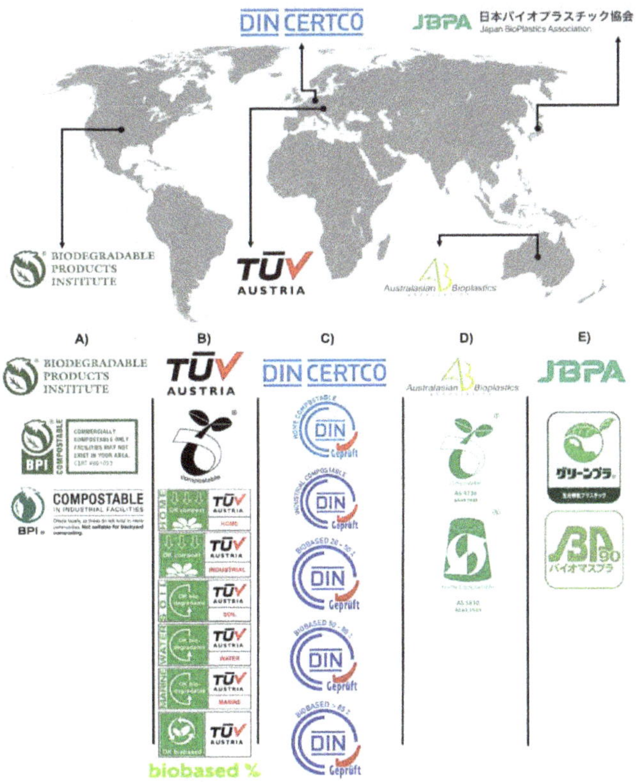

Figure 6. Bioplastic certification world map. (**A**) Biodegradable Product Institute; (**B**) Tüv Austria; (**C**) DIN Certco; (**D**) Australasian Bioplastics Association; (**E**) Japan BioPlastics Association.

Notwithstanding the tendency to use bioplastics for their wide-ranging properties, functionalities, and applications, the scheme of regulations and certifications is changing towards more sustainable approaches. Alliances to join the governmental, industrial, and academic sectors to establish new regulations and improve the existing ones are needed. In this sense, it should be a reality in the near future that there are more initiatives to certify the characteristics of industrial products which are standardized worldwide and that there are key standardization bodies to legislate and certify products with logos that describe the properties of the family of bioplastics.

8. Conclusions

The field of plastics has been changing over the years, with biobased molecules originally made to be synthetic, complex, and nonbiodegradable. The stability conferred by the polymeric nature of traditional plastics hinders their degradation in the environment, resulting in the accumulation of plastics in the ecosystem instead of their chemical degradation, macroscopic fragmentation, and biological assimilation. Research in the plastic field is overwhelmingly focused on tackling its environmental and economic problems. Trends in the use of oxodegradable additives and their potential degradation by microbial and physical means are an insight into these topics. Nevertheless, the expectation to use different raw materials derived from natural sources and create a circular economy to make biopolymers that substitute for fossil resources to produce plastics is on the rise. Although the physicochemical properties of the materials could be exceptional in relation to biodegradability, challenges derived from their production, implementation, and recyclability have been identified. The relationship between the activities of (1) governments with public policy, (2) academics with research on new materials and their properties, (3) the industry with the implementation and products, (4) international organizations and nonprofit organizations that supply regulations and certifications, and (5) the social culture regarding the use of plastics and recyclability will bring better initiatives for the future of plastics.

Author Contributions: Conceptualization, E.M.M.-M. and R.P.-S.; software, R.M.-G.; investigation, A.G.A., R.P.B.M. and K.A.S.C.; writing—original draft preparation, E.M.M.-M., R.M.-G., L.A.-R., D.R.-G. and L.P.-A. writing—review and editing, R.G.A., J.E.S.-H. and A.R.-M.; visualization, M.Y.L.; supervision, H.M.N.I.; funding acquisition, R.P.-S. All authors have read and agreed to the published version of the manuscript.

Funding: This research was funded by Cadena Comercial OXXO, S.A de C.V, grant number DJ-PI-JMC-DO-ITESM-EIC-2021/12.

Institutional Review Board Statement: Not applicable.

Informed Consent Statement: Not applicable.

Data Availability Statement: Not applicable.

Acknowledgments: The authors thankfully acknowledge Altera (Ricardo Alberto Ruiz Tapia and Cassandra Aguilar Figueroa); Qualyplast (Lorenia Miranda López, José Puig Escamilla, and Leobardo Burboa Maldonado); and Kristafilms for their support in this work. Consejo Nacional de Ciencia y Tecnología (CONACYT) is thankfully acknowledged for partially supporting this work under the Sistema Nacional de Investigadores (SNI) program awarded to Elda M. Melchor-Martínez (CVU: 230784), Juan Eduardo Sosa-Hernández (CVU: 375202), Rafael Gomes Araújo (CVU:714118), Adriana Reyes-Mayer (CVU: 288227), Hafiz M.N. Iqbal (CVU:735340), and Roberto Parra-Saldívar (CVU: 35753), and the PhD scholarship awarded to Lynette Alvarado Ramírez (CVU: 486638). The authors thank Tecnologico de Monterrey for providing access to scientific literature.

Conflicts of Interest: The authors declare no conflict of interest.

References

1. Geyer, R.; Jambeck, J.R.; Law, K.L. Production, use, and fate of all plastics ever made. *Sci. Adv.* **2017**, *3*, 3–8. [CrossRef] [PubMed]
2. Thakur, S.; Chaudhary, J.; Sharma, B.; Verma, A.; Tamulevicius, S.; Thakur, V.K. Sustainability of bioplastics: Opportunities and challenges. *Curr. Opin. Green Sustain. Chem.* **2018**, *13*, 68–75. [CrossRef]
3. Ramkumar, M.; Balasubramani, K.; Santosh, M.; Nagarajan, R. The plastisphere: A morphometric genetic classification of plastic pollutants in the natural environment. *Gondwana Res.* 2021; *in press*. [CrossRef]

4. Nagalakshmaiah, M.; Afrin, S.; Malladi, R.P.; Elkoun, S.; Robert, M.; Ansari, M.A.; Svedberg, A.; Karim, Z. Biocomposites: Present trends and challenges for the future. In *Green Composites for Automotive Applications*; Koronis, G., Silva, A., Eds.; Woodhead Publishing Series in Composites Science and Engineering; Woodhead Publishing: Cambridge, UK, 2019; pp. 197–215. ISBN 978-0-08-102177-4.
5. Chen, Y.; Awasthi, A.K.; Wei, F.; Tan, Q.; Li, J. Single-use plastics: Production, usage, disposal, and adverse impacts. *Sci. Total Environ.* **2021**, *752*, 141772. [CrossRef] [PubMed]
6. Mittal, M.; Mittal, D.; Aggarwal, N.K. Plastic accumulation during COVID-19: Call for another pandemic; bioplastic a step towards this challenge? *Environ. Sci. Pollut. Res.* **2022**, *29*, 11039–11053. [CrossRef]
7. de Sousa, F.D.B. Plastic and its consequences during the COVID-19 pandemic. *Environ. Sci. Pollut. Res.* **2021**, *28*, 46067–46078. [CrossRef]
8. Ali, S.S.; Elsamahy, T.; Koutra, E.; Kornaros, M.; El-Sheekh, M.; Abdelkarim, E.A.; Zhu, D.; Sun, J. Degradation of conventional plastic wastes in the environment: A review on current status of knowledge and future perspectives of disposal. *Sci. Total Environ.* **2021**, *771*, 144719. [CrossRef]
9. William, R. Goal 14. Conserve and Sustainably Use the Oceans, Seas, and Marine Resources for Sustainable Development. In *A New Era in Global Health Nursing and the United Nations 2030 Agenda for Sustainable Development*; William, R., Ed.; Springer Publishing: Berlin/Heidelberg, Germany, 2015.
10. Walker, T.R. (Micro)plastics and the UN Sustainable Development Goals. *Curr. Opin. Green Sustain. Chem.* **2021**, *30*, 100497. [CrossRef]
11. Arikan, E.B.; Ozsoy, H.D. A review: Investigation of bioplastics. *J. Civ. Eng. Arch.* **2015**, *9*, 188–192.
12. Narancic, T.; Cerrone, F.; Beagan, N.; O'Connor, K.E. Recent advances in bioplastics: Application and biodegradation. *Polymers* **2020**, *12*, 920. [CrossRef]
13. Bhagwat, G.; Gray, K.; Wilson, S.P.; Muniyasamy, S.; Vincent, S.G.T.; Bush, R.; Palanisami, T. Benchmarking Bioplastics: A Natural Step Towards a Sustainable Future. *J. Polym. Environ.* **2020**, *28*, 3055–3075. [CrossRef]
14. Pellis, A.; Malinconico, M.; Guarneri, A.; Gardossi, L. Renewable polymers and plastics: Performance beyond the green. *New Biotechnol.* **2021**, *60*, 146–158. [CrossRef] [PubMed]
15. The BPF—A History. Available online: https://www.bpf.co.uk/about_the_bpf/The_BPF_A_History.aspx (accessed on 3 February 2022).
16. Gilbert, M. Plastics Materials: Introduction and Historical Development. In *Brydson's Plastics Materials*, 8th ed.; Elsevier Ltd.: Amsterdam, The Netherlands, 2017; pp. 2–18. [CrossRef]
17. Zhang, G.; Yin, T.; Nian, G.; Suo, Z. Fatigue-resistant polyurethane elastomer composites. *Extrem. Mech. Lett.* **2021**, *48*, 101434. [CrossRef]
18. Das, A.; Mahanwar, P. A brief discussion on advances in polyurethane applications. *Adv. Ind. Eng. Polym. Res.* **2020**, *3*, 93–101. [CrossRef]
19. Jassal, M.; Agrawal, A.K.; Gupta, D.; Panwar, K. Aramid Fibers. In *Handbook of Fibrous Materials*; Wiley-VCH Verlag GmbH & Co. KGaA: Weinheim, Germany, 2020; pp. 207–231.
20. Jethy, B.; Paul, S.; Das, S.K.; Adesina, A.; Mustakim, S.M. Critical review on the evolution, properties, and utilization of plasticwastes for construction applications. *J. Mater. Cycles Waste Manag.* **2022**, *24*, 435–451. [CrossRef]
21. Kalai Thendral, K.; Abraham Antony, D. Ballistic Impact analysis of 2D woven Kevlar/Basalt reinforced composite. *IOP Conf. Ser. Mater. Sci. Eng.* **2020**, *912*, 052023. [CrossRef]
22. Pathak, S.; Sneha, C.; Mathew, B.B. Bioplastics: Its Timeline Based Scenario & Challenges. *J. Polym. Biopolym. Phys. Chem.* **2014**, *2*, 84–90. [CrossRef]
23. Kabasci, S. *Bio-Based Plastics Materials and Applications*; Wiley: Hoboken, NJ, USA, 2014; ISBN 9781118676646.
24. Bakar, N.F.A.; Othman, S.A. Corn Bio-plastics for Packaging Application. *J. Des. Sustain. Environ.* **2019**, *1*, 1–3.
25. *Plasctic Europe—Association of Plastics Manufactures*; Plastic Europe: Brussels, Belgium, 2020; pp. 1–64.
26. da Costa, J.P. The 2019 global pandemic and plastic pollution prevention measures: Playing catch-up. *Sci. Total Environ. J.* **2021**, *774*, 145806. [CrossRef]
27. Li, P.; Wang, X.; Su, M.; Zou, X.; Duan, L.; Zhang, H. Characteristics of Plastic Pollution in the Environment: A Review. *Bull. Environ. Contam. Toxicol.* **2021**, *107*, 577–584. [CrossRef]
28. Liang, Y.; Tan, Q.; Song, Q.; Li, J. An analysis of the plastic waste trade and management in Asia. *Waste Manag.* **2021**, *119*, 242–253. [CrossRef] [PubMed]
29. Li, L.; Zuo, J.; Duan, X.; Wang, S.; Hu, K.; Chang, R. Impacts and mitigation measures of plastic waste: A critical review. *Environ. Impact Assess. Rev.* **2021**, *90*, 106642. [CrossRef]
30. Zhang, K.; Hamidian, A.H.; Tubić, A.; Zhang, Y.; Fang, J.K.H.; Wu, C.; Lam, P.K.S. Understanding plastic degradation and microplastic formation in the environment: A review. *Environ. Pollut.* **2021**, *274*, 116554. [CrossRef] [PubMed]
31. Cooper, D.A.; Corcoran, P. Effects of Chemical and Mechanical Weathering Processes on the Degradation of Plastic Debris on Marine Beaches Graduate Program in Geology. Ph.D. Thesis, Western University, London, ON, Canada, 2012.
32. Ahmed, T.; Shahid, M.; Azeem, F.; Rasul, I.; Shah, A.A.; Noman, M.; Hameed, A.; Manzoor, N.; Manzoor, I.; Muhammad, S. Biodegradation of plastics: Current scenario and future prospects for environmental safety. *Environ. Sci. Pollut. Res.* **2018**, *25*, 7287–7298. [CrossRef] [PubMed]

33. Blettler, M.C.M.; Ulla, M.A.; Rabuffetti, A.P.; Garello, N. Plastic pollution in freshwater ecosystems: Macro-, meso-, and microplastic debris in a floodplain lake. *Environ. Monit. Assess.* **2017**, *189*, 581. [CrossRef]
34. Helm, P.A. Occurrence, Sources, Transport, and Fate of Microplastics in the Great Lakes–St. Lawrence River Basin. In *Contaminants in the Great Lakes: An Introduction*; Springer: Cham, Switzerland, 2020; pp. 15–47.
35. Pojar, I.; Stănică, A.; Stock, F.; Kochleus, C.; Schultz, M.; Bradley, C. Sedimentary microplastic concentrations from the Romanian Danube River to the Black Sea. *Sci. Rep.* **2021**, *11*, 2000. [CrossRef] [PubMed]
36. Napper, I.E.; Baroth, A.; Barrett, A.C.; Bhola, S.; Chowdhury, G.W.; Davies, B.F.R.; Duncan, E.M.; Kumar, S.; Nelms, S.E.; Hasan Niloy, M.N.; et al. The abundance and characteristics of microplastics in surface water in the transboundary Ganges River. *Environ. Pollut.* **2021**, *274*, 116348. [CrossRef]
37. Whitehead, P.G.; Bussi, G.; Hughes, J.M.R.; Castro-Castellon, A.T.; Norling, M.D.; Jeffers, E.S.; Rampley, C.P.N.; Read, D.S.; Horton, A.A. Modelling microplastics in the river thames: Sources, sinks and policy implications. *Water* **2021**, *13*, 861. [CrossRef]
38. Chae, Y.; Kim, D.; Kim, S.W.; An, Y.J. Trophic transfer and individual impact of nano-sized polystyrene in a four-species freshwater food chain. *Sci. Rep.* **2018**, *8*, 284. [CrossRef]
39. Shen, M.; Zeng, Z.; Wen, X.; Ren, X.; Zeng, G.; Zhang, Y.; Xiao, R. Presence of microplastics in drinking water from freshwater sources: The investigation in Changsha, China. *Environ. Sci. Pollut. Res.* **2021**, *28*, 42313–42324. [CrossRef]
40. Zhang, Q.; Xu, E.G.; Li, J.; Chen, Q.; Ma, L.; Zeng, E.Y.; Shi, H. A Review of Microplastics in Table Salt, Drinking Water, and Air: Direct Human Exposure. *Environ. Sci. Technol.* **2020**, *54*, 3740–3751. [CrossRef] [PubMed]
41. Bhatt, P.; Pathak, V.M.; Bagheri, A.R.; Bilal, M. Microplastic contaminants in the aqueous environment, fate, toxicity consequences, and remediation strategies. *Environ. Res.* **2021**, *200*, 111762. [CrossRef] [PubMed]
42. Lebreton, L.; Slat, B.; Ferrari, F.; Sainte-Rose, B.; Aitken, J.; Marthouse, R.; Hajbane, S.; Cunsolo, S.; Schwarz, A.; Levivier, A.; et al. Evidence that the Great Pacific Garbage Patch is rapidly accumulating plastic. *Sci. Rep.* **2018**, *8*, 4666. [CrossRef] [PubMed]
43. Provencher, J.F.; Ammendolia, J.; Rochman, C.M.; Mallory, M.L. Assessing plastic debris in aquatic food webs: What we know and don't know about uptake and trophic transfer. *Environ. Rev.* **2019**, *27*, 304–317. [CrossRef]
44. Mbachu, O.; Jenkins, G.; Kaparaju, P.; Pratt, C. The rise of artificial soil carbon inputs: Reviewing microplastic pollution effects in the soil environment. *Sci. Total Environ.* **2021**, *780*, 146569. [CrossRef]
45. Rillig, M.C.; Ziersch, L.; Hempel, S. Microplastic transport in soil by earthworms. *Sci. Rep.* **2017**, *7*, 1362. [CrossRef]
46. Huerta Lwanga, E.; Mendoza Vega, J.; Ku Quej, V.; Chi, J.D.L.A.; Sanchez del Cid, L.; Chi, C.; Escalona Segura, G.; Gertsen, H.; Salánki, T.; van der Ploeg, M.; et al. Field evidence for transfer of plastic debris along a terrestrial food chain. *Sci. Rep.* **2017**, *7*, 14071. [CrossRef]
47. Sun, Y.; Duan, C.; Cao, N.; Li, X.; Li, X.; Chen, Y.; Huang, Y.; Wang, J. Effects of microplastics on soil microbiome: The impacts of polymer type, shape, and concentration. *Sci. Total Environ.* **2022**, *806*, 150516. [CrossRef]
48. Webb, H.K.; Arnott, J.; Crawford, R.J.; Ivanova, E.P. Plastic degradation and its environmental implications with special reference to poly(ethylene terephthalate). *Polymers* **2013**, *5*, 1–18. [CrossRef]
49. Verma, R.; Vinoda, K.S.; Papireddy, M.; Gowda, A.N.S. Toxic Pollutants from Plastic Waste—A Review. *Procedia Environ. Sci.* **2016**, *35*, 701–708. [CrossRef]
50. Wright, S.L.; Ulke, J.; Font, A.; Chan, K.L.A.; Kelly, F.J. Atmospheric microplastic deposition in an urban environment and an evaluation of transport. *Environ. Int.* **2020**, *136*, 105411. [CrossRef] [PubMed]
51. Chen, X.; Yan, N. A brief overview of renewable plastics. *Mater. Today Sustain.* **2020**, *7–8*, 100031. [CrossRef]
52. Huang, W.; Song, B.; Liang, J.; Niu, Q.; Zeng, G.; Shen, M.; Deng, J.; Luo, Y.; Wen, X.; Zhang, Y. Microplastics and associated contaminants in the aquatic environment: A review on their ecotoxicological effects, trophic transfer, and potential impacts to human health. *J. Hazard. Mater.* **2021**, *405*, 124187. [CrossRef] [PubMed]
53. Fanini, L.; Defeo, O.; Elliott, M.; Paragkamian, S.; Pinna, M.; Salvo, V.-S. Coupling beach ecology and macroplastics litter studies: Current trends and the way ahead. *Mar. Pollut. Bull.* **2021**, *173*, 112951. [CrossRef]
54. Parolini, M.; De Felice, B.; Lamonica, C.; Cioccarelli, S.; Crosta, A.; Diolaiuti, G.; Ortenzi, M.A.; Ambrosini, R. Macroplastics contamination on glaciers from Italian Central-Western Alps. *Environ. Adv.* **2021**, *5*, 100084. [CrossRef]
55. Pietz, O.; Augenstein, M.; Georgakakos, C.B.; Singh, K.; McDonald, M.; Walter, M.T. Macroplastic accumulation in roadside ditches of New York State's Finger Lakes region (USA) across land uses and the COVID-19 pandemic. *J. Environ. Manag.* **2021**, *298*, 113524. [CrossRef]
56. Patrício Silva, A.L.; Prata, J.C.; Walker, T.R.; Duarte, A.C.; Ouyang, W.; Barcelò, D.; Rocha-Santos, T. Increased plastic pollution due to COVID-19 pandemic: Challenges and recommendations. *Chem. Eng. J.* **2021**, *405*, 126683. [CrossRef]
57. Pascuta, M.S.; Vodnar, D.C. Nanocarriers for Sustainable Active Packaging: An Overview during and Post COVID-19. *Coatings* **2022**, *12*, 102. [CrossRef]
58. Blettler, M.C.M.; Mitchell, C. Dangerous traps: Macroplastic encounters affecting freshwater and terrestrial wildlife. *Sci. Total Environ.* **2021**, *798*, 149317. [CrossRef]
59. Law, K.L.; Narayan, R. Reducing environmental plastic pollution by designing polymer materials for managed end-of-life. *Nat. Rev. Mater.* **2021**, *7*, 104–116. [CrossRef]
60. Rech, S. Marine Plastic Pollution as a Vector for Non-Native Species Transport. Los Plásticos Contaminantes Marinos Como Vector de Transporte para Especies Exóticas. Ph.D. Thesis, Universidad de Oviedo, Oviedo, Spain, 2018. Available online: https://digibuo.uniovi.es/dspace/handle/10651/50378 (accessed on 11 March 2022).

61. Naidoo, T.; Rajkaran, A.; Sershen. Impacts of plastic debris on biota and implications for human health: A South African perspective. *S. Afr. J. Sci.* **2020**, *116*, 1–8. [CrossRef]
62. Machovsky-Capuska, G.E.; Amiot, C.; Denuncio, P.; Grainger, R.; Raubenheimer, D. A nutritional perspective on plastic ingestion in wildlife. *Sci. Total Environ.* **2019**, *656*, 789–796. [CrossRef] [PubMed]
63. Santos, R.G.; Machovsky-Capuska, G.E.; Andrades, R. Plastic ingestion as an evolutionary trap: Toward a holistic understanding. *Science* **2021**, *373*, 56–60. [CrossRef]
64. Honingh, D.; van Emmerik, T.; Uijttewaal, W.; Kardhana, H.; Hoes, O.; van de Giesen, N. Urban River Water Level Increase Through Plastic Waste Accumulation at a Rack Structure. *Front. Earth Sci.* **2020**, *8*, 28. [CrossRef]
65. Umer, M.; Abbasi, B.N.; Sohail, A.; Tang, J.; Ullah, I.; Abbasi, H. Determinants of the Usage of Plastic Bags. *Int. J. Bus. Econ. Manag. Works* **2018**, *5*, 8–22.
66. Superio, M.D.A.; Abreo, N.A.S. Plastic in Freshwater Ecosytems: A looming crisis in the Philippines. *Philipp. Sci. Lett.* **2020**, *13*, 1–5.
67. Olofinnade, O.; Chandra, S.; Chakraborty, P. Recycling of high impact polystyrene and low-density polyethylene plastic wastes in lightweight based concrete for sustainable construction. *Mater. Today Proc.* **2020**, *38*, 2151–2156. [CrossRef]
68. Piehl, S.; Leibner, A.; Löder, M.G.J.; Dris, R.; Bogner, C.; Laforsch, C. Identification and quantification of macro- and microplastics on an agricultural farmland. *Sci. Rep.* **2018**, *8*, 17950. [CrossRef]
69. Suyadi; Manullang, C.Y. Distribution of plastic debris pollution and it is implications on mangrove vegetation. *Mar. Pollut. Bull.* **2020**, *160*, 111642. [CrossRef]
70. Leonov, V.D.; Tiunov, A.V. Interaction of Invertebrates and Synthetic Polymers in Soil: A Review. *Russ. J. Ecol.* **2020**, *51*, 503–517. [CrossRef]
71. Lehmann, A.; Leifheit, E.F.; Gerdawischke, M.; Rillig, M.C. Microplastics have shape- and polymer-dependent effects on soil aggregation and organic matter loss—An experimental and meta-analytical approach. *Microplast. Nanoplast.* **2021**, *1*, 1–14. [CrossRef]
72. Rochman, C.M.; Brookson, C.; Bikker, J.; Djuric, N.; Earn, A.; Bucci, K.; Athey, S.; Huntington, A.; McIlwraith, H.; Munno, K.; et al. Rethinking microplastics as a diverse contaminant suite. *Environ. Toxicol. Chem.* **2019**, *38*, 703–711. [CrossRef]
73. Shi, W.; Cui, T.; Wu, H.; LeBlanc, G.A.; Wang, F.; An, L. A proposed nomenclature for microplastic contaminants. *Mar. Pollut. Bull.* **2021**, *172*, 112960. [CrossRef] [PubMed]
74. Wang, C.; Zhao, J.; Xing, B. Environmental source, fate, and toxicity of microplastics. *J. Hazard. Mater.* **2021**, *407*, 124357. [CrossRef] [PubMed]
75. Hidayaturrahman, H.; Lee, T.-G. A study on characteristics of microplastic in wastewater of South Korea: Identification, quantification, and fate of microplastics during treatment process. *Mar. Pollut. Bull.* **2019**, *146*, 696–702. [CrossRef]
76. Hebner, T.S.; Maurer-Jones, M.A. Characterizing microplastic size and morphology of photodegraded polymers placed in simulated moving water conditions. *Environ. Sci. Process. Impacts* **2020**, *22*, 398–407. [CrossRef]
77. Burrows, S.D.; Frustaci, S.; Thomas, K.V.; Galloway, T. Expanding exploration of dynamic microplastic surface characteristics and interactions. *TrAC-Trends Anal. Chem.* **2020**, *130*, 115993. [CrossRef]
78. Rosal, R. Morphological description of microplastic particles for environmental fate studies. *Mar. Pollut. Bull.* **2021**, *171*, 112716. [CrossRef]
79. Vivekanand, A.C.; Mohapatra, S.; Tyagi, V.K. Microplastics in aquatic environment: Challenges and perspectives. *Chemosphere* **2021**, *282*, 131151. [CrossRef]
80. Xiang, Y.; Jiang, L.; Zhou, Y.; Luo, Z.; Zhi, D.; Yang, J.; Lam, S.S. Microplastics and environmental pollutants: Key interaction and toxicology in aquatic and soil environments. *J. Hazard. Mater.* **2022**, *422*, 126843. [CrossRef]
81. Muñiz-González, A.B.; Silva, C.J.M.; Patricio Silva, A.L.; Campos, D.; Pestana, J.L.T.; Martínez-Guitarte, J.L. Suborganismal responses of the aquatic midge Chironomus riparius to polyethylene microplastics. *Sci. Total Environ.* **2021**, *783*, 146981. [CrossRef] [PubMed]
82. Zandaryaa, S. Freshwater Microplastic Pollution: The State of Knowledge and Research. In *Plastics in the Aquatic Environment—Part* I: Current Status and Challenges; Stock, F., Reifferscheid, G., Brennholt, N., Kostianaia, E., Eds.; The Handbook of Environmental Chemistry; Springer International Publishing: Cham, Switzerland, 2022; pp. 255–272. ISBN 978-3-030-84118-8.
83. Fajardo, C.; Martín, C.; Costa, G.; Sánchez-Fortún, S.; Rodríguez, C.; de Lucas Burneo, J.J.; Nande, M.; Mengs, G.; Martín, M. Assessing the role of polyethylene microplastics as a vector for organic pollutants in soil: Ecotoxicological and molecular approaches. *Chemosphere* **2022**, *288*, 132460. [CrossRef] [PubMed]
84. Godoy, V.; Blázquez, G.; Calero, M.; Quesada, L.; Martín-Lara, M.A. The potential of microplastics as carriers of metals. *Environ. Pollut.* **2019**, *255*, 113363. [CrossRef] [PubMed]
85. Tang, Y.; Liu, Y.; Chen, Y.; Zhang, W.; Zhao, J.; He, S.; Yang, C.; Zhang, T.; Tang, C.; Zhang, C.; et al. A review: Research progress on microplastic pollutants in aquatic environments. *Sci. Total Environ.* **2021**, *766*, 142572. [CrossRef] [PubMed]
86. Bakir, A.; Rowland, S.J.; Thompson, R.C. Transport of persistent organic pollutants by microplastics in estuarine conditions. *Estuar. Coast. Shelf Sci.* **2014**, *140*, 14–21. [CrossRef]
87. Campanale, C.; Massarelli, C.; Savino, I.; Locaputo, V.; Uricchio, V.F. A detailed review study on potential effects of microplastics and additives of concern on human health. *Int. J. Environ. Res. Public Health* **2020**, *17*, 1212. [CrossRef]
88. Prata, J.C. Airborne microplastics: Consequences to human health? *Environ. Pollut.* **2018**, *234*, 115–126. [CrossRef]

89. Vianello, A.; Jensen, R.L.; Liu, L.; Vollertsen, J. Simulating human exposure to indoor airborne microplastics using a Breathing Thermal Manikin. *Sci. Rep.* **2019**, *9*, 8670. [CrossRef]
90. Carr, S.A. Sources and dispersive modes of micro-fibers in the environment. *Integr. Environ. Assess. Manag.* **2017**, *13*, 466–469. [CrossRef]
91. Hahladakis, J.N.; Velis, C.A.; Weber, R.; Iacovidou, E.; Purnell, P. An overview of chemical additives present in plastics: Migration, release, fate and environmental impact during their use, disposal and recycling. *J. Hazard. Mater.* **2018**, *344*, 179–199. [CrossRef]
92. Briou, B.; Caillol, S.; Robin, J.; Lapinte, V.; Briou, B.; Caillol, S.; Robin, J.; Non-endocrine, V.L. Non-endocrine disruptor effect for cardanol based plasticizer. *Ind. Crops Prod.* **2019**, *130*, 1–8. [CrossRef]
93. Jamarani, R.; Erythropel, H.C.; Nicell, J.A.; Leask, R.L.; Marić, M. How green is your plasticizer? *Polymers* **2018**, *10*, 834. [CrossRef] [PubMed]
94. Gao, M.; Xu, Y.; Dong, Y.; Song, Z.; Liu, Y. Accumulation and metabolism of di(n-butyl) phthalate (DBP) and di(2-ethylhexyl) phthalate (DEHP) in mature wheat tissues and their effects on detoxification and the antioxidant system in grain. *Sci. Total Environ.* **2019**, *697*, 133981. [CrossRef] [PubMed]
95. Rodrigues, M.O.; Abrantes, N.; Gonçalves, F.J.M.; Nogueira, H.; Marques, J.C.; Gonçalves, A.M.M. Impacts of plastic products used in daily life on the environment and human health: What is known? *Environ. Toxicol. Pharmacol.* **2019**, *72*, 103239. [CrossRef] [PubMed]
96. Chen, X.; Xu, S.; Tan, T.; Lee, S.T.; Cheng, S.H.; Lee, F.W.F.; Xu, S.J.L.; Ho, K.C. Toxicity and estrogenic endocrine disrupting activity of phthalates and their mixtures. *Int. J. Environ. Res. Public Health* **2014**, *11*, 3156–3168. [CrossRef]
97. Sandeep, S.; Rowdhwal, S. Toxic Effects of Di-2-ethylhexyl Phthalate: An Overview. *Biomed Res. Int.* **2018**, *2018*, 1750368.
98. Kotowska, U.; Kapelewska, J.; Sawczuk, R. Occurrence, removal, and environmental risk of phthalates in wastewaters, landfill leachates, and groundwater in Poland. *Environ. Pollut.* **2020**, *267*, 115643. [CrossRef]
99. Xia, B.; Zhu, Q.; Zhao, Y.; Ge, W.; Zhao, Y.; Song, Q.; Zhou, Y.; Shi, H.; Zhang, Y. Phthalate exposure and childhood overweight and obesity: Urinary metabolomic evidence. *Environ. Int.* **2018**, *121*, 159–168. [CrossRef]
100. Wen, Y.; Rattan, S.; Flaws, J.A.; Irudayaraj, J. Multi and transgenerational epigenetic effects of di-(2-ethylhexyl) phthalate (DEHP) in liver. *Toxicol. Appl. Pharmacol.* **2020**, *402*, 115123. [CrossRef]
101. Fred-Ahmadu, O.H.; Bhagwat, G.; Oluyoye, I.; Benson, N.U.; Ayejuyo, O.O.; Palanisami, T. Interaction of chemical contaminants with microplastics: Principles and perspectives. *Sci. Total Environ.* **2020**, *706*, 135978. [CrossRef]
102. Alp, A.C.; Yerlikaya, P. Phthalate ester migration into food: Effect of packaging material and time. *Eur. Food Res. Technol.* **2020**, *246*, 425–435. [CrossRef]
103. Novotna, K.; Cermakova, L.; Pivokonska, L.; Cajthaml, T.; Pivokonsky, M. Microplastics in drinking water treatment—Current knowledge and research needs. *Sci. Total Environ.* **2019**, *667*, 730–740. [CrossRef] [PubMed]
104. Abdelmoez, W.; Dahab, I.; Ragab, E.M.; Abdelsalam, O.A.; Mustafa, A. Bio- and oxo-degradable plastics: Insights on facts and challenges. *Polym. Adv. Technol.* **2021**, *32*, 1981–1996. [CrossRef]
105. Schiavo, S.; Oliviero, M.; Chiavarini, S.; Manzo, S. Adverse effects of oxo-degradable plastic leachates in freshwater environment. *Environ. Sci. Pollut. Res.* **2020**, *27*, 8586–8595. [CrossRef] [PubMed]
106. Markowicz, F.; Szymańska-Pulikowska, A. Analysis of the possibility of environmental pollution by composted biodegradable and oxobiodegradable plastics. *Geosciences* **2019**, *9*, 460. [CrossRef]
107. Vázquez-Morillas, A.; Beltrán-Villavicencio, M.; Alvarez-Zeferino, J.C.; Osada-Velázquez, M.H.; Moreno, A.; Martínez, L.; Yañez, J.M. Biodegradation and Ecotoxicity of Polyethylene Films Containing Pro-Oxidant Additive. *J. Polym. Environ.* **2016**, *24*, 221–229. [CrossRef]
108. Foschi, E.; Bonoli, A. The commitment of packaging industry in the framework of the european strategy for plastics in a circular economy. *Adm. Sci.* **2019**, *9*, 18. [CrossRef]
109. Goel, V.; Luthra, P.; Kapur, G.S.; Ramakumar, S.S.V. Biodegradable/Bio-plastics: Myths and Realities. *J. Polym. Environ.* **2021**, *29*, 3079–3104. [CrossRef]
110. Tabba, H.D.; Hijji, Y.M.; Abu-Surrah, A.S. Olefin Polymerization. In *Polyolefin Compounds and Materials: Fundamentals and Industrial Applications*; AlMa'adeed, M.A.-A., Krupa, I., Eds.; Springer Series on Polymer and Composite Materials; Springer: Cham, Switzerland, 2016; pp. 51–77. ISBN 978-3-319-25982-6.
111. Zeenat; Elahi, A.; Bukhari, D.A.; Shamim, S.; Rehman, A. Plastics degradation by microbes: A sustainable approach. *J. King Saud Univ.-Sci.* **2021**, *33*, 101538. [CrossRef]
112. Eyheraguibel, B.; Traikia, M.; Fontanella, S.; Sancelme, M.; Bonhomme, S.; Fromageot, D.; Lemaire, J.; Lauranson, G.; Lacoste, J.; Delort, A.M. Characterization of oxidized oligomers from polyethylene films by mass spectrometry and NMR spectroscopy before and after biodegradation by a Rhodococcus rhodochrous strain. *Chemosphere* **2017**, *184*, 366–374. [CrossRef]
113. Dang, T.C.H.; Nguyen, D.T.; Thai, H.; Nguyen, T.C.; Hien Tran, T.T.; Le, V.H.; Nguyen, V.H.; Tran, X.B.; Thao Pham, T.P.; Nguyen, T.G.; et al. Plastic degradation by thermophilic Bacillus sp. BCBT21 isolated from composting agricultural residual in Vietnam. *Adv. Nat. Sci. Nanosci. Nanotechnol.* **2018**, *9*, 015014. [CrossRef]
114. Rujnić-Sokele, M.; Pilipović, A. Challenges and opportunities of biodegradable plastics: A mini review. *Waste Manag. Res.* **2017**, *35*, 132–140. [CrossRef] [PubMed]
115. Siracusa, V.; Blanco, I. Polymers Analogous to Petroleum-Derived Ones for Packaging and Engineering Applications. *Polymers* **2020**, *12*, 1641. [CrossRef]

116. Thiruchelvi, R.; Das, A.; Sikdar, E. Bioplastics as better alternative to petro plastic. *Mater. Today Proc.* **2021**, *37*, 1634–1639. [CrossRef]
117. European Bioplastics. Available online: https://www.european-bioplastics.org (accessed on 12 September 2021).
118. Sid, S.; Mor, R.S.; Kishore, A.; Sharanagat, V.S. Bio-sourced polymers as alternatives to conventional food packaging materials: A review. *Trends Food Sci. Technol.* **2021**, *115*, 87–104. [CrossRef]
119. Garrido, R.; Cabeza, L.F.; Falguera, V. An overview of bioplastic research on its relation to national policies. *Sustainability* **2021**, *13*, 7848. [CrossRef]
120. Fredi, G.; Dorigato, A. Recycling of bioplastic waste: A review. *Adv. Ind. Eng. Polym. Res.* **2021**, *4*, 159–177. [CrossRef]
121. Jian, J.; Xiangbin, Z.; Xianbo, H. An overview on synthesis, properties and applications of poly(butylene-adipate-co-terephthalate)-PBAT. *Adv. Ind. Eng. Polym. Res.* **2020**, *3*, 19–26. [CrossRef]
122. Kim, S.Y. Application of the three-dimensionally printed biodegradable polycaprolactone (PCL) mesh in repair of orbital wall fractures. *J. Cranio-Maxillofac. Surg.* **2019**, *47*, 1065–1071. [CrossRef]
123. Atiwesh, G.; Mikhael, A.; Parrish, C.C.; Banoub, J.; Le, T.A.T. Environmental impact of bioplastic use: A review. *Heliyon* **2021**, *7*, e07918. [CrossRef]
124. Montava-Jordà, S.; Torres-Giner, S.; Ferrandiz-Bou, S.; Quiles-Carrillo, L.; Montanes, N. Development of sustainable and cost-competitive injection-molded pieces of partially bio-based polyethylene terephthalate through the valorization of cotton textile waste. *Int. J. Mol. Sci.* **2019**, *20*, 1378. [CrossRef]
125. Shafqat, A.; Tahir, A.; Mahmood, A.; Tabinda, A.B.; Yasar, A.; Pugazhendhi, A. A review on environmental significance carbon foot prints of starch based bio-plastic: A substitute of conventional plastics. *Biocatal. Agric. Biotechnol.* **2020**, *27*, 101540. [CrossRef]
126. Thakur, R.; Pristijono, P.; Scarlett, C.J.; Bowyer, M.; Singh, S.P.; Vuong, Q.V. Starch-based films: Major factors affecting their properties. *Int. J. Biol. Macromol.* **2019**, *132*, 1079–1089. [CrossRef] [PubMed]
127. Zoungranan, Y.; Lynda, E.; Dobi-Brice, K.K.; Tchirioua, E.; Bakary, C.; Yannick, D.D. Influence of natural factors on the biodegradation of simple and composite bioplastics based on cassava starch and corn starch. *J. Environ. Chem. Eng.* **2020**, *8*, 104396. [CrossRef]
128. Srinivasa Rao, L.; Naidu, C.D.; Tiwari, S. Investigation on synthesis, structure and degradability of starch based bioplastics. *Mater. Today Proc.* **2022**, *49*, 257–261. [CrossRef]
129. Asrofi, M.; Sapuan, S.M.; Ilyas, R.A.; Ramesh, M. Characteristic of composite bioplastics from tapioca starch and sugarcane bagasse fiber: Effect of time duration of ultrasonication (Bath-Type). *Mater. Today Proc.* **2020**, *46*, 1626–1630. [CrossRef]
130. Astuti, P.; Erprihana, A.A. Antimicrobial Edible Film from Banana Peels as Food Packaging. Abstract. *Am. J. Oil Chem. Technol.* **2014**, *2*, 65–70.
131. Sujuthi, R.A.F.M.; Liew, K.C. Properties of bioplastic sheets made from different types of starch incorporated with recycled newspaper pulp. *Trans. Sci. Technol.* **2016**, *3*, 257–264.
132. Maulida; Siagian, M.; Tarigan, P. Production of Starch Based Bioplastic from Cassava Peel Reinforced with Microcrystalline Celllulose Avicel PH101 Using Sorbitol as Plasticizer. *J. Phys. Conf. Ser.* **2016**, *710*, 012012. [CrossRef]
133. Tsou, C.H.; Suen, M.C.; Yao, W.H.; Yeh, J.T.; Wu, C.S.; Tsou, C.Y.; Chiu, S.H.; Chen, J.C.; Wang, R.Y.; Lin, S.M.; et al. Preparation and characterization of Bioplastic-Based green renewable composites from tapioca with acetyl tributyl citrate as a plasticizer. *Materials* **2014**, *7*, 5617–5632. [CrossRef]
134. Fabra, M.J.; Martínez-Sanz, M.; Gómez-Mascaraque, L.G.; Gavara, R.; López-Rubio, A. Structural and physicochemical characterization of thermoplastic corn starch films containing microalgae. *Carbohydr. Polym.* **2018**, *186*, 184–191. [CrossRef] [PubMed]
135. Zhang, C.; Wang, C.; Cao, G.; Wang, D.; Ho, S.H. A sustainable solution to plastics pollution: An eco-friendly bioplastic film production from high-salt contained Spirulina sp. residues. *J. Hazard. Mater.* **2020**, *388*, 121773. [CrossRef] [PubMed]
136. Mathiot, C.; Ponge, P.; Gallard, B.; Sassi, J.F.; Delrue, F.; Le Moigne, N. Microalgae starch-based bioplastics: Screening of ten strains and plasticization of unfractionated microalgae by extrusion. *Carbohydr. Polym.* **2019**, *208*, 142–151. [CrossRef] [PubMed]
137. Mostafa, N.A.; Farag, A.A.; Abo-dief, H.M.; Tayeb, A.M. Production of biodegradable plastic from agricultural wastes. *Arab. J. Chem.* **2018**, *11*, 546–553. [CrossRef]
138. Nigam, S.; Das, A.K.; Patidar, M.K. Synthesis, characterization and biodegradation of bioplastic films produced from Parthenium hysterophorus by incorporating a plasticizer (PEG600). *Environ. Chall.* **2021**, *5*, 100280. [CrossRef]
139. Abdul Khalil, H.P.S.; Tye, Y.Y.; Saurabh, C.K.; Leh, C.P.; Lai, T.K.; Chong, E.W.N.; Nurul Fazita, M.R.; Hafiidz, J.M.; Banerjee, A.; Syakir, M.I. Biodegradable polymer films from seaweed polysaccharides: A review on cellulose as a reinforcement material. *Express Polym. Lett.* **2017**, *11*, 244–265. [CrossRef]
140. Velásquez-Riaño, M.; Bojacá, V. Production of bacterial cellulose from alternative low-cost substrates. *Cellulose* **2017**, *24*, 2677–2698. [CrossRef]
141. Soemphol, W.; Hongsachart, P.; Tanamool, V. Production and characterization of bacterial cellulose produced from agricultural by-product by Gluconacetobacter strains. *Mater. Today Proc.* **2018**, *5*, 11159–11168. [CrossRef]
142. Gayathri, G.; Srinikethan, G. Bacterial Cellulose production by K. saccharivorans BC1 strain using crude distillery effluent as cheap and cost effective nutrient medium. *Int. J. Biol. Macromol.* **2019**, *138*, 950–957. [CrossRef]
143. Revin, V.; Liyaskina, E.; Nazarkina, M.; Bogatyreva, A.; Shchankin, M. Cost-effective production of bacterial cellulose using acidic food industry by-products. *Braz. J. Microbiol.* **2018**, *49*, 151–159. [CrossRef]

144. Sadalage, P.S.; Pawar, K.D. Production of microcrystalline cellulose and bacterial nanocellulose through biological valorization of lignocellulosic biomass wastes. *J. Clean. Prod.* **2021**, *327*, 129462. [CrossRef]
145. Huang, C.; Guo, H.J.; Xiong, L.; Wang, B.; Shi, S.L.; Chen, X.F.; Lin, X.Q.; Wang, C.; Luo, J.; Chen, X. De Using wastewater after lipid fermentation as substrate for bacterial cellulose production by Gluconacetobacter xylinus. *Carbohydr. Polym.* **2016**, *136*, 198–202. [CrossRef] [PubMed]
146. Li, Z.; Wang, L.; Hua, J.; Jia, S.; Zhang, J.; Liu, H. Production of nano bacterial cellulose from waste water of candied jujube-processing industry using Acetobacter xylinum. *Carbohydr. Polym.* **2015**, *120*, 115–119. [CrossRef]
147. Zhao, H.; Xia, J.; Wang, J.; Yan, X.; Wang, C.; Lei, T.; Xian, M.; Zhang, H. Production of bacterial cellulose using polysaccharide fermentation wastewater as inexpensive nutrient sources. *Biotechnol. Biotechnol. Equip.* **2018**, *32*, 350–356. [CrossRef]
148. Vytejčková, S.; Vápenka, L.; Hradecký, J.; Dobiáš, J.; Hajšlová, J.; Loriot, C.; Vannini, L.; Poustka, J. Testing of polybutylene succinate based films for poultry meat packaging. *Polym. Test.* **2017**, *60*, 357–364. [CrossRef]
149. de la Torre, I.; Acedos, M.G.; Ladero, M.; Santos, V.E. On the use of resting L. delbrueckii spp. delbrueckii cells for D-lactic acid production from orange peel wastes hydrolysates. *Biochem. Eng. J.* **2019**, *145*, 162–169. [CrossRef]
150. Kim, J.W.; Jang, J.H.; Yeo, H.J.; Seol, J.; Kim, S.R.; Jung, Y.H. Lactic Acid Production from a Whole Slurry of Acid-Pretreated Spent Coffee Grounds by Engineered Saccharomyces cerevisiae. *Appl. Biochem. Biotechnol.* **2019**, *189*, 206–216. [CrossRef]
151. Lian, T.; Zhang, W.; Cao, Q.; Wang, S.; Dong, H. Enhanced lactic acid production from the anaerobic co-digestion of swine manure with apple or potato waste via ratio adjustment. *Bioresour. Technol.* **2020**, *318*, 124237. [CrossRef]
152. Alves de Oliveira, R.; Schneider, R.; Vaz Rossell, C.E.; Maciel Filho, R.; Venus, J. Polymer grade L-lactic acid production from sugarcane bagasse hemicellulosic hydrolysate using Bacillus coagulans. *Bioresour. Technol. Rep.* **2019**, *6*, 26–31. [CrossRef]
153. Bai, J.; Xu, D.; Xie, D.; Wang, M.; Li, Z.; Guo, X. Effects of antibacterial peptide-producing Bacillus subtilis and Lactobacillus buchneri on fermentation, aerobic stability, and microbial community of alfalfa silage. *Bioresour. Technol.* **2020**, *315*, 123881. [CrossRef]
154. Jiang, S.; Xu, P.; Tao, F. L-Lactic acid production by Bacillus coagulans through simultaneous saccharification and fermentation of lignocellulosic corncob residue. *Bioresour. Technol. Rep.* **2019**, *6*, 131–137. [CrossRef]
155. Mulla, M.Z.; Rahman, M.R.T.; Marcos, B.; Tiwari, B.; Pathania, S. Poly lactic acid (Pla) nanocomposites: Effect of inorganic nanoparticles reinforcement on its performance and food packaging applications. *Molecules* **2021**, *26*, 1967. [CrossRef] [PubMed]
156. Rajesh Banu, J.; Kavitha, S.; Yukesh Kannah, R.; Poornima Devi, T.; Gunasekaran, M.; Kim, S.H.; Kumar, G. A review on biopolymer production via lignin valorization. *Bioresour. Technol.* **2019**, *290*, 121790. [CrossRef] [PubMed]
157. Albuquerque, P.B.S.; Malafaia, C.B. Perspectives on the production, structural characteristics and potential applications of bioplastics derived from polyhydroxyalkanoates. *Int. J. Biol. Macromol.* **2018**, *107*, 615–625. [CrossRef]
158. Rahman, A.; Miller, C.D. *Microalgae as a Source of Bioplastics*; Elsevier B.V.: Amsterdam, The Netherlands, 2017; ISBN 9780444640413.
159. Andler, R.; Valdés, C.; Urtuvia, V.; Andreeßen, C.; Díaz-Barrera, A. Fruit residues as a sustainable feedstock for the production of bacterial polyhydroxyalkanoates. *J. Clean. Prod.* **2021**, *307*, 127236. [CrossRef]
160. Borrero-de Acuña, J.M.; Aravena-Carrasco, C.; Gutierrez-Urrutia, I.; Duchens, D.; Poblete-Castro, I. Enhanced synthesis of medium-chain-length poly(3-hydroxyalkanoates) by inactivating the tricarboxylate transport system of Pseudomonas putida KT2440 and process development using waste vegetable oil. *Process Biochem.* **2019**, *77*, 23–30. [CrossRef]
161. Ruiz, C.; Kenny, S.T.; Narancic, T.; Babu, R.; Connor, K.O. Conversion of waste cooking oil into medium chain polyhydroxyalkanoates in a high cell density fermentation. *J. Biotechnol.* **2019**, *306*, 9–15. [CrossRef]
162. Kulkarni, S.O.; Kanekar, P.P.; Jog, J.P.; Sarnaik, S.S.; Nilegaonkar, S.S. Production of copolymer, poly (hydroxybutyrate-co-hydroxyvalerate) by Halomonas campisalis MCM B-1027 using agro-wastes. *Int. J. Biol. Macromol.* **2015**, *72*, 784–789. [CrossRef]
163. Valdez-Calderón, A.; Barraza-Salas, M.; Quezada-Cruz, M.; Islas-Ponce, M.A.; Angeles-Padilla, A.F.; Carrillo-Ibarra, S.; Rodríguez, M.; Rojas-Avelizapa, N.G.; Garrido-Hernández, A.; Rivas-Castillo, A.M. Production of polyhydroxybutyrate (PHB) by a novel Klebsiella pneumoniae strain using low-cost media from fruit peel residues. *Biomass Convers. Biorefin.* **2020**. [CrossRef]
164. Jahan, F.; Kumar, V.; Saxena, R.K. Distillery effluent as a potential medium for bacterial cellulose production: A biopolymer of great commercial importance. *Bioresour. Technol.* **2018**, *250*, 922–926. [CrossRef]
165. Acedos, M.G.; Moreno-Cid, J.; Verdú, F.; González, J.A.; Tena, S.; López, J.C. Exploring the potential of slaughterhouse waste valorization: Development and scale-up of a new bioprocess for medium-chain length polyhydroxyalkanoates production. *Chemosphere* **2022**, *287*, 132401. [CrossRef] [PubMed]
166. Pereira, J.R.; Araújo, D.; Freitas, P.; Marques, A.C.; Alves, V.D.; Sevrin, C.; Grandfils, C.; Fortunato, E.; Reis, M.A.M.; Freitas, F. Production of medium-chain-length polyhydroxyalkanoates by Pseudomonas chlororaphis subsp. aurantiaca: Cultivation on fruit pulp waste and polymer characterization. *Int. J. Biol. Macromol.* **2021**, *167*, 85–92. [CrossRef] [PubMed]
167. Rao, A.; Haque, S.; El-Enshasy, H.A.; Singh, V.; Mishra, B.N. RSM–GA based optimization of bacterial PHA production and In Silico modulation of citrate synthase for enhancing PHA production. *Biomolecules* **2019**, *9*, 872. [CrossRef] [PubMed]
168. Rebocho, A.T.; Pereira, J.R.; Neves, L.A.; Alves, V.D.; Sevrin, C.; Grandfils, C.; Freitas, F.; Reis, M.A.M. Preparation and characterization of films based on a natural p(3hb)/mcl-pha blend obtained through the co-culture of cupriavidus necator and pseudomonas citronellolis in apple pulp waste. *Bioengineering* **2020**, *7*, 34. [CrossRef]
169. Wang, Z.; Li, B.; Zhou, L.; Yu, S.; Su, Z.; Song, J.; Sun, Q.; Sha, O.; Wang, X.; Jiang, W.; et al. Prodigiosin inhibits Wnt/β-catenin signaling and exerts anticancer activity in breast cancer cells. *Proc. Natl. Acad. Sci. USA* **2016**, *113*, 13150–13155. [CrossRef]

170. Hoheneder, R.; Fitz, E.; Bischof, R.H.; Russmayer, H.; Ferrero, P.; Peacock, S.; Sauer, M. Efficient conversion of hemicellulose sugars from spent sulfite liquor into optically pure L-lactic acid by Enterococcus mundtii. *Bioresour. Technol.* **2021**, *333*, 125215. [CrossRef]
171. Bernardo, M.P.; Coelho, L.F.; Sass, D.C.; Contiero, J. L-(+)-Lactic acid production by Lactobacillus rhamnosus B103 from dairy industry waste. *Braz. J. Microbiol.* **2016**, *47*, 640–646. [CrossRef]
172. Lim, C.; Yusoff, S.; Ng, C.G.; Lim, P.E.; Ching, Y.C. Bioplastic made from seaweed polysaccharides with green production methods. *J. Environ. Chem. Eng.* **2021**, *9*, 105895. [CrossRef]
173. Sankaranarayanan, S.; Hermosilla, J.; Acevedo, F.; Navia, R. The influences of solvents on the electrospun of whole Scenedesmus almeriensis and poly(ethylene oxide) for the preparation of composite nanofibers. *Compos. Commun.* **2018**, *10*, 18–24. [CrossRef]
174. Kim, Y.; Ruedy, D. Mushroom Packages an Ecovative Approach in Packaging Industry. In *Handbook of Engaged Sustainability*; Springer: Berlin/Heidelberg, Germany, 2019; pp. 1–25.
175. Wu, J. Extraction of Chitin Nanofibers and Utilization for Sustainable Composites and Foams. Ph.D. Thesis, Georgia Institute of Technology, Atlanta, GA, USA, 2014.
176. Nandakumar, A.; Chuah, J.A.; Sudesh, K. Bioplastics: A boon or bane? *Renew. Sustain. Energy Rev.* **2021**, *147*, 111237. [CrossRef]
177. Kawaguchi, H.; Takada, K.; Elkasaby, T.; Pangestu, R.; Toyoshima, M.; Kahar, P.; Ogino, C.; Kaneko, T.; Kondo, A. Recent advances in lignocellulosic biomass white biotechnology for bioplastics. *Bioresour. Technol.* **2022**, *344*, 126165. [CrossRef] [PubMed]
178. Xiao, F.; Fontaine, G.; Bourbigot, S. Recent developments in fire retardancy of polybutylene succinate. *Polym. Degrad. Stab.* **2021**, *183*, 109466. [CrossRef]
179. QMILK. The Material of the Future. Available online: https://www.qmilkfiber.eu/?lang=en (accessed on 3 February 2022).
180. Bioplástico–Biofase. Available online: https://biofase.com.mx/bioplastico (accessed on 3 February 2022).
181. NatureWorks | What Is Ingeo. Available online: https://www.natureworksllc.com/What-is-Ingeo (accessed on 3 February 2022).
182. Thiounn, T.; Smith, R.C. Advances and approaches for chemical recycling of plastic waste. *J. Polym. Sci.* **2020**, *58*, 1347–1364. [CrossRef]
183. Meys, R.; Frick, F.; Westhues, S.; Sternberg, A.; Klankermayer, J.; Bardow, A. Towards a circular economy for plastic packaging wastes—The environmental potential of chemical recycling. *Resour. Conserv. Recycl.* **2020**, *162*, 105010. [CrossRef]
184. Wei, R.; Tiso, T.; Bertling, J.; O'Connor, K.; Blank, L.M.; Bornscheuer, U.T. Possibilities and limitations of biotechnological plastic degradation and recycling. *Nat. Catal.* **2020**, *3*, 867–871. [CrossRef]
185. Hatti-Kaul, R.; Nilsson, L.J.; Zhang, B.; Rehnberg, N.; Lundmark, S. Designing Biobased Recyclable Polymers for Plastics. *Trends Biotechnol.* **2020**, *38*, 50–67. [CrossRef]
186. Tuominen, J.; Kylma, J.; Kapanen, A.; Venelampi, O.; Itävaara, M.; Seppälä, J. Biodegradation of lactic acid based polymers under controlled composting conditions and evaluation of the ecotoxicological impact. *Biomacromolecules* **2002**, *3*, 445–455. [CrossRef]
187. Hoffmann, J.; Řezníčková, I.; Kozáková, J.; Růžička, J.; Alexy, P.; Bakoš, D.; Precnerová, L. Assessing biodegradability of plastics based on poly(vinyl alcohol) and protein wastes. *Polym. Degrad. Stab.* **2003**, *79*, 511–519. [CrossRef]
188. Muenmee, S.; Chiemchaisri, W.; Chiemchaisri, C. Microbial consortium involving biological methane oxidation in relation to the biodegradation of waste plastics in a solid waste disposal open dump site. *Int. Biodeterior. Biodegrad.* **2015**, *102*, 172–181. [CrossRef]
189. Massardier-Nageotte, V.; Pestre, C.; Cruard-Pradet, T.; Bayard, R. Aerobic and anaerobic biodegradability of polymer films and physico-chemical characterization. *Polym. Degrad. Stab.* **2006**, *91*, 620–627. [CrossRef]
190. Tournier, V.; Topham, C.M.; Gilles, A.; David, B.; Folgoas, C.; Moya-Leclair, E.; Kamionka, E.; Desrousseaux, M.L.; Texier, H.; Gavalda, S.; et al. An engineered PET depolymerase to break down and recycle plastic bottles. *Nature* **2020**, *580*, 216–219. [CrossRef] [PubMed]
191. Wei, R.; Breite, D.; Song, C.; Gräsing, D.; Ploss, T.; Hille, P.; Schwerdtfeger, R.; Matysik, J.; Schulze, A.; Zimmermann, W. Biocatalytic Degradation Efficiency of Postconsumer Polyethylene Terephthalate Packaging Determined by Their Polymer Microstructures. *Adv. Sci.* **2019**, *6*, 1900491. [CrossRef] [PubMed]
192. Marcilla, A.; Beltrán, M.I.; Navarro, R. Evolution of products during the degradation of polyethylene in a batch reactor. *J. Anal. Appl. Pyrolysis* **2009**, *86*, 14–21. [CrossRef]
193. Onwudili, J.A.; Insura, N.; Williams, P.T. Composition of products from the pyrolysis of polyethylene and polystyrene in a closed batch reactor: Effects of temperature and residence time. *J. Anal. Appl. Pyrolysis* **2009**, *86*, 293–303. [CrossRef]
194. Artetxe, M.; Lopez, G.; Amutio, M.; Barbarias, I.; Arregi, A.; Aguado, R.; Bilbao, J.; Olazar, M. Styrene recovery from polystyrene by flash pyrolysis in a conical spouted bed reactor. *Waste Manag.* **2015**, *45*, 126–133. [CrossRef] [PubMed]
195. Ahmad, I.; Ismail Khan, M.; Khan, H.; Ishaq, M.; Tariq, R.; Gul, K.; Ahmad, W. Pyrolysis study of polypropylene and polyethylene into premium oil products. *Int. J. Green Energy* **2015**, *12*, 663–671. [CrossRef]
196. López-Mondéjar, R.; Zühlke, D.; Becher, D.; Riedel, K.; Baldrian, P. Cellulose and hemicellulose decomposition by forest soil bacteria proceeds by the action of structurally variable enzymatic systems. *Sci. Rep.* **2016**, *6*, 25279. [CrossRef]
197. Nawaz, A.; Hasan, F.; Shah, A.A. Degradation of poly(ε-caprolactone) (PCL) by a newly isolated Brevundimonas sp. strain MRL-AN1 from soil. *FEMS Microbiol. Lett.* **2015**, *362*, 1–7. [CrossRef]
198. Dobrucka, R. Bioplastic packaging materials in circular economy. *Logforum* **2019**, *15*, 129–137. [CrossRef]
199. Galgani, L.; Loiselle, S.A. Plastic pollution impacts on marine carbon biogeochemistry. *Environ. Pollut.* **2021**, *268*, 115598. [CrossRef]

200. Alves de Oliveira, R.; Komesu, A.; Vaz Rossell, C.E.; Maciel Filho, R. Challenges and opportunities in lactic acid bioprocess design—From economic to production aspects. *Biochem. Eng. J.* **2018**, *133*, 219–239. [CrossRef]
201. Otoni, C.G.; Azeredo, H.M.C.; Mattos, B.D.; Beaumont, M.; Correa, D.S.; Rojas, O.J. The Food–Materials Nexus: Next Generation Bioplastics and Advanced Materials from Agri-Food Residues. *Adv. Mater.* **2021**, *33*, 2102520. [CrossRef]
202. Asgher, M.; Qamar, S.A.; Bilal, M.; Iqbal, H.M.N. Bio-based active food packaging materials: Sustainable alternative to conventional petrochemical-based packaging materials. *Food Res. Int.* **2020**, *137*, 109625. [CrossRef]
203. Briassoulis, D.; Pikasi, A.; Hiskakis, M. Recirculation potential of post-consumer /industrial bio-based plastics through mechanical recycling—Techno-economic sustainability criteria and indicators. *Polym. Degrad. Stab.* **2021**, *183*, 109217. [CrossRef]
204. Jõgi, K.; Bhat, R. Valorization of food processing wastes and by-products for bioplastic production. *Sustain. Chem. Pharm.* **2020**, *18*, 100326. [CrossRef]
205. Dilkes-Hoffman, L.; Ashworth, P.; Laycock, B.; Pratt, S.; Lant, P. Public attitudes towards bioplastics—Knowledge, perception and end-of-life management. *Resour. Conserv. Recycl.* **2019**, *151*, 104479. [CrossRef]
206. Geueke, B.; Groh, K.; Muncke, J. Food packaging in the circular economy: Overview of chemical safety aspects for commonly used materials. *J. Clean. Prod.* **2018**, *193*, 491–505. [CrossRef]
207. Pelamatti, T.; Rios-Mendoza, L.M.; Hoyos-Padilla, E.M.; Galván-Magaña, F.; De Camillis, R.; Marmolejo-Rodríguez, A.J.; González-Armas, R. Contamination knows no borders: Toxic organic compounds pollute plastics in the biodiversity hotspot of Revillagigedo Archipelago National Park, Mexico. *Mar. Pollut. Bull.* **2021**, *170*, 112623. [CrossRef]
208. Meereboer, K.W.; Misra, M.; Mohanty, A.K. Review of recent advances in the biodegradability of polyhydroxyalkanoate (PHA) bioplastics and their composites. *Green Chem.* **2020**, *22*, 5519–5558. [CrossRef]
209. Di Bartolo, A.; Infurna, G.; Dintcheva, N.T. A review of bioplastics and their adoption in the circular economy. *Polymers* **2021**, *13*, 1229. [CrossRef] [PubMed]
210. Talan, A.; Pokhrel, S.; Tyagi, R.D.; Drogui, P. Biorefinery strategies for microbial bioplastics production: Sustainable pathway towards Circular Bioeconomy. *Bioresour. Technol. Rep.* **2022**, *17*, 100875. [CrossRef]
211. Devadas, V.V.; Khoo, K.S.; Chia, W.Y.; Chew, K.W.; Munawaroh, H.S.H.; Lam, M.K.; Lim, J.W.; Ho, Y.C.; Lee, K.T.; Show, P.L. Algae biopolymer towards sustainable circular economy. *Bioresour. Technol.* **2021**, *325*, 124702. [CrossRef] [PubMed]
212. Teleky, B.E.; Vodnar, D.C. Recent advances in biotechnological itaconic acid production, and application for a sustainable approach. *Polymers* **2021**, *13*, 3574. [CrossRef]
213. Filiciotto, L.; Rothenberg, G. Biodegradable Plastics: Standards, Policies, and Impacts. *ChemSusChem* **2021**, *14*, 56–72. [CrossRef] [PubMed]
214. OK Compost & Seedling. Available online: https://www.tuv-at.be/green-marks/certifications/ok-compost-seedling/ (accessed on 3 February 2022).
215. OK Biodegradable. Available online: https://www.tuv-at.be/green-marks/certifications/ok-biodegradable/ (accessed on 3 February 2022).
216. OK Biobased. Available online: https://www.tuv-at.be/green-marks/certifications/ok-biobased/ (accessed on 3 February 2022).
217. NEN Bio-Based Content. Available online: https://www.tuv-at.be/green-marks/certifications/nen-bio-based-content/ (accessed on 3 February 2022).

polymers

MDPI

Article

Effect of Prosopis Juliflora Thorns on Mechanical Properties of Plastic Waste Reinforced Epoxy Composites

Sakthi Balan Ganapathy [1], Aravind Raj Sakthivel [1,*], Mohamed Thariq Hameed Sultan [2,3,4,*], Farah Syazwani Shahar [2], Ain Umaira Md Shah [2], Tabrej Khan [5] and Tamer A. Sebaey [5,6,*]

1 Department of Manufacturing Engineering, School of Mechanical Engineering (SMEC), Vellore Institute of Technology, Vellore 632014, Tamil Nadu, India; sakthibala33@gmail.com
2 Department of Aerospace, Faculty of Engineering, Universiti Putra Malaysia, Serdang 43400, Malaysia; farahsyaz93@yahoo.com (F.S.S.); ainumaira91@gmail.com (A.U.M.S.)
3 Laboratory of Biocomposite Technology, Institute of Tropical Forestry and Forest Products (INTROP), Universiti Putra Malaysia, Serdang 43400, Malaysia
4 Aerospace Malaysia Innovation Centre (944751-A), Prime Minister's Department, MIGHT Partnership Hub, Jalan Impact, Cyberjaya 63000, Malaysia
5 Engineering Management Department, College of Engineering, Prince Sultan University, Riyadh 12435, Saudi Arabia; tkhan@psu.edu.sa
6 Mechanical Design and Production Department, Faculty of Engineering, Zagazig University, Zagazig 44519, Sharkia, Egypt
* Correspondence: aravindsakthivel@hotmail.com (A.R.S.); thariq@upm.edu.my (M.T.H.S.); tsebaey@psu.edu.sa (T.A.S.)

Citation: Ganapathy, S.B.; Sakthivel, A.R.; Sultan, M.T.H.; Shahar, F.S.; Shah, A.U.M.; Khan, T.; Sebaey, T.A. Effect of Prosopis Juliflora Thorns on Mechanical Properties of Plastic Waste Reinforced Epoxy Composites. *Polymers* **2022**, *14*, 1278. https://doi.org/10.3390/polym14071278

Academic Editors: Maya Jacob John and Sabu Thomas

Received: 1 March 2022
Accepted: 16 March 2022
Published: 22 March 2022

Abstract: Plastics are unavoidable at this times, putting our planet in danger. The Prosopis juliflora (PJ) thorns are collected, processed, and powdered. The mechanical characteristics of these powders are examined when combined with polymer composites. Pores are the main cause of moisture input, hence using powder filler materials reduces the number of pores in the composite, increasing water resistance. The composites are made by altering three parameters: waste plastic content, filler powder composition, and chemical treatment. It was discovered that the integration of thorn powder increased the wear resistance. The composites were tested in accordance with ASTM standards, and the results were optimized. Based on the results, composite specimens were created and tested for validation.

Keywords: waste plastics; fillers; Prosopis juliflora; thorn powder; abrasive wear resistance

1. Introduction

Due to the plastic thrash, the land, water, and air are polluted [1]. If the plastics are dumped in the landscapes, they will block the percolation of rainwater into the land which leads to a decrease in the groundwater levels. If the plastics are thrown into water bodies, it leads to leaching, and the water gets polluted. We cannot burn the plastics, as it would release harmful gases into the atmosphere. To avoid the degradation of the earth, the used plastics have to be reused in some way. In the growing world, the need for alternate materials is increasing. At the same time, lots of landscapes are occupied by Prosopis juliflora plants, which makes it difficult to cultivate crops and also it absorbs the water and nutrients from the land and makes it uncultivable [2,3]. A lot of research works are undergoing to use these plants effectively for various applications. Kailappan et al. produced the activated carbon from Prosopis juliflora using a chemical method and proved it can be used in oil, food, and pharma industries [4]. As the plastics are used in this composite, it can be used in automobile, marine applications where lightweight and less corrosive materials are required. Plastics have moisture repelling property and the Prosopis juliflora thorn powder provides hardness to the composites and also it fills the minute pores produced during the production of composites.

When Saravanakumar and his colleagues analysed the parameters of the fibres, they found that they had an average microfibril angle of 10.64 and utmost elongation strength of 558 MPa. It was found that the aspect ratio of the PJ fibres had a significant impact on the mechanical qualities of the finished product. A fibre aspect ratio of 136 and fibre loading of 23.53 wt. percent were found to yield the best mechanical properties. [5]. The tensile and thermal properties of treated PJ fibres were shown to be superior to those of untreated fibres in a study by Madhu et al. [6].

Structural applications may benefit from the properties of PJ fiber- and glass-fiber reinforced polymer composites, according to Manoj Kumar et al. [7]. Due to the presence of lignin, researchers led by Luis Valencia found functional nanomaterials with a mean radius of 10 nm and a length of 150 nm in the PJ [8]. Filler-reinforced epoxy composites can improve mechanical properties by as much as 45 percent, according to research by Santhosh et al., who studied the morphology and properties of PJ and RH—reinforced composites [9]. PJ ash powder can be used as a substitute for cement up to 20%, according to Parthiban Kathirvel et al., and they were able to attain the same strength for their newly designed concrete [10]. An onion-like porous carbon made from the PJ has been advocated as an efficient electrode material by Sathyanarayanan Shanmugapriya and his colleagues [11].

The novel composites' mechanical, tribological, and water absorption properties must be thoroughly investigated. It was discovered by Sakthi Balan et al. that adding 30 wt percent of waste plastic particle to jute fibre and waste plastic-filled composites resulted in high resistance to water absorption [12]. Water absorption and tribological properties were increased by the inclusion of filler and fibres to the composite, which was made of plastic waste, fiber glass, and silica sand fillers. It has been reported that the thermal and mechanical characteristics of epoxy composites manufactured with 20% fibres and NaOH treatment have been improved [13] by Arthanarieswaran and colleagues. Using dates palm seeds and glass fibre reinforced polymer composite, Heba I. Elkhouly et al. demonstrated an increase in the composite's wear resistance and toughness. Tapas by Priyadarshi Tapas Ranjan Swain et al. made a composite out of jute fibres and studied how it wore. The wear resistance has been modified by the chemical treatment [14]. They observed that the abrading distance was the most important element in determining the wear of waste silk fiber-reinforced epoxy composites [15], followed by the loading of fibers.

Taguchi is an effective strategy for designing experiments. Materials scientists use it to their advantage when examining the effects of various process variables. Polymer-based composites are being made using the Taguchi method. Boron nitride reinforcement of Nylon composite was studied by Shiva Kumar and Chennakesava Reddy using the Taguchi approach and it was found that the composite's wear resistance was improved by the addition of boron nitride [16]. It was proposed by Wahid Ferdous et al. that the bond length and thickness for higher strength be studied using Taguchi design in polymer-based composites [17]. Polymer composites with various reinforcements such kevlar, carbon, and glass fibers were subjected to the Taguchi technique by Karthik et al. in order to optimize the wear parameters [18]. They found that the hybrid composites had improved wear behaviors. As a result of the Taguchi approach and ANOVA, Siva Prasad and Chaitanya were able to optimize the drilling parameters for the GFRP composites [19].

Natural fibers have gained popularity and are beginning to supplant synthetic fibers, owing to their contribution to sustainable practices. As a result of environmental, social, and economic development, numerous industries have altered their manufacturing processes, materials, and procedures in order to ensure a sustainable future. While natural fibers have significant disadvantages, they can be overcome with appropriate chemical treatments and fiber processing. Numerous goods composed of natural fibers are developed and used in sports, electronics, and musical instrument manufacturing [20]. Natural fibers exhibit comparable wear resistance to synthetic fibers. The wear resistance of natural fibers can be increased by reinforcing them with synthetic fibers [21]. Recent research has concentrated on green fillers such as date seed powders, coconut and cashew nut shell powders, and rice and maize husks. Natural fibers and fillers are used in composites because they are inexpensive, widely available, biodegradable, recyclable, and lightweight [22]. Researchers are becoming more interested in starch-based bioplastics due to their environmentally beneficial characteristics. Starches have been isolated from a variety of plants, including PJ plants, and bio composites have been constructed and tested for mechanical properties and biodegradability. The results indicated that composites might be used as a substitute material in the packaging industry [1].

Plastic trash and PJ thorn granules are both included into the polymer matrix in this study. The motive of this work is to investigate the influence of chemical treatment, the amount of waste plastics, and PJ thorn powders on the composite's hardness and moisture absorption and wear capabilities using the Taguchi technique.

2. Experimental Details

2.1. Materials

The scraped compact discs are employed as reinforcement in this composite fabrication, and most of them are constructed of polycarbonate. The decomposition of polycarbonate can take hundreds of years, yet it is possible. The reinforcements are placed in a matrix made of epoxy resin. Araldite AW106 grade resin and HV 953U hardener are utilized in this. Epoxies have features that distinguish them from other resins, and adjustments can be made to meet our specifications. Epoxies have superior mechanical qualities, including increased thermal stability, wear and chemical resistance, and resistance to ageing caused by environmental factors [22]. In order to improve a composite's specific characteristics and traits, fillers are incorporated into the material. Fillers in this work include PJ thorn powders. In general, PJ plants cause a wide range of issues for both humans and the environment. As a result, the PJ plant's thorns are harvested, dried for a period of time, and ground into a powder. The waste compact discs were collected from Vellore, Tamilnadu, India, the resin and hardener were purchased from Ayishwarya polymers, Coimbatore, Tamilnadu, India, and the PJ thorn powders were made manually by collecting the PJ thorns from nearby areas in and around Vellore, Tamilnadu, India, drying them in the sun, and then powdering them using a mixer grinder. Finally, before use, the PJ powder is sieved and processed. Once these powders have been chemically treated, they can be used as reinforcements and the matrix, as they have been neutralized of their characteristics.

2.2. Fabrication Method

Initially, the raw materials are collected like waste used plastics. The collected plastics are then separated according to their grades, cleaned, dried, and then they are crushed into particulates. Then the filler powder is made by collecting the PJ thorns, drying and then they are chemically treated with alkalis and silanes for one hour. The NaOH solution (5% w/v) is used for alkali treatment and Triethoxy vinyl silane (5% v/v) is used for silane treatment [23]. In both cases, the thorn powders are dipped into the chemicals for an immersion time of one hour. After the alkali treatment, the thorns are washed with HCL solution to make their pH value-neutral and the silane treated thorns are washed with de-ionized water and then dried. Then they are crushed and made into powders. The resin and the hardener are mixed in the proper ratios and the composite is made by the spray layup method. Spray layup has some added advantages to the hand layup method and the defects are low when compared to other manufacturing techniques. As the plastic particulates and PJ thorn powder can be dispersed through air medium, spray layup is preferred. Initially, the resin was applied and then the reinforcement and the fillers are loaded in the spray gun, and with the help of compressed air as a medium, the particles are sprayed evenly on the resin surface. This ensures the even spread of the fillers.

2.3. Mechanical Properties Examination Methods

The Vickers micro hardness tester manufactured by Fuel Instruments & Engineers Pvt. Ltd., Ichalkaranji, Maharashtra, India, is used to measure micro hardness, and the ASTM E10 standards are followed for the method. Test specimens are imprinted in Vickers hardness testers with 2.5-mm diameter balls of 10 kg of force. The specimen was indented in numerous places, and the average value is used to get the end result. To compute the average values, the diagonals of the cavity generated by pressing a diamond pyramid, are used.

The elongation of specimens on a UTM was measured using the ASTM D 3039-76 standard testing procedure. In this experiment, we used a UK-made H10KS model from Tinius Olsen, Redhill, UK. Crosshead speed and strain rate are maintained at 5 mm/min. Water jet cutting minimizes the formation of micro cracks compared to other cutting procedures, the specimens are cut using this method.

ASTM D570 standards were used to conduct the water intake examination for composites. As a precautionary measure, sticky tape is used to cover the specimen's sides. A scale is used to weigh the samples in advance of testing, and the results are recorded. Test pieces are then submerged in water for 24 h for examination. Afterward, the test pieces are taken out, patted dry, and weighed to determine the weight gain that occurred during the testing period. The percentage of water consumed will be determined by comparing the pre- and post-testing results. The moisture percentage is calculated using the following equation [12].

$$\text{MA\%} = (W_2 - W_1)/W_1 \times 100, \quad (1)$$

where,

MA% = Moisture absorption percentage
W_1 = Sample weight before the experiment in gms
W_2 = Sample weight after the experiment in gms

The abrasive wear of the composite is measured using a pin on a disc wear testing equipment. The specimen was cut in accordance with the specifications of ASTM G99-05. Wear testing equipment supplied by DUCOM instruments, Peenya Industrial Area, Bengaluru, Karnataka, India, has a maximum wear track diameter of 135 mm and a maximum disc speed of 2000 rpm, which was used for the testing. A frictional force of 200 N and a load of 20 Kg are the maximums that can be applied. An abrading distance of 420 m and a load of 10 Newtons are applied to the specimen, which measures 5930 mm in size. With the use of a cantilever unit, the specimen is put into the holder and held there by the disc. Before loading the specimen, it is weighed, and then the test is performed. The samples will be weighed before and after the tests. The density is used to compute the wear volume. In order to calculate the wear rates, Equation (2) is used [24].

$$\text{Wear rate } (K_s) = \text{Differences in wear volumes in mm}^3/(\text{Normal load in N}) \times (\text{Abrading distance in m}) \quad (2)$$

2.4. Optimization Technique

Optimization was done on three concepts based on our requirements. Larger is better, nominal is better, and smaller is better are the three concepts under which the optimization was done [5]. In our case for water absorption, the percentages must be low, so smaller is better concept is used [25]. For micro hardness, larger is better concept is used as normally polymer composites have good tensile and bending characteristics [6]. The hardness is required in some cases where it is subjected to wear and abrasion applications. The influencing parameters such as the composition of waste plastics, PJ thorn powder, and type of chemical treatment are chosen and the most influencing parameter which affects the hardness and the water intake properties are found out. The optimum values of these parameters are also found and the validations were done to find out the nearness of the results with the predicted results. The design for specimen production is obtained through Taguchi's L27 full factorial experimental design, then the prepared specimens are tested and the test results are fed to the software and run for getting the optimum values [21].

In full factorial design the number of experiments is more, so that the results of the full factorial will be more accurate than the fractional factorial. The process parameter chosen and its stages were indicated in Table 1. The experimental results are given in Table 2.

Table 1. Process variables and their various stages.

S. No.	Variables	Units	Stages		
			1	2	3
1	Particulates of waste plastics	Wt%	10	20	30
2	Prosopis juliflora thorn powder	Wt%	5	10	15
3	Chemical treatment	-	Nil	NaOH (Alkali)	Triethoxy vinyl silane

Table 2. Full factorial inputs and outcomes.

Trail No	Process Parameters			Hardness (Hv)	Tensile Strength (MPa)	Water Intake Percentage for 24 h	Wear Rate × 10^{-5} mm^3/Nm
	Plastic Particulates (Wt%) (A)	Thorn Powder (Wt%) (B)	Type of Chemical Treatment (C)				
1	10	5	1	23.6	63.56	1.35	9.55
2	10	5	1	23.5	63.55	1.32	9.55
3	10	5	1	23.7	63.56	1.33	9.58
4	10	10	2	27.2	62.44	1.44	8.6
5	10	10	2	27.5	62.44	1.41	8.62
6	10	10	2	27.3	62.46	1.43	8.61
7	10	15	3	32.1	62.12	1.53	7.2
8	10	15	3	32.4	62.11	1.54	7.22
9	10	15	3	32	62.09	1.54	7.2
10	20	5	2	24.8	65.22	1.15	9.2
11	20	5	2	24.6	65.22	1.12	9.21
12	20	5	2	24.1	65.21	1.13	9.21
13	20	10	3	27.8	64.87	1.23	8.33
14	20	10	3	27.9	64.85	1.24	8.32
15	20	10	3	27.8	64.86	1.26	8.33
16	20	15	1	32.5	64.15	1.36	7.01
17	20	15	1	32.6	64.12	1.35	7.02
18	20	15	1	32.8	64.13	1.37	7
19	30	5	3	25.9	70.54	0.8	8.91
20	30	5	3	26.1	70.51	0.79	8.91
21	30	5	3	26.2	70.52	0.75	8.92
22	30	10	1	29.1	69.3	0.97	8
23	30	10	1	29.2	69.31	0.96	8.01
24	30	10	1	29.5	69.31	1.04	8.02
25	30	15	2	34.1	68.79	1.18	6.5
26	30	15	2	34.2	68.8	1.15	6.49
27	30	15	2	34.05	68.79	1.1	6.5

3. Results and Discussion

3.1. Hardness Analysis

The primary effect map for the composite samples' hardness may be seen in Figure 1 with mean of the output factor in the Y axis and values of the levels in the X axis. Figure 1 shows that the amount of plastic and the type of chemical treatment have the less impact on hardness, than composition of PJ thorn powder. In this case, the chemical therapy has the least impact. The harder the final product is obtained as a result of the increased filler content. The fillers in the matrix provide resistance, which enhances hardness [26]. Another component that improves the composite's properties is the bonding between the matrix and filler [27]. Not all materials will adhere to the resin. Fillers are distributed uniformly throughout the composites; however there is some variance in hardness due to the substance with which the indenter makes contact. At times, the indenter comes into touch with the filler, and at other times, it comes into contact with the resin or plastic particles, resulting in a range of hardness values. To ensure accuracy, readings are taken at multiple locations and the average values are used as the final result. Due to the hydrophobic nature of natural fillers and fibers, they will have low wettability. This renders them incompatible with resins, and following a chemical reaction, the hydrophobic nature of the fillers is converted to hydrophilic, increasing their wettability. Figures 2 and 3 illustrate the interaction and contour plot for the hardness of the material. Using a contour map, hardness enhancement by the plastics and thorn powder composition inclusion is clearly seen. It is possible to see the ANOVA findings in Table 3. Thorn powder composition is determined to have the greatest impact on hardness properties, according to ANOVA results.

In addition, the composition of plastics as well as the chemical treatment has least effect on the hardness of the material. According to the model summary table, the R-square number indicates how near the findings are to the mean values; an R-square value of 98.53 percent indicates that the results are extremely close to the mean values. The contribution plot for the hardness is depicted in Figure 4. The contribution plot is created by plotting the F-values from the ANOVA results together. The F-value indicates that the addition of thorn powder has the most influence on hardness among all of the components, and the contribution plot suggests that the addition of plastics has the second greatest influence on hardness. As seen in Equation (3), the regression equation for hardness is a linear relationship.

Regression equation for Hardness (HV) = 18.057 + 0.1058 Composition of plastics + 0.8250 Composition of thorn powder + 0.094 type of chemical treatment (3)

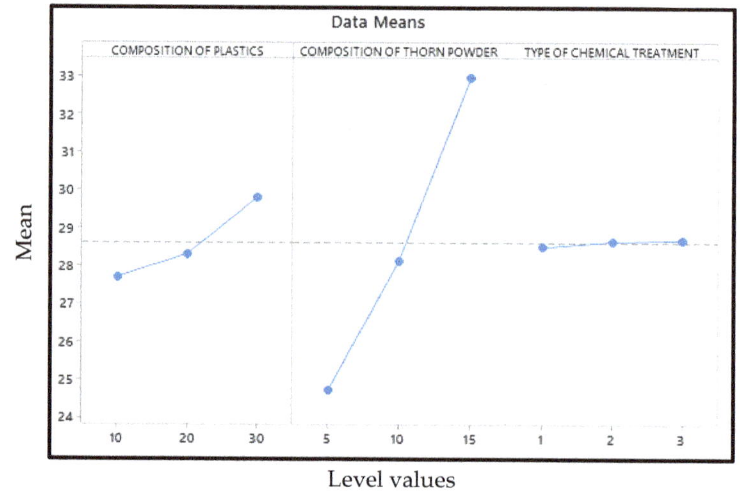

Figure 1. Effects chart for Hardness.

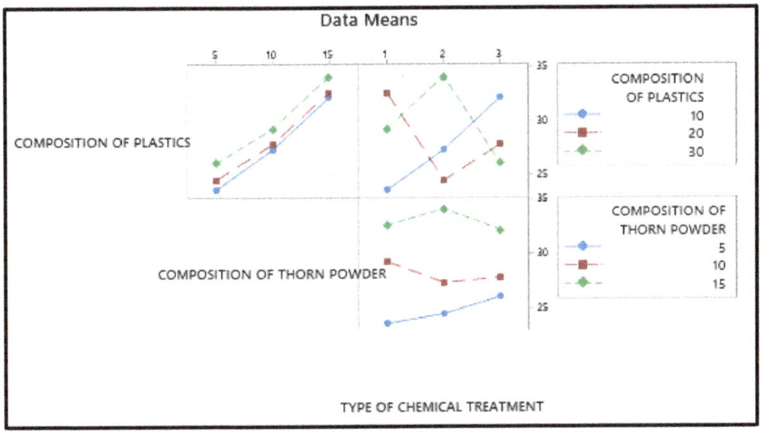

Figure 2. Interaction plot for SN ratios for Hardness.

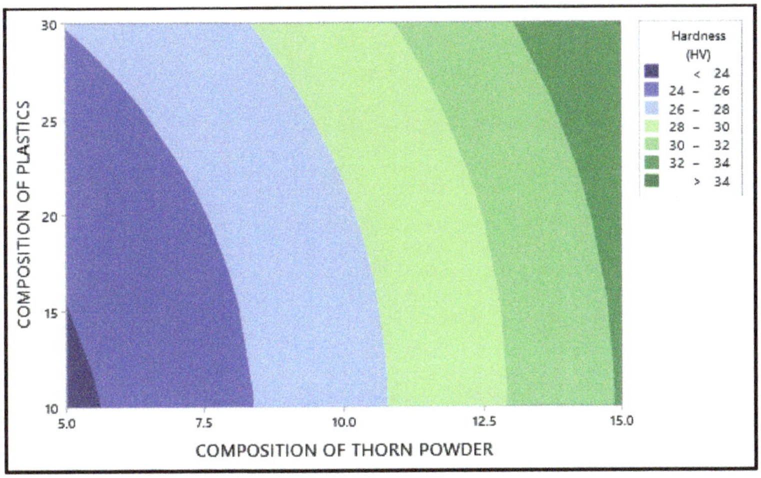

Figure 3. Contour plot for Maximum Hardness.

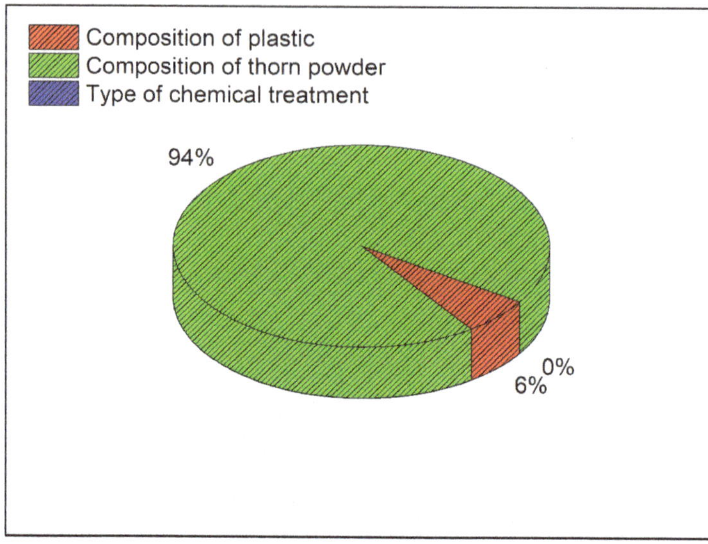

Figure 4. Contribution plot for Maximum Hardness.

Table 3. Hardness ANOVA.

Source	DOF	Adjacent SS	Adjacent MS	F-Values	*p*-Values
Regression	3	326.603	108.868	514.7	0
Composition Of Plastics	1	20.161	20.161	95.32	0
Composition Of Thorn Powder	1	306.281	306.281	1448.02	0
Type Of Chemical Treatment	1	0.161	0.161	0.76	0.393
Error	23	4.865	0.212		
Lack-Of-Fit	5	4.253	0.851	25.03	0
Pure Error	18	0.612	0.034		
Total	26	331.468			

S: 0.459911, R^2 (Adjacent): 98.53%, R^2 (Predicted): 98.03%.

3.2. Water Absorption Property

Figure 5 shows that the plastic composition should be 30%, the inclusion of thorn powder should be 5%, and it should be silane-treated. In Figure 5, the mean values of the output factor (water absorption percentage) is in the Y axis and the level values are in the X axis. The water absorption property of waste plastic is more strongly influenced by its composition. As a general rule, water cannot be absorbed by plastic. As a result, adding plastic to the composite makes it more water resistant [28]. Additionally, the chemical bonds that bind the monomers are strong enough that plastics take years to breakdown. Plastics are constructed in such a way that water cannot percolate into them. Additionally, the filler powders aid in resisting water percolation, albeit only to a certain amount [29]. If they come into direct touch with water, they dissolve and, in certain situations, cause the composites to bulge. Figure 6 depicts the interdependence of the variables, showing how they interact with one another. It demonstrates that the addition of waste plastics and the addition of thorn powder are not interdependent. Figure 7 shows the contour plot, which shows that for the least amount of water absorption, 30 weight percent of plastics and 5 weight percent of thorn powder are added. Table 4 shows that the addition of waste plastic particles has the greatest impact on the composites' ability to absorb moisture, followed by the composition of thorn powders. The water intake qualities of PJ thorn powders are unaffected by the chemical treatment that was performed on them. Nearness to the mean is indicated by the R-Square value of 97.06. The F-value from the ANOVA table was used to generate the contribution plot in Figure 8.

$$\text{The regression equation for minimum water intake in 24 h} = 1.45 - 0.023056\ \text{Composition of plastics} + 0.02644\ \text{Composition of thorn powder} - 0.02056\ \text{types of chemical treatment} \quad (4)$$

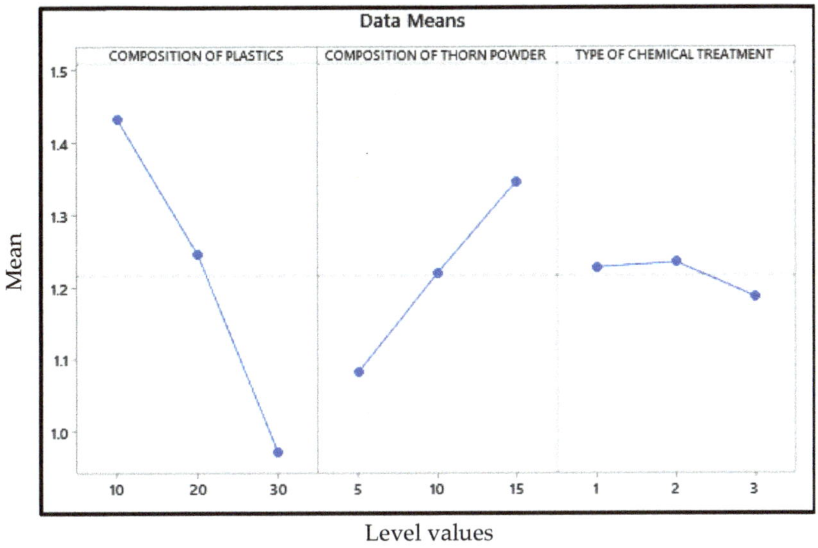

Figure 5. Effects plot for water absorption.

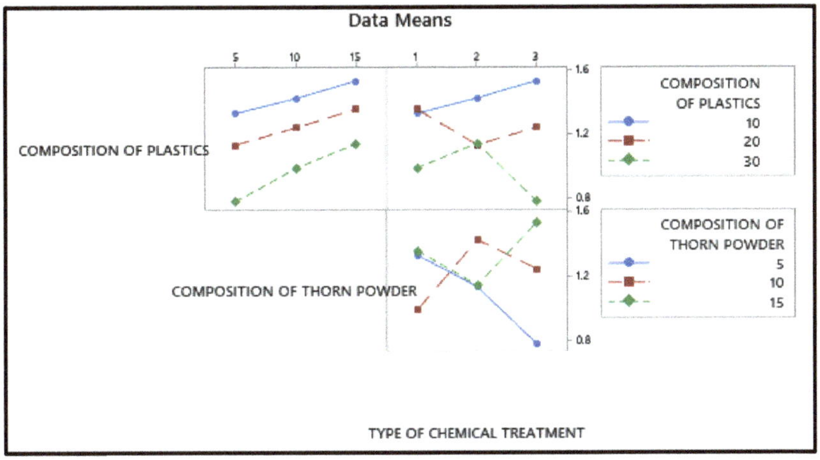

Figure 6. Interaction plot for water absorption.

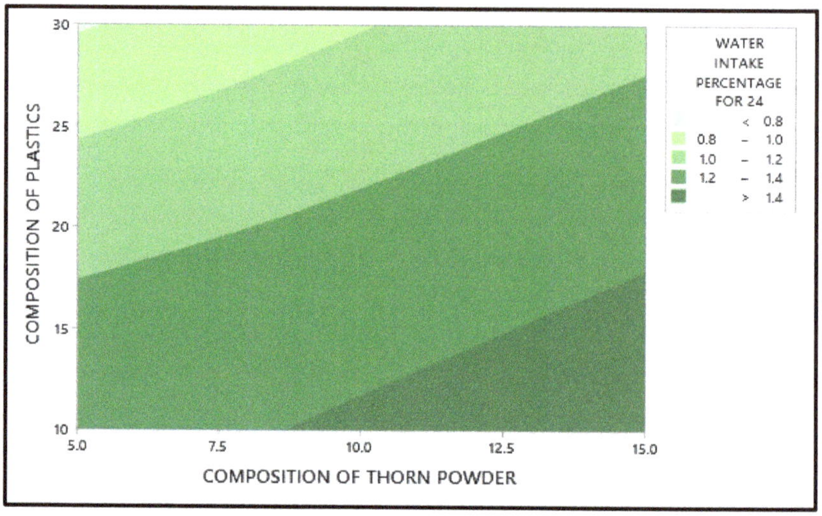

Figure 7. Water intake test result through contour graph.

Table 4. Water intake results from DOE.

Variables	D.F	Adj SS	Adj MS	F-Value	*p*-Value
Regression	3	1.2791	0.426367	253.2	0
Composition of Plastics	1	0.95681	0.956806	568.21	0
Composition of Thorn Powder	1	0.31469	0.314689	186.88	0
Chemical Treatment	1	0.00761	0.007606	4.52	0.045
Error	23	0.03873	0.001684		
Lack-of-Fit	5	0.02813	0.005626	9.55	0
Error	18	0.0106	0.000589		
Total	26	1.31783			

S: 0.041035, R-Sq: 97.06%, R-Squared (Adj): 96.68%, R-Squared (Predicated): 95.98%.

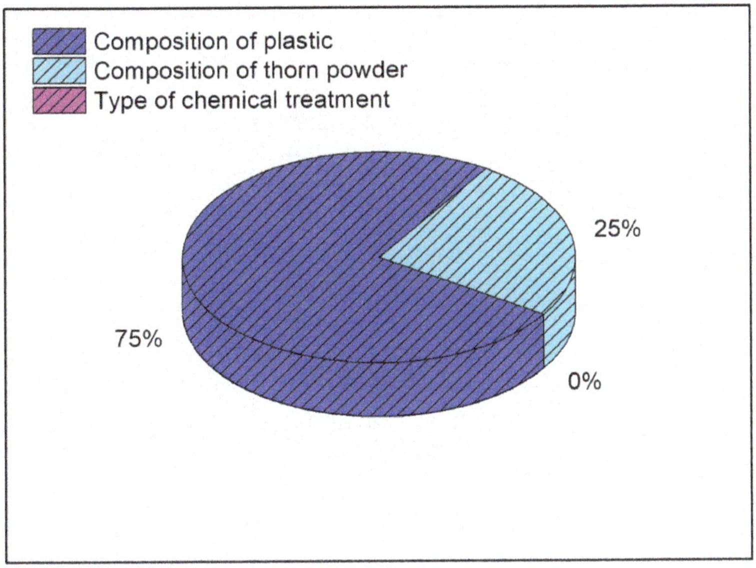

Figure 8. Contribution plot for water absorption.

3.3. Tensile Strength

According to the results of the optimization, in order to achieve optimum tensile strength, the amount of plastics added must be maximized, and the amount of thorn powder added must be 5 weight percent, with the treatment of the thorn powder having little effect. The addition of greater volumes of waste plastic improved the elongation strength. It is possible to achieve an improvement in tensile strength by improving the bonding between the matrix and the reinforcement [30]. The tensile results demonstrate that the plastics introduced as reinforcement have a stronger bonding to the matrix. There will be gaps between the reinforcements and the matrix in some locations due to the resin shrinking during curing, which will result in the composite failing when the load is applied. However, in this scenario, the problem is resolved, since the fillers cover the gaps and improve the composite's performance under pressure. As shown in Figure 9, the scanning electron microscope pictures were captured after the sample had ruptured, with the emphasis being on the ruptured area of the sample. SEM images will show the wear and erosion mechanisms that happened during wear testing [21]. The key effects plot for tensile strength is depicted in Figure 10 in which the tensile strength mean is mentioned in the Y axis and their levels are indicated in X axis.

Particulate plastics are visible; thorn powder is equally dispersed on top of the composite. The fiber-matrix interface reveals improved adhesion between plastics and fibers. Due to the detachment of plastic particles, pits can form on the surface. Figure 11 shows the relationship between the input and output parameters through an interaction plot. A3B1C3 are the input elements needed to achieve maximal tensile strength. In Figure 12, contour plot likewise mirrors the main effect plot's results. To see the tensile strength of the plastic and thorn powder mixture, look at the graph in Figure 13. Plastic particles make up most, while thorn powder has little effect on tensile characteristics which is seen in Table 5. There is almost a 95 percent confidence level in the model is seen based on the R-Squared value. The results are in agreement with the findings of earlier researchers [6,7]. Figure 13 displays the tensile strength contribution plot.

$$\text{Regression equation for Maximum tensile strength} = 60.079 + 0.3419\ \text{Composition of plastics} - 0.1421\ \text{Composition of thorn powder} + 0.082\ \text{type of chemical treatment} \quad (5)$$

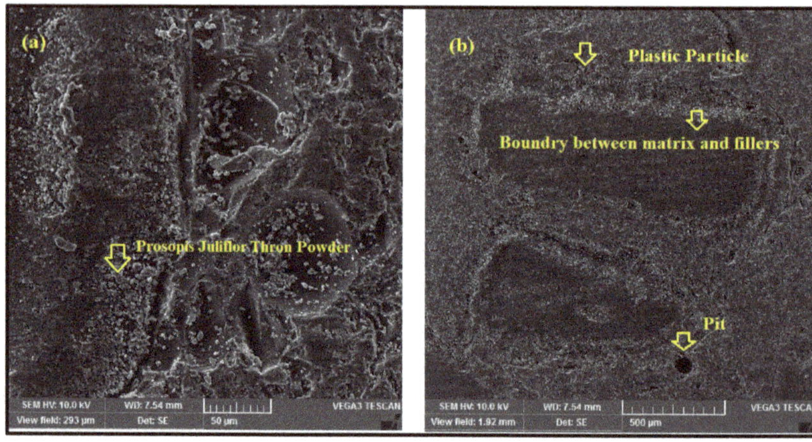

Figure 9. (**a**). SEM image with 50 µm magnification (**b**). SEM image with 500 µm magnification.

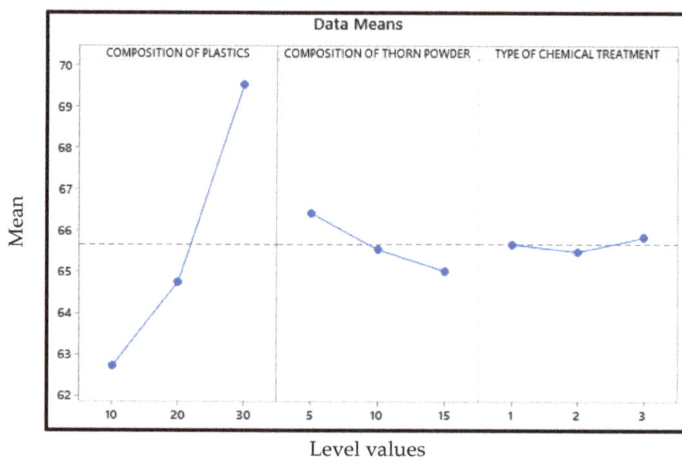

Figure 10. Effects graph for tensile strength.

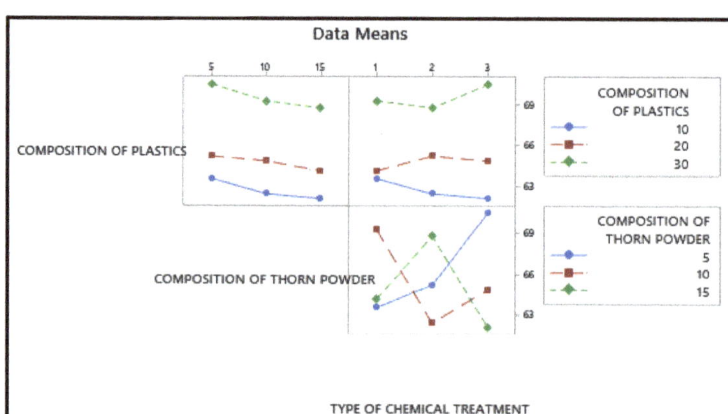

Figure 11. Interaction plot for Maximum tensile strength.

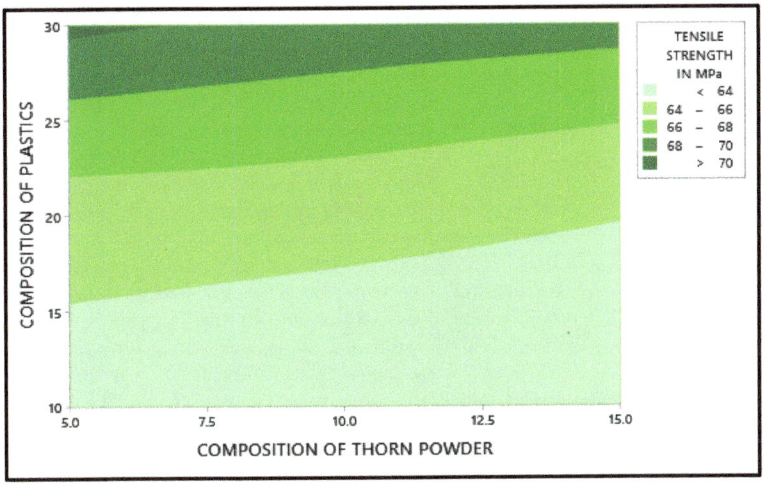

Figure 12. Tensile strength results through contour graph.

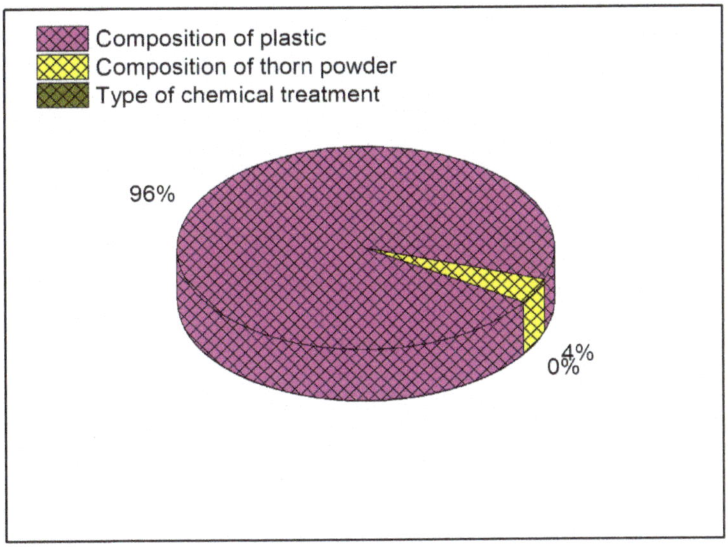

Figure 13. Contribution plot for Maximum tensile strength.

Table 5. Tensile strength results from DOE.

Variables	D.F	Adj SS	Adj MS	F-Value	*p*-Value
Regression	3	219.608	73.203	136.59	0
Composition Of Plastics	1	210.398	210.398	392.57	0
Composition Of Thorn Powder	1	9.088	9.088	16.96	0
Type Of Chemical Treatment	1	0.122	0.122	0.23	0.638
Error	23	12.327	0.536		
Lack-of-Fit	5	12.325	2.465	20797.84	0
Pure Error	18	0.002	0		
Total	26	231.935			

S: 0.732084, R^2: 94.69%, R^{-2} (Adj): 93.99%, R^{-2} (Pred): 92.84%.

3.4. Abrasion Wear Behavior

For the lowest wear rate, the main effects plot is shown in Figure 14 in which the mean value of the wear rate is mentioned in Y axis with their levels in X axis. Porosity reduction due to the inclusion of thorn powder is the important factor in deciding composite's wear resistance. However, the amount of waste plastic and the type of chemical treatment do not have a significant impact. The increase in hardness is due to the increased amount of thorn powder [14]. Hardness has a direct correlation to wear resistance [31]. Wear resistance is improved by increasing the amount of hardness. The wear-resistant is boosted by the thorn powder. Results for several polymer-based composite fillers have been published by others [14,15,32,33]. The interaction plot for the lowest wear rate can be seen in Figure 15. Plastic and thorn powder have the strongest interactions. For a clearer picture of the wide range of possible reactions, see the contour plot in Figure 16. There must be 30 percent plastics and 15 percent thorn powder added in order to achieve a minimum wear rate of 0.05%. The main effect plot for minimal wear rate is shown in ANOVA Table 6, thorn powder contributes more to wear resistance than other additives. Thorn powder contributes 93 percent of the total contribution, while plastics supply 7 percent. The wear resistance of the composite is not affected by the chemical treatment. The composites' wear rate contribution plot is shown in Figure 17. Equation (4) contains the regression equation, and an R-Squared value of 98.39% shows that the experimental responses are closer to the mean values.

$$\text{Regression equation for Minimum rate of wear} = 11.168 - 0.03261\ \text{composition of plastics} - 0.232\ \text{composition of thorn powder} - 0.0222\ \text{type of chemical treatment} \quad (6)$$

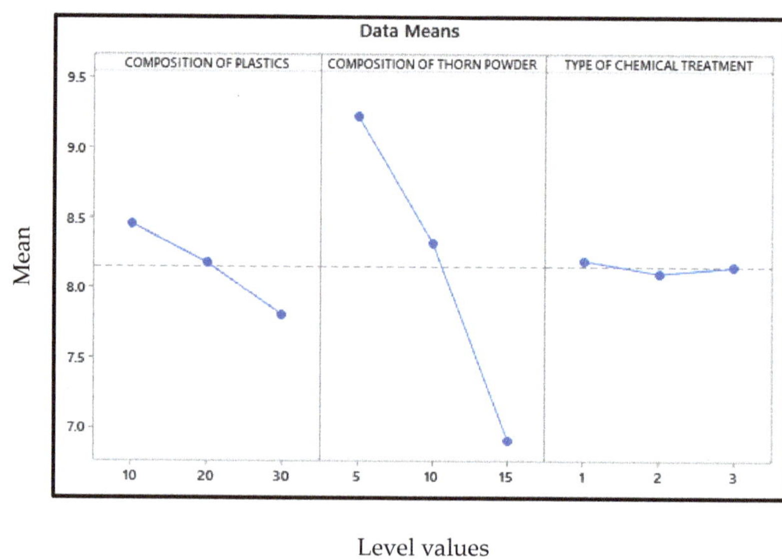

Figure 14. Main effect plot for minimum wear rate.

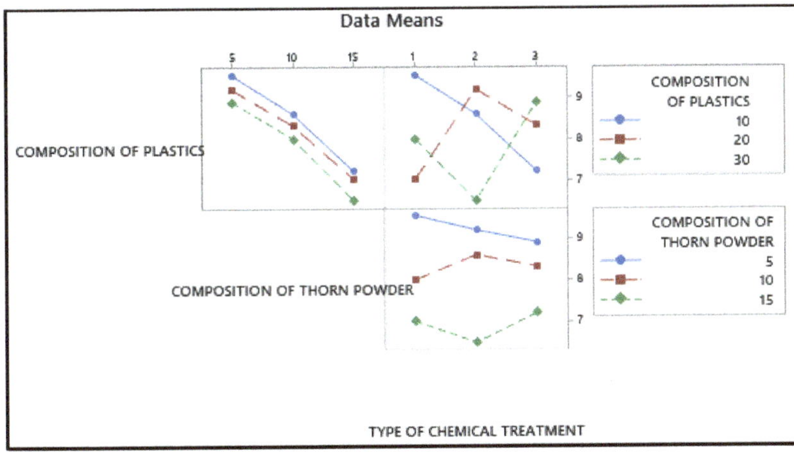

Figure 15. Interaction plot for minimum wear rate.

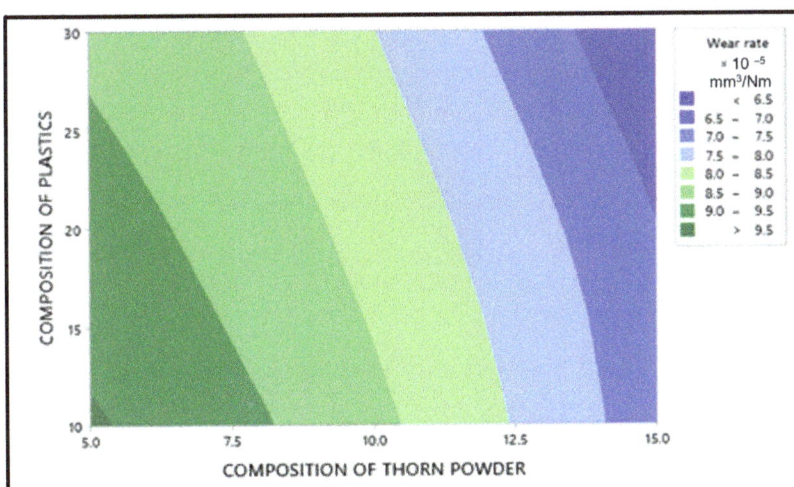

Figure 16. Wear rate results through contour graph.

Table 6. Wear rate results from DOE.

Variables	D.F	Adjacent SS	Adjacent MS	F-Value	*p*-Value
Regression	3	26.1904	8.7301	467.52	0
Composition Of Plastics	1	1.9143	1.9143	102.51	0
Composition Of Thorn Powder	1	24.2672	24.2672	1299.58	0
Type Of Chemical Treatment	1	0.0089	0.0089	0.48	0.497
Error	23	0.4295	0.0187		
Lack-of-Fit	5	0.4278	0.0856	888.4	0
Pure Error	18	0.0017	0.0001		
Total	26	26.6199			

S: 0.13665, R^{-2}: 98.39%, R^{-2} (Adj): 98.18%, R^{-2} (Pred): 97.86%.

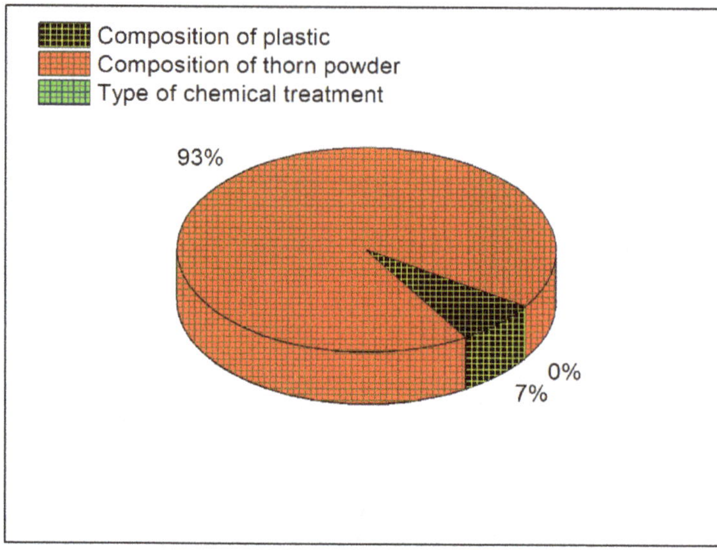

Figure 17. Contribution plot for Minimum wear rate.

4. Conclusions

PJ fibers are mixed with a variety of natural and synthetic fibers to create enhanced-property hybrid composites. Natural fibers' high strength-to-weight ratio, longevity, and inexpensive cost make them an excellent choice for polymer composites. Natural fiber composites are widely used in defense, automotive, and marine applications. A composite is made and tested using glass fiber and PJ in powder form. The results indicate that adding 6% PJ powder to glass fiber composites leads in increased impact and compressive strengths, as well as increased hardness [34].

In this study, the composite was created by combining waste plastic particles with Prosopis juliflora thorn powder according to Taguchi's full factorial design and laying it out by spraying. The trials are carried out in accordance with ASTM standards, and the results are entered into a software programme for further optimization. According to the optimal values, the additions of thorn powder improves the hardness and wear resistance property, and the inclusion of waste plastics improves the resistance to moisture absorption and the tensile properties of the material. A material with a high hardness will have a higher resistance to wear due to friction and abrasion. The use of fillers increases the composite's hardness, which is reflected in the composite's wear rate [35].

It is necessary to have a 30 weight percent composition of plastics, a 15 weight percent addition of thorn powder, and it must be silane treated in order to get maximum hardness. In order to get the lowest possible water intake while maintaining the highest possible tensile property, 30 weight percent waste plastic particles and 5 weight percent thorn powders must be incorporated. Due to the natural nature of the filler powder, its inclusion must be kept to a minimum to ensure optimal resistance to moisture absorption. Even if chemically treated, they will lose their hydrophobic properties in extreme circumstances and for extended periods of time, allowing water to permeate through the composites. Overall, chemical treatment had little effect on the hardness, tensile strength, or moisture intake characteristics of the material. Finally, in order to get the lowest possible abrasion wear, the maximum amount of polymers and thorn powder should be used. The validation tests are carried out for hardness in accordance with the projected levels, and it is discovered that the error value falls below the acceptability criteria.

Author Contributions: Original draft preparation, S.B.G. and A.R.S.; Investigation and supervision, A.R.S.; Final manuscript preparation, M.T.H.S., F.S.S., T.A.S. and A.U.M.S.; Funding acquisition, T.A.S. and T.K. All authors have read and agreed to the published version of the manuscript.

Funding: The APC was funded by Ministry of Higher Education, Malaysia, through the Fundamental Research Grant Scheme FGRS/1/2019/STG07/UPM/02/2, grant number 5540320.

Institutional Review Board Statement: Not applicable.

Informed Consent Statement: Not applicable.

Data Availability Statement: Not applicable.

Acknowledgments: The authors would like to thank the Department of Aerospace Engineering, Faculty of Engineering, Universiti Putra Malaysia and Laboratory of Bio-composite Technology, Institute of Tropical Forestry and Forest Product (INTROP), Universiti Putra Malaysia (HICOE) for the close collaboration in this research. The authors would like to acknowledge the support of Prince Sultan University for paying the Article Processing Charge (APC) of this publication.

Conflicts of Interest: The authors declare no conflict of interest.

Abbreviations

PJ	Prosopis juliflora
RH	Rice husk
NaOH	Sodium hydroxide
ANOVA	Analysis of Variance
GFRP	Glass fiber reinforced polymer
W/V	Weight by volume
V/V	Volume by volume
HCL	Hydrochloric acid
pH	Potential of Hydrogen
UTM	Universal testing machine
ASTM	American society for testing and materials
UK	United Kingdom
DOE	Design of Experiments
DOF	Degrees of freedom
SS	Sum of squares
MS	Mean sum of squares
F-value	Fisher's value
p-value	Probability value
R^2	Regression square
SEM	Scanning electron microscopy

References

1. Marichelvam, M.K.; Manimaran, P.; Sanjay, M.R.; Siengchin, S.; Geetha, M.; Kandakodeeswaran, K.; Boonyasopon, P.; Gorbatyuk, S. Extraction and Development of Starch-Based Bioplastics from Prosopis Juliflora Plant: Eco-Friendly and Sustainability Aspects. *Curr. Res. Green Sustain. Chem.* **2022**, *5*, 100296. [CrossRef]
2. Wakie, T.T.; Laituri, M.; Evangelista, P.H. Assessing the Distribution and Impacts of Prosopis Juliflora through Participatory Approaches. *Appl. Geogr.* **2016**, *66*, 132–143. [CrossRef]
3. Sato, T. Beyond Water-Intensive Agriculture: Expansion of Prosopis Juliflora and Its Growing Economic Use in Tamil Nadu, India. *Land Use Policy* **2013**, *35*, 283–292. [CrossRef]
4. Kailappan, R.; Gothandapani, L.; Viswanathan, R. Production of Activated Carbon from Prosopis (Prosopis Juliflora). *Bioresour. Technol.* **2000**, *75*, 241–243. [CrossRef]
5. Surya Rajan, B.; Balaji, M.A.S.; Saravanakumar, S.S. Effect of Chemical Treatment and Fiber Loading on Physico-Mechanical Properties of Prosopis Juliflora Fiber Reinforced Hybrid Friction Composite. *Mater. Res. Express* **2019**, *6*, 035302. [CrossRef]
6. Madhu, P.; Pradeep, S.; Sanjay, M.R.; Siengchin, S. Characterization of Raw and Alkali Treated Prosopis Juliflora Fibers for Potential Polymer Composite Reinforcement. *IOP Conf. Ser. Mater. Sci. Eng.* **2019**, *653*, 012016. [CrossRef]
7. Manoj Kumar, G.; Uthranarayan, C.; Joseph Jebaraj, D.J.; Keerthana, S.; Ganesh, N. Exploration of Tensile, Flexural and Hardness Test Properties of Prosopis Juliflora/Glass/Epoxy Hybrid Composite Laminates. *J. Phys. Conf. Ser.* **2019**, *1362*, 012015. [CrossRef]

8. Valencia, L.; Arumughan, V.; Jalvo, B.; Maria, H.J.; Thomas, S.; Mathew, A.P. Nanolignocellulose Extracted from Environmentally Undesired Prosopis Juliflora. *ACS Omega* **2019**, *4*, 4330–4338. [CrossRef]
9. Santhosh, M.S.; Karthikeyan, G.; Sasikumar, R.; Hariharan, R.; Mohanraj, R. Mechanical and Morphological Behaviour of Rice Husk/Prosopis Juliflora Reinforced Bio Composites. *Mater. Today Proc.* **2020**, *27*, 556–560. [CrossRef]
10. Kathirvel, P.; Anik, G.A.; Kaliyaperumal, S.R.M. Effect of Partial Replacement of Cement with Prosopis Juliflora Ash on the Strength and Microstructural Characteristics of Cement Concrete. *Constr. Build. Mater.* **2019**, *225*, 273–282. [CrossRef]
11. Shanmugapriya, S.; Surendran, S.; Lee, Y.S.; Selvan, R.K. Improved Surface Charge Storage Properties of Prosopis Juliflora (Pods) Derived Onion-like Porous Carbon through Redox-Mediated Reactions for Electric Double Layer Capacitors. *Appl. Surf. Sci.* **2019**, *492*, 896–908. [CrossRef]
12. Balan, G.S.; Ravichandran, M. Study of Moisture Absorption Characteristics of Jute Fiber Reinforced Waste Plastic Filled Polymer Composite. *Mater. Today Proc.* **2020**, *27*, 712–717. [CrossRef]
13. Arthanarieswaran, V.P.; Kumaravel, A.; Kathirselvam, M.; Saravanakumar, S.S. Mechanical and Thermal Properties of Acacia Leucophloea Fiber/Epoxy Composites: Influence of Fiber Loading and Alkali Treatment. *Int. J. Polym. Anal. Charact.* **2016**, *21*, 571–583. [CrossRef]
14. Swain, P.T.R.; Biswas, S. Abrasive Wear Behaviour of Surface Modified Jute Fiber Reinforced Epoxy Composites. *Mater. Res.* **2017**, *20*, 661–671. [CrossRef]
15. Darshan, S.M.; Suresha, B. Mechanical and Abrasive Wear Behaviour of Waste Silk Fiber Reinforced Epoxy Biocomposites Using Taguchi Method. *Mater. Sci. Forum* **2019**, *969*, 787–793. [CrossRef]
16. Shiva Kumar, K.; Chennakesava Reddy, A. Investigation on Mechanical Properties and Wear Performance of Nylon-6/Boron Nitride Polymer Composites by Using Taguchi Technique. *Results Mater.* **2020**, *5*, 100070. [CrossRef]
17. Ferdous, W.; Manalo, A.; Aravinthan, T. Bond Behaviour of Composite Sandwich Panel and Epoxy Polymer Matrix: Taguchi Design of Experiments and Theoretical Predictions. *Constr. Build. Mater.* **2017**, *145*, 76–87. [CrossRef]
18. Karthik, K.; Rajamani, D.; Manimaran, A.; Udaya Prakash, J. Wear Behaviour of Hybrid Polymer Matrix Composites Using Taguchi Technique. *Mater. Today Proc.* **2020**, *33*, 3186–3190. [CrossRef]
19. Siva Prasad, K.; Chaitanya, G. Optimization of Process Parameters on Surface Roughness during Drilling of GFRP Composites Using Taguchi Technique. *Mater. Today Proc.* **2020**, *39*, 1553–1558. [CrossRef]
20. Azman, M.A.; Asyraf, M.R.M.; Khalina, A.; Petrů, M.; Ruzaidi, C.M.; Sapuan, S.M.; Wan Nik, W.B.; Ishak, M.R.; Ilyas, R.A.; Suriani, M.J. Natural Fiber Reinforced Composite Material for Product Design: A Short Review. *Polymers* **2021**, *13*, 1917. [CrossRef]
21. Khare, J.M.; Dahiya, S.; Gangil, B.; Ranakoti, L.; Sharma, S.; Huzaifah, M.R.M.; Ilyas, R.A.; Dwivedi, S.P.; Chattopadhyaya, S.; Kilinc, H.C.; et al. Comparative Analysis of Erosive Wear Behaviour of Epoxy, Polyester and Vinyl Esters Based Thermosetting Polymer Composites for Human Prosthetic Applications Using Taguchi Design. *Polymers* **2021**, *13*, 3607. [CrossRef]
22. Sienkiewicz, N.; Dominic, M.; Parameswaranpillai, J. Natural Fillers as Potential Modifying Agents for Epoxy Composition: A Review. *Polymers* **2022**, *14*, 265. [CrossRef]
23. Sakthi Balan, G.; Ravichandran, M.; Kumar, V.S. Study of ageing effect on mechanical properties of prosopis juliflora fibre reinforced palm seed powder filled polymer composite. *Aust. J. Mech. Eng.* **2020**, *18*, 1–13. [CrossRef]
24. Agunsoye, J.O.; Isaac, T.S.; Awe, O.I.; Onwuegbuzie, A.T. Effect of Silicon Additions on the Wear Properties of Grey Cast Iron. *J. Miner. Mater. Charact. Eng.* **2013**, *2013*, 61–67. [CrossRef]
25. Sujin Jose, A.; Athijayamani, A.; Ramanathan, K.; Sidhardhan, S. Effects of Aspect Ratio and Loading on the Mechanical Properties of Prosopis Juliflora Fibre-Reinforced Phenol Formaldehyde Composites. *Fibres Text. East. Eur.* **2017**, *25*, 59–64. [CrossRef]
26. Stalin, B.; Nagaprasad, N.; Vignesh, V.; Ravichandran, M.; Rajini, N.; Ismail, S.O.; Mohammad, F. Evaluation of Mechanical, Thermal and Water Absorption Behaviors of Polyalthia Longifolia Seed Reinforced Vinyl Ester Composites. *Carbohydr. Polym.* **2020**, *248*, 116748. [CrossRef] [PubMed]
27. Gopal Krishna, U.B.; Srinivasa, C.S.; Amara, N.S.; Gudoor, S. Processing, Characterization and Property Evaluation of Seashell and Glass Fibre Added Epoxy Based Polymer Matrix Composite. *Mater. Today Proc.* **2019**, *35*, 417–422. [CrossRef]
28. Saba, N.; Paridah, M.T.; Jawaid, M. Mechanical Properties of Kenaf Fibre Reinforced Polymer Composite: A Review. *Constr. Build. Mater.* **2015**, *76*, 87–96. [CrossRef]
29. Nagaraj, N.; Balasubramaniam, S.; Venkataraman, V.; Manickam, R.; Nagarajan, R.; Sikiru Oluwarotimi, I. Effect of Cellulosic Filler Loading on Mechanical and Thermal Properties of Date Palm Seed/Vinyl Ester Composites. *Int. J. Biol. Macromol.* **2020**, *147*, 53–66. [CrossRef]
30. Vigneshwaran, K.; Venkateshwaran, N.; Srinivasan, S.P. Mechanical, Thermal and Vibration Characteristics of Dosinia Exoleta Dispersed Polymer Composites. *Int. J. Polym. Anal. Charact.* **2018**, *23*, 646–656. [CrossRef]
31. Elkhouly, H.I.; Abdel-Magied, R.K.; Aly, M.F. Date Palm Seed as Suitable Filler Material in Glass–Epoxy Composites. *Iran. Polym. J. Engl. Ed.* **2019**, *28*, 65–73. [CrossRef]
32. Balan, G.S.; Ganesh, N.; Ravichandran, M. Study of Tribological and Water Intake Characteristics of Epoxy Based Hybrid Composite. *Mater. Today Proc.* **2020**, *27*, 729–735. [CrossRef]
33. Ananthu, M.; Shamnadh, M.; Dileep, P.N. Experimental Evaluation on Mechanical Properties and Wear Resistance in PMMA Seashell Bionanocomposite for Medical Application. *Mater. Today Proc.* **2018**, *5*, 25657–25666. [CrossRef]

34. Palanivendhan, M.; Chandradass, J.; Kaviyarasu, T.; Philip, J. Fabrication and Characteristics of Hybrid Glass Fiber/Prosopis Juliflora Reinforced Epoxy Composite. *Mater. Today Proc.* **2021**, *45*, 6833–6837. [CrossRef]
35. Balasundaram, R.; Devi, S.S.; Balan, G.S. Machine learning approaches for prediction of properties of natural fiber composites: Apriori algorithm. *Aust. J. Mech. Eng.* **2022**, *20*, 1–16. [CrossRef]

polymers

MDPI

Article

Mechanical and Dielectric Properties of Fly Ash Geopolymer/Sugarcane Bagasse Ash Composites

Nattapong Chuewangkam [1], Theeranuch Nachaithong [1], Narong Chanlek [2], Prasit Thongbai [1,3] and Supree Pinitsoontorn [1,3,*]

1 Department of Physics, Faculty of Science, Khon Kaen University, Khon Kaen 40002, Thailand; nattapong.ch@kkumail.com (N.C.); theeranuch.tn@gmail.com (T.N.); pthongbai@kku.ac.th (P.T.)
2 Synchrotron Light Research Institute (Public Organization), 111 University Avenue, Muang District, Nakhon Ratchasima 30000, Thailand; narong@slri.or.th
3 Institute of Nanomaterials Research and Innovation for Energy (IN-RIE), Khon Kaen University, Khon Kaen 40002, Thailand
* Correspondence: psupree@kku.ac.th

Abstract: Fly ash (FA) and sugarcane bagasse ash (SCBA) are the wastes from lignite power plants and sugar industries, usually disposed of as landfills. In this research, these wastes were effectively utilized as a construction material, namely geopolymer. The effect of the SCBA (0–40 wt.%) addition to the FA geopolymers was investigated. The compressive strength of the FA geopolymers was reduced with the SCBA addition. The reduction was mainly due to the presence of the highly stable and non-reactive quartz (SiO_2) phase in SCBA. The SCBA was not dissolved in the alkaline activated solution and hence did not contribute to the geopolymerization process. The unreacted SCBA particles remained in the geopolymer matrix but did not provide strength. However, if the amount of SCBA was about 10 wt.% or less, the impact on the characteristics and properties of FA geopolymers was minimal. Furthermore, this research also studied the dielectric properties of the FA geopolymer/SCBA composites. The relatively large dielectric constant ($\varepsilon' = 3.6 \times 10^3$) was found for the pristine geopolymer. The addition of SCBA decreased the ε' slightly due to high carbon content in SCBA. Nevertheless, the variation in ε' was mainly controlled by the geopolymerization process to form the aluminosilicate gel structure.

Keywords: geopolymer; fly ash; sugarcane bagasse ash; mechanical properties; dielectric properties

Citation: Chuewangkam, N.; Nachaithong, T.; Chanlek, N.; Thongbai, P.; Pinitsoontorn, S. Mechanical and Dielectric Properties of Fly Ash Geopolymer/Sugarcane Bagasse Ash Composites. *Polymers* **2022**, *14*, 1140. https://doi.org/10.3390/polym14061140

Academic Editors: Sabu Thomas and Maya Jacob John

Received: 17 February 2022
Accepted: 10 March 2022
Published: 12 March 2022

1. Introduction

The Mae Moh power plant in Lampang province is Thailand's largest lignite electricity generating station. The process uses 45 thousand tons of lignite per day, or 16 million tons per year, and emits 4.4 million tons of fly ash (FA) per year [1]. On the other hand, in the sugar industry, sugarcane is crushed to extract the juice. The fibrous residue, called bagasse, is used as a fuel source for feeding a boiler. Sugarcane bagasse ash (SCBA) is thus a residue obtained from the burning of bagasse in the sugar industry [2]. Thailand is the world's fourth-largest sugar producer and the second-largest exporter. In 2019, the sugar production capacity was 14.58 million tons, which used sugarcane of 125 million tons, consequently producing 800,000 tons of SCBA [3]. In general, both FA and SCBA are usually disposed of as landfills. Thus, vast space is necessary to dispose of these wastes, which generate major environmental issues, harming flora, animals, and people's health. Recycling and reusing waste should be seen from both ecological and economic standpoints. Therefore, several research studies have investigated waste utilization for various technological purposes [4–6]. For example, FA has been utilized in construction as a geopolymer-based material due to its high silica and alumina, low cost, and is highly reactive for geopolymerization.

Geopolymer is an inorganic binding material, mainly composed of alumina (Al_2O_3) and silica (SiO_2). Polymeric structures of Al–O–Si form the fundamental building blocks of a geopolymer. Many types of materials can be used as raw materials in geopolymer production. Most of them are recycled or by-product materials, such as FA, rice husk ash, or metakaolin [7]. Alkali metal salts and/or hydroxide are usually required for dissolving silica and alumina from raw materials [8]. Generally, geopolymer provides excellent mechanical and thermal properties, high durability, high initial strength, and environmental greenness [9,10]. Geopolymer manufacturing emits 80% less CO_2 than OPC production processes.

On the other hand, the SCBA contains a large amount of silica (62%) and some Al_2O_3, CaO, Fe_2O_3, and K_2O. Loss of ignition (LOI) of about 10% implies the high content of unburnt organic matter [11]. Many researchers have used SCBA as pozzolanic materials, a broad class of SiO_2 and Al_2O_3 materials possessing cementitious properties. SCBA was initially utilized in construction materials in 1998 [12]. The research studied the reaction between limestone and SCBA with pozzolanic characteristics by analyzing the mechanical properties of hardened cement pastes. It was found that adding a suitable amount of SCBA could enhance the compressive strength of the cement pastes [12]. According to most research, SCBA of 5–15 wt.% could be added to cement paste, mortar, and concrete to enhance its strength [5,13,14]. However, when a higher amount of SCBA was added, the mechanical properties dropped because of insufficient cement to bind aggregates and the presence of incompletely burned sugarcane bagasse and amorphous carbon in SCBA with poor strength [11,13–15]. However, Bahurudeen et al. reported that sieving SCBA before mixing with cement concrete increased the strength due to the complete removal of incompletely burned sugarcane bagasse and amorphous carbon [5]. As a result, SCBA could be explored as pozzolanic materials in geopolymer for improving mechanical properties.

Moreover, while SCBA could improve the strength of construction materials, it could act as a dielectric material due to its high silica content. A dielectric material is an electrical insulator that can be polarized by an applied electric field. When an electric field is applied to a dielectric material, electric charges do not flow through it as they do in an electrical conductor. Instead, they move slightly from their normal equilibrium positions, resulting in dielectric polarization [16]. Recently, many studies have been carried out on the dielectric properties of geopolymer [17–21]. Geopolymer is a cross-link long-chain inorganic polymer of AlO_4 and SiO_4 so that it requires charge balancing from alkali cations, such as Na^+ and K^+. The dielectric constant (ε') of the geopolymer paste and mortar at 24 h after mixing were found as 3.5 and 7–10, respectively [22,23]. At room temperature, the most important variables influencing the electrical conductivity and dielectric property of geopolymers are water molecules and hydroxide [24].

In this research, we investigated the effect of SCBA addition in FA-based geopolymer paste. The mechanical strengths, dielectric properties, microstructure, and functional groups of the geopolymer/SCBA composites were studied and discussed. We demonstrated that the waste from a lignite power plant (FA) and the waste from a sugar industry (SCBA) could be turned into useful construction materials. However, the addition of SCBA had a strong effect on the microstructure and properties of FA geopolymer. Therefore, the added amount of SCBA should not be too high. Our research suggested that the SCBA concentration should be limited to about 10 wt.% of the total FA + SCBA weight. By appropriately controlling the FA and SCBA, this research has the innovation to utilize industrial wastes into more value-added products. The geopolymer/SCBA composites in this work can be used as pre-formed bricks with dielectric properties that make them very useful as functional and smart construction materials.

2. Experimental Section/Methods

2.1. Raw Materials

The source material for geopolymer preparation was high calcium lignite fly ash (FA) from the Mae Moh power plant in Lampang Province, Thailand. Sugarcane bagasse ash

(SCBA) was donated from the sugar plant in Thailand (Thai Roong Ruang Research & Development Co., Ltd., Uthai Thani, Thailand). The alkaline-activated solutions were sodium hydroxide (NaOH) and sodium silicate (Na_2SiO_3) solutions. The 10 M NaOH solution was prepared by diluting NaOH flake (98% purity, AGC Chemicals (Thailand Co., Ltd., Samut Prakan, Thailand) in DI water. The Na_2SiO_3 solution was purchased from Eastern Silicate Co., Ltd., Chonburi, Thailand (12.53 wt.% Na_2O, 30.24 wt.% SiO_2, and 57.23 wt.% H_2O). The mole ratio of the Na_2SiO_3 solution was 2.49.

2.2. Sample Preparation

Process of treated SCBA as shown in Figure 1: the as-received SCBA was sieved to 75–600 μm (200–30 mesh) to remove unwanted fragile residues, such as incompletely burned sugarcane bagasse and other impurities. It was further treated by oven drying at 120 °C for 24 h to eliminate moisture. To prepare the FA-based geopolymer/SCBA composite, firstly, FA and SCBA were blended into a bowl for 5 min. The SCBA content was set to be 0, 10, 20, 30, and 40 wt.% of the powders' total weight. Then, the NaOH solution was added to the mixture and stirred by a mechanical blender at 285 rpm for 5 min. Subsequently, the Na_2SiO_3 solution was added and mixed for another 5 min at the same speed. The ratio of Na_2SiO_3 to NaOH was 1.0, and a liquid to ash ratio (L/A) of 0.45 was used. After that, the geopolymer paste was poured into a $2.5 \times 2.5 \times 2.5$ cm^3 acrylic mold for the mechanical property test. It was also poured into a disk-shaped mold (2.4 cm diameter and 0.6 cm thick) for the dielectric properties measurement. Finally, the geopolymer paste was left at the ambient temperature for 1 h, before being wrapped with a plastic film and cured in an oven at 60 °C for 24 h. This way of curing is useful for the pre-formed geopolymer bricks but may not be suitable for manufacturing at the construction sites. The sample was kept in a control room (25 °C and 50% RH) before the tests at 7 and 28 days.

Figure 1. A process for treatment of the raw sugarcane bagasse ash (SCBA).

2.3. Characterization Techniques

The rheological behavior of the geopolymer/SCBA paste right after mixing was tested using a miniature slump (mini-slump) cone test. [25,26]. The paste was injected into a truncated conical mold, which was then lifted to a vertical position in accordance with ASTM C143. The mixture was allowed to spread after the conical mold was removed, and the diameters of the paste were measured. The workability (%W) was calculated as:

$$\%W = \left(\frac{d - d_0}{d_0}\right) \times 100 \quad (1)$$

where d is the spread-out diameter of geopolymer paste, and d_0 is the original diameter.

X-ray fluorescence spectroscopy (XRF, Rigaku, ZSX Primus, Texas, USA) was used to evaluate the chemical composition of the as-received FA ash and the treated SCBA. The phase and crystal structure of the raw materials and geopolymer composites were examined by X-ray diffraction (XRD, Panalytical, Empyrean, Worcester, UK). Scanning electron microscopy (SEM, SEC, and SNE-4500M) was used to assess the microstructure of the geopolymer/SCBA composite. The fractured specimen after the mechanical test was gold-coated before being subjected to SEM investigation. The functional groups of the geopolymer were identified using a Fourier transform infrared spectroscope (FTIR, Bruker, TENSOR27, Boston, MA, USA). Moreover, the surface composition of the as-received FA ash and the treated SCBA were measured using X-ray photoelectron spectroscopy (XPS, PHI5000 VersaProbe II, ULVAC-PHI), at the SUT-NANOTEC- SLRI research facility, Synchrotron Light Research Institute (SLRI), Thailand.

The compressive strength was conducted on the geopolymer/SCBA specimens cured for 7 and 28 days using a universal testing equipment (Chun yen, CY-6040A12). Compressive load was applied at a rate of 50 kN/min until the specimen fractured, according to ASTM C109/C109M-20b [27]. Six samples were used for each experiment, and the average value was calculated for the compressive strength.

For dielectric measurement, the disk-shaped sample was used. The silver paste was painted at the top and bottom surfaces of the samples (diameter of 2.4 cm), before heating in air at 60 °C for 15 min to make good electrode contact. The impedance analyzer (Keysight E4990A) was used for dielectric measurement over the frequency range from 40 to 10^7 Hz using an oscillation voltage of 0.5 V at room temperature. The relative permittivity or dielectric constant (ε') was calculated from:

$$\varepsilon' = \frac{Ct}{\varepsilon_0 A} \quad (2)$$

where C is the sample's capacitance, t is the sample's thickness, ε_0 is the permittivity of free space (8.854×10^{-12} F/m), and A is the electrode area.

3. Results and Discussion

The raw SCBA was treated, as explained in Section 2, before mixing with FA and other chemicals to form a geopolymer. The chemical compositions of the FA and treated SCBA were determined using XRF, as shown in Table 1. The major oxides of FA are SiO_2, CaO, Al_2O_3, and Fe_2O_3. The high CaO content indicates the FA as Class C fly ash. In contrast, the main composition of SCBA is SiO_2 (66.9 wt.%) with relatively low other oxides. The microstructure of raw FA and treated SCBA was observed under scanning electron microscopy (SEM), as shown in Figure 2, along with the particle size distribution curves. The FA was spherical particles with a smooth surface. The particle size was around 2–20 µm, with the average particle size of 2.23 µm. On the other hand, the treated SCBA was of irregular shape with a rough surface. The size of the SCBA particle was more than an order of magnitude larger than the FA particle, with the average particle size of 105.22 µm.

Table 1. Chemical composition of as-received FA and treated SCBA.

Oxide Compound	Fly Ash (wt.%)	Treated SCBA (wt.%)
Silicon dioxide (SiO_2)	28.54	66.91
Calcium oxide (CaO)	26.37	9.48
Aluminum oxide (Al_2O_3)	14.94	6.66
Ferric oxide (Fe_2O_3)	18.14	8.11
Sulfur trioxide (SO_3)	4.56	0.32
Potassium oxide (K_2O)	2.80	3.49
Magnesium oxide (MgO)	2.01	1.62
Sodium oxide (Na_2O)	1.05	0.40

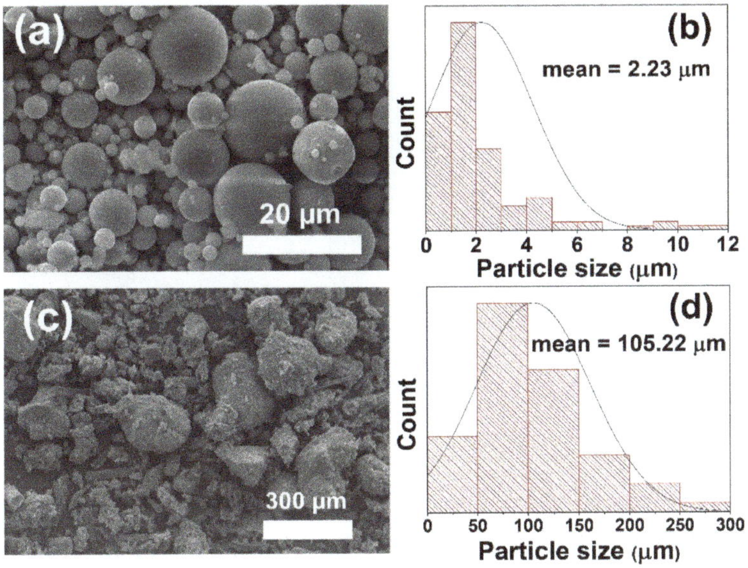

Figure 2. SEM micrographs and particle size distribution curves for (**a**,**b**) raw FA, and (**c**,**d**) treated SCBA.

The workability of the geopolymer composite as a function of SCBA concentration is presented in Figure 3. Workability significantly deteriorates at the SCBA levels greater than 10 wt.%. Figure 3b–d show the pictures of the geopolymer pastes during the workability test. For the geopolymer paste without SCBA, the paste spreads readily and exhibits good rheological flow. However, adding SCBA into the paste increases its viscosity, especially the geopolymer/SCBA-40 wt.% paste, which becomes so viscous that it hardly flows. This result could be due to the SCBA's high porousness and high rate of solution absorption. Workability between 150–250% is suitable for the casting and drying processes [28,29]. As a result, the optimal concentration of the added in SCBA in the geopolymer paste was between 10 to 30 wt.%.

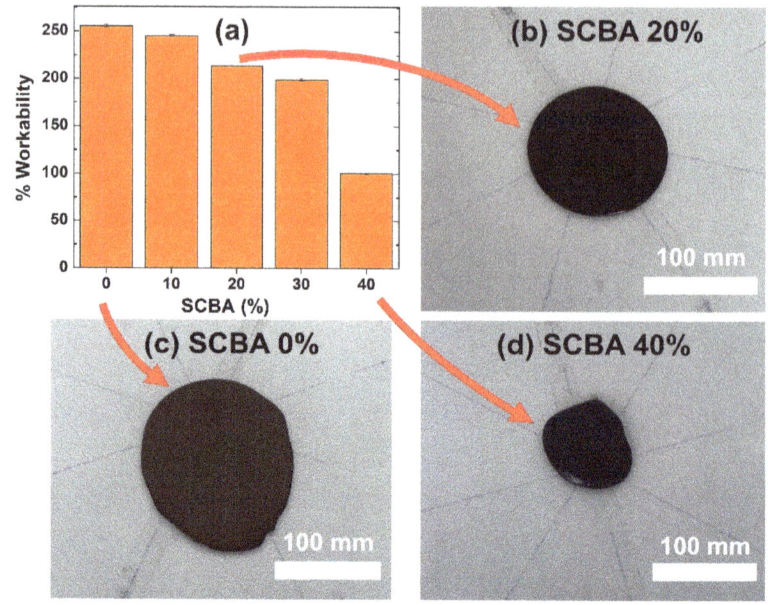

Figure 3. (**a**) Workability of the geopolymer/SCBA pastes; (**b–d**) the spread of geopolymer pastes during workability tests.

The X-ray diffraction (XRD) patterns of raw FA, treated SCBA, geopolymer paste, and geopolymer/SCBA composite are shown in Figure 4. The raw FA shows a broad XRD hump around 2θ = 20–40°, a common characteristic of an amorphous phase. Apart from that, other crystalline peaks are observed for the raw FA, namely Anhydrite (A), quartz (Q), hematite (F), and calcium oxide (C), as indicated in Figure 4. For the treated SCBA, the XRD pattern clearly shows the high crystalline peak of the quartz phase, similar to previous reports [30]. The FA geopolymer exhibits mostly an amorphous phase with minor quartz peaks. The amorphous phase is the characteristic of geopolymeric gel as observed elsewhere [17], whereas the minor quartz peaks are due to the quartz crystal in FA. When SCBA was added to the geopolymer, the XRD pattern became the combination between the FA geopolymer and the raw SCBA. In other words, it shows the feature of the broad hump from the amorphous phase of geopolymeric gel and also the sharp quartz peak from SCBA. The intensity of the quartz peak increased proportionally with the amount of SCBA addition. It implies that the highly crystalline quartz phase of SCBA did not interact or form chemical bonds with geopolymeric gel during the geopolymerization process [31]. Thus, the SCBA did not contribute to the aluminosilicate building block of geopolymer.

The compressive strengths of the geopolymer/SCBA composite pastes at 7 and 28 days are presented in Figure 5. The obvious point from the figure is that the strength at 28 days is much higher than at 7 days. In addition, the general trend shows that adding SCBA reduces the strength of geopolymer both at 7 and 28 days. The explanation for both observations is as follows. Geopolymerization consists of three stages: deconstruction, polymerization, and stabilization [32]. Deconstruction dissolves alumina and silica from FA in the alkaline-activated solution. The alumina and silica then form aluminosilicate geopolymeric gel during the polymerization stage. For stabilization, the gels are interconnected to form more extensive networks, and the strength of the geopolymer paste develops. However, the stabilization is slow and requires many days for the strength to be fully developed [33]. That is why the overall strengths are lower for the 7-day sample.

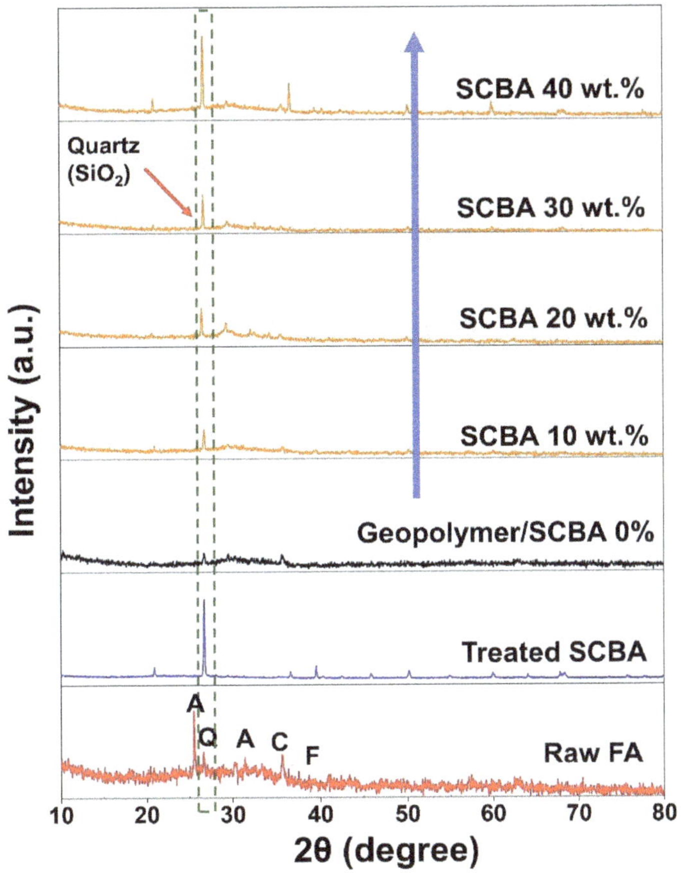

Figure 4. XRD patterns of the raw fly ash, treated SCBA, geopolymer paste, and geopolymer/SCBA composite. Denote the initial of phases: Q = quartz, A = anhydrite, C = calcium oxide, F = hematite.

The reduced strength with higher SCBA could be due to the highly crystalline quartz (SiO_2) in SCBA. Although the basis of the geopolymer structure is an Al-O-Si polymeric chain [34], the initial alumina and silica sources need to be firstly dissolved in an alkaline activated solution before proceeding to the subsequent geopolymerization stages. Thus, they are better in an amorphous form for ease of dissolution. This is what happens to the raw FA for their geopolymerization. However, since SCBA contains a significant amount of quartz, which is very stable and nearly undissolved in the alkaline activated solution, the deconstruction step was not achieved, and the geopolymerization never occurred. Therefore, SCBA did not contribute to the formation of aluminosilicate gels, as evidenced from XRD in Figure 4. Adding more SCBA means there is less FA geopolymer, which provides the strength. The other reason is attributed to the different sizes of the raw FA and SCBA, as shown in Figure 2. As the size of SCBA particles is more than 10 times larger than the FA, the SCBA is less reactive and thus had a negative impact on the geopolymer strength.

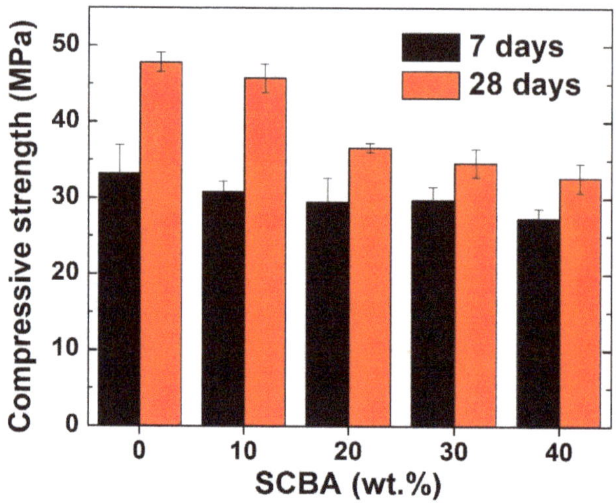

Figure 5. Compressive strength of geopolymer paste and geopolymer/SCBA composite pastes.

Furthermore, the effect of strength reduction with SCBA is more prominent for the geopolymer pastes aged for 28 days. The strength at 28 days reduced from 47.8 MPa for the geopolymer paste without SCBA to 32.7 MPa for the geopolymer/SCBA-40 wt.% composite, which accounts for >30% in strength reduction. On the other hand, the strength of the geopolymer/SCBA-10 wt.% composite is 45.7 MPa, almost unchanged compared to the pristine paste. Thus, it can be concluded that a small amount of SCBA (up to 10 wt.%) can be added in FA geopolymer, as an effective way for utilizing SCBA wastes, without sacrificing its mechanical property.

The microstructures of geopolymer/SCBA pastes were examined, as illustrated in Figure 6. The fractured surfaces were examined after mechanical tests at 7 and 28 days. The distinct features from the SEM images are that the geopolymer paste without SCBA is fully dense with a smooth surface due to the formation of aluminosilicate gels from geopolymerization. However, few unreacted FA particles could still be observed (red arrows). In contrast, the surface of the composite pastes is rougher and is covered by flaky unreacted particles. These particles are partly unreacted FA but mostly are unreacted SCBA, which mainly consists of stable crystalline SiO_2. As more SCBA was added to the geopolymer composites, a higher fraction of unreacted SCBA particles is observed (yellow circles in Figure 6i). It was reported that the number of unreacted particles and the contact between them and the geopolymer matrix had a substantial negative impact on the overall strength of the material [35]. Thus, the observation from SEM also supports the changes in compressive strengths. As more SCBA particles were added to the geopolymer composite, these SCBA particles did not involve geopolymerization. Instead, it resulted in more unreacted particles left in the geopolymer matrix, which led to decreased mechanical property.

It should be noted that the SEM images in Figure 6 are not significantly different between the 7-day and 28-day. This implies that the microstructure of the geopolymer paste did not change much by aging. However, the geopolymerization process still goes on at the chemical bonding level. To prove this, the FTIR spectra were measured, as shown in Figure 7. Figure 7a shows a broad scan spectrum of the geopolymer paste. Several absorption bands are observed, for instance, the symmetric stretching of Al-O at 684 cm^{-1} [36], the asymmetric stretching vibrations of Si-O-Si or Al-O-Si at 948 cm^{-1}, [36,37], and the stretching vibration of O-C-O at 1416 cm^{-1} [38]. Moreover, water absorption on the geopolymer surface resulted in the bending vibration of H-O-H at 1648 cm^{-1} [39], and the hydroxyl (–OH) functional groups at 3356 cm^{-1} [40].

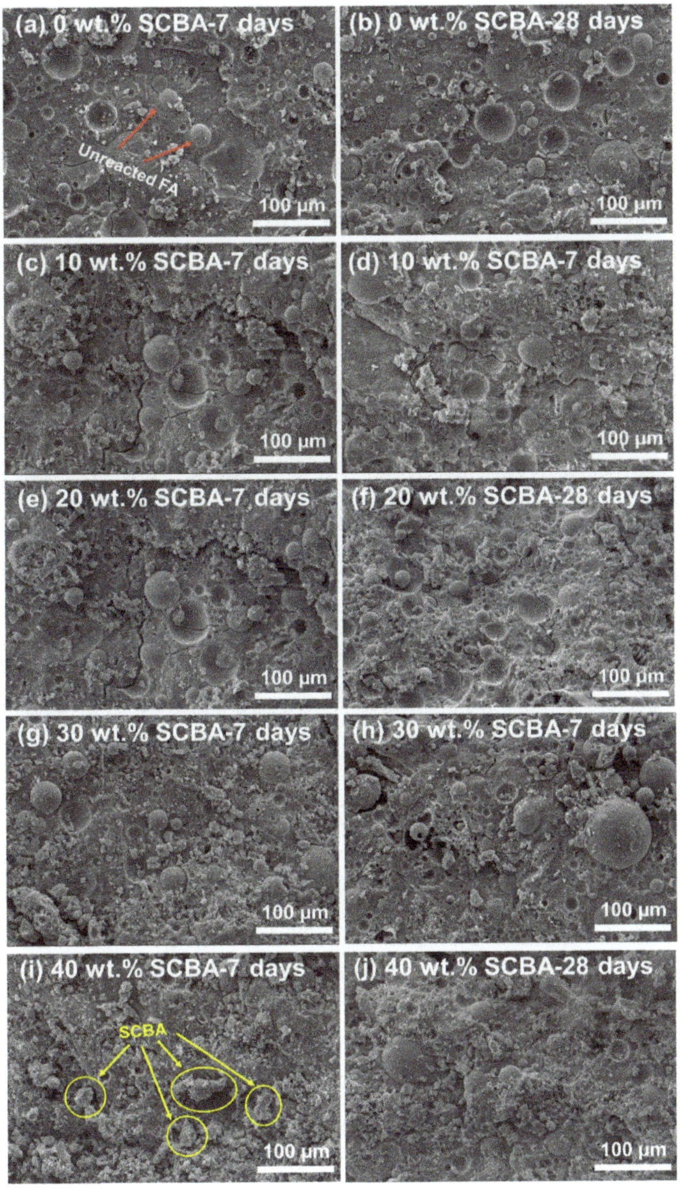

Figure 6. SEM micrographs of the geopolymer pastes at 7 and 28 days with SCBA of (**a**,**b**) 0 wt.%, (**c**,**d**) 10 wt.%, (**e**,**f**) 20 wt.%, (**g**,**h**) 30 wt.%, and (**i**,**j**) 40 wt.%.

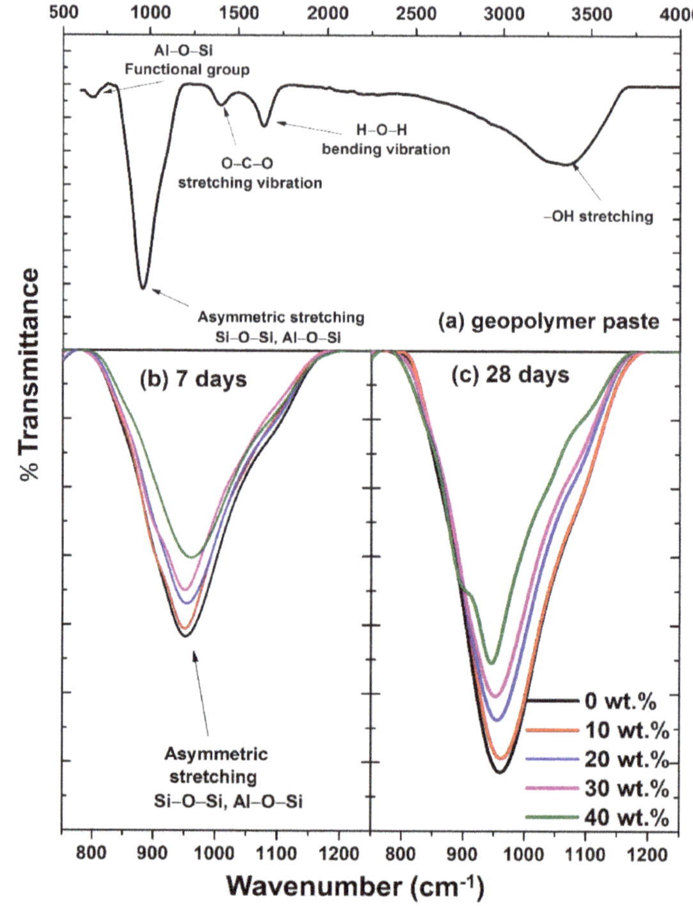

Figure 7. FTIR spectra of the geopolymer/SCBA composite pastes: (**a**) wide scan from 500 to 4000 cm^{-1}; (**b**,**c**) expanded views from 850 to 1250 cm^{-1} for determining the change in Si-O-Si and Al-O-Si band as a function of SCBA wt.%.

Geopolymerzation is related mainly to the asymmetric stretching vibrations of Si-O-Si or Al-O-Si at 948 cm^{-1}, as they indicate the formation of aluminosilicate building blocks. Therefore, we expanded that band spectra and compared them amongst different samples, as shown in Figure 7b,c. The intensity of the Si-O-Si (Al) band decreases as higher SCBA (0–40 wt.%) is added in the geopolymer composite. The same trend is observed for the samples aged 7 and 28 days. This indicates that the addition of SCBA suppresses the formation of aluminosilicate geopolymeric gels. Thus, the FTIR result is another key evidence to support the changes in mechanical properties.

The dielectric properties of the geopolymer/SCBA composites were measured, as shown in Figure 8. The dielectric constants (ε') for all samples decrease with increasing frequency. The dielectric response in a low-frequency region is usually caused by the interfacial polarization of composite materials or sample–electrode contact [41,42]. At low frequencies, the molecules are given sufficient time to spin and orient themselves in the direction of the applied AC at low frequencies [43]. However, at high frequencies, the time for re-orientation is not sufficient, resulting in the decreased ε' from the relaxation of a polarization process within the system [44].

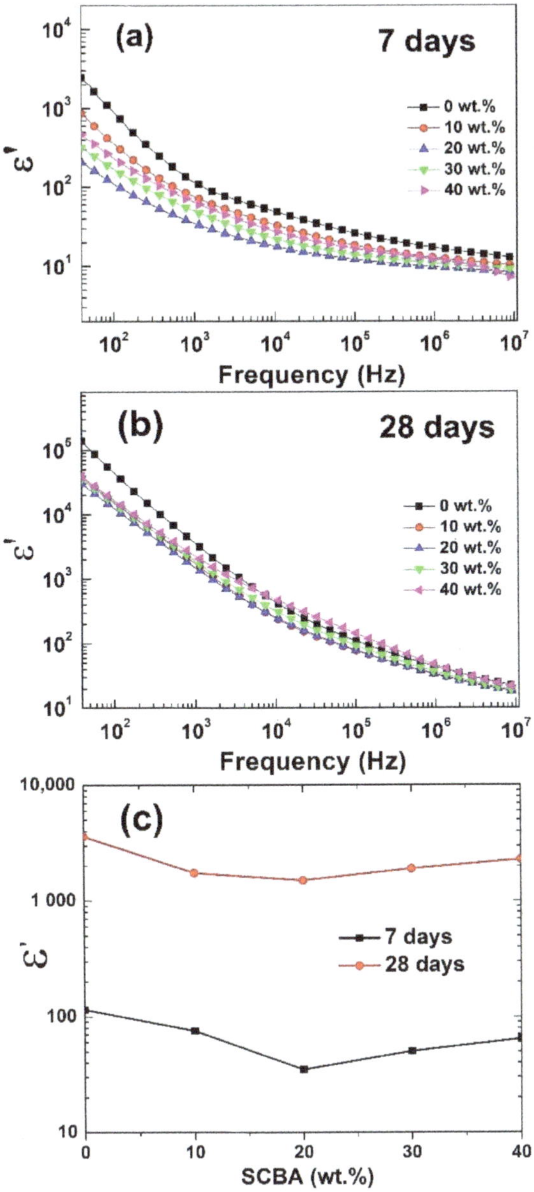

Figure 8. The dielectric constant ε' of geopolymer/SCBA pastes at (**a**) 7 days and (**b**) 28 days. (**c**) The variation of ε' at 1 kHz as a function of SCBA wt.% in geopolymer.

Furthermore, compared between the samples aged 7 and 28 days, the ε' constants of 28-day geopolymers are one order of magnitude larger. For example, the ε' values at 1 kHz of the geopolymer paste without SCBA were 1.2×10^2 and 3.6×10^3 for a 7-day and 28-day age, respectively. The increased ε' with the geopolymer age could be attributed to aluminosilicate gel and relative humidity. As mentioned earlier, geopolymerization is a slow process, and the aluminosilicate geopolymeric structure continues to develop with time. Thus, the extensive network of geopolymeric gel is more developed at 28 days, leading to larger ε', in an agreement in previous reports [45]. Moreover, the dielectric property is strongly affected by humidity in the samples—the higher the humidity, the lower ε' [17]. Therefore, as the aluminosilicate structure is in the more advanced stage at 28 days, the relative humidity in the geopolymer composite is decreased, contributing to the higher ε'.

The dielectric constants were also affected by the addition of SCBA. The ε' values decreased with SCBA inclusion in the FA geopolymer at 7 days (Figure 8a) and 28 days (Figure 8b). The variation of ε' at 1 kHz as a function of SCBA wt.% in geopolymer is plotted in Figure 8c. The decreased ε' could be attributed to an electrically conductive phase presented in SCBA. Table 2 shows the chemical compositions of the raw FA and the treated SCBA determined by X-ray photoelectron spectroscopy (XPS) analysis. Obviously, the carbon content of the SCBA is significantly higher than that of the FA. Figure 9 compares the XPS carbon peaks between FA and SCBA. It shows that the intensity of the carbon for SCBA is much higher. As carbon is a conductive material, the higher carbon fraction in SCBA reduces the sample's insulative property and suppresses the dielectric constants. However, the effect of SCBA addition was minor when compared to the geopolymer curing age (Figure 8c). In other words, the geopolymer/SCBA composites at 28 days still exhibit much higher ε' with respect to the pristine geopolymer paste at 7 days. It infers that a crucial factor in controlling the dielectric constant is the geopolymerization process to form the aluminosilicate gel structure.

Table 2. Chemical compositions of raw FA and treated SCBA determined by XPS analysis.

Compound	Fly Ash (at.%)	SCBA Treated (at.%)
C 1s	34.26	41.14
O 1s	52.66	46.75
Si 2p	7.92	10.88
Ca 2p	5.16	1.23

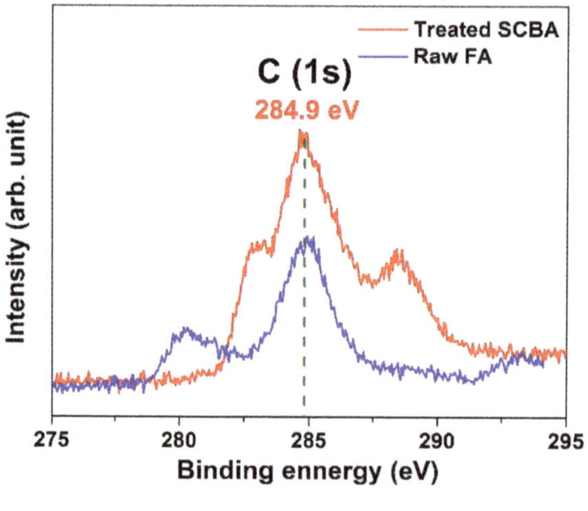

Figure 9. The carbon (1S) XPS spectra of the raw FA and the treated SCBA powders.

4. Conclusions

In this research, we have utilized the waste from a lignite power plant (FA) and the waste from a sugar industry (SCBA) to fabricate a construction material, the geopolymer. The FA geopolymer without SCBA shows excellent properties, such as good rheological flow of the paste after mixing (workability = 255%), high compressive strength (47.8 MPa), and relatively large dielectric constant (ε' = 3.6 $\times$ 10^3). These excellent properties were due to the formation of the aluminosilicate gels from the geopolymerization process of FA when dissolved in an alkaline-activated solution. The amorphous phase of geopolymeric gel was detected by XRD. The SEM images showed the fully dense samples with a smooth surface of aluminosilicate gels. Moreover, the intensity of the asymmetric stretching vibrations of the Si-O-Si (Al) observed from FTIR was very strong, supporting the formation of aluminosilicate building blocks.

On the contrary, the FA geopolymer composited with SCBA showed inferior characteristics. The workability, compressive strength, and dielectric properties deteriorated as compared to the pristine geopolymer. The main reason is due to the highly crystalline quartz (SiO_2) phase in SCBA, which is very stable and not reactive. Thus, the SCBA did not dissolve in the alkaline-activated solution and did not take part in the geopolymerization process. This led to the unreacted SCBA particles leftover in the geopolymer paste. These particles did not provide strength to the geopolymer and thus led to decreased mechanical properties. Furthermore, the high carbon content in SCBA contributed to an electrically conductive phase in the geopolymer composite, which in turn reduced the dielectric constant. However, our results suggested that if the amount of SCBA was about 10 wt.% or less, the impact on the characteristics and properties of FA geopolymers was minimal. Therefore, the FA with approximately 10 wt.% SCBA could be utilized to fabricate geopolymer composites.

Author Contributions: Conceptualization, S.P. and N.C. (Nattapong Chuewangkam); writing—original draft preparation, N.C. (Nattapong Chuewangkam); methodology, N.C. (Nattapong Chuewangkam), T.N. and N.C. (Narong Chanlek); formal analysis, N.C. (Nattapong Chuewangkam), T.N., P.T. and N.C. (Narong Chanlek); supervision, S.P.; project administration, S.P. and P.T.; writing—review and editing, S.P. and N.C. (Nattapong Chuewangkam). All authors have read and agreed to the published version of the manuscript.

Funding: This research was funded by the Thailand Research Fund (TRF) in cooperation with Synchrotron Light Research Institute (public organization) and Khon Kaen University (RSA6280020), the Research and Graduate Studies of Khon Kaen University.

Institutional Review Board Statement: Not applicable.

Informed Consent Statement: Not applicable.

Data Availability Statement: The data and the code that support the results within this paper and other findings of this study are available from the corresponding author upon reasonable request.

Conflicts of Interest: The authors declare no conflict of interest.

Abbreviations

DI	Deionization
FA	Fly ash
FTIR	Fourier transform infrared
LOI	Loss of ignition
RH	Relative humidity
SCBA	Sugarcane bagasse ash
SEM	Scanning electron microscope
XRF	X-ray fluorescence
XRD	X-ray diffraction
XPS	X-ray photoelectron spectroscopy
ε'	Dielectric constant
C	Sample's capacitance
t	Sample's thickness
ε_0	Permittivity of free space (8.854×10^{-12} F/m)
$\%W$	Workability
d	Spread-out diameter of geopolymer paste
d_0	Original diameter

References

1. Supasri, T.; Vorayos, N.; Thongchiew, P. Performance Analysis Model Development for Mae Moh Coal-Fired Power Plant. *Int. J. Mech.* **2013**, *7*, 1882–1886.
2. Phonphuak, N.; Chindaprasirt, P. Types of waste, properties, and durability of pore-forming waste-based fired masonry bricks. In *Eco-Efficient Masonry Bricks and Blocks*; Elsevier: Amsterdam, The Netherlands, 2015; pp. 103–127.
3. Apisitniran, L. Drought conditions and low cane supply are pushing production way down. In *Bitter Outlook for Sugar Industry*; Bangkok Post Public Company Limited: Bangkok, Thailand, 2020.
4. Alves, H.P.; Silva, J.B.; Campos, L.F.; Torres, S.M.; Dutra, R.P.; Macedo, D.A. Preparation of mullite based ceramics from clay–kaolin waste mixtures. *Ceram. Int.* **2016**, *42*, 19086–19090. [CrossRef]
5. Bahurudeen, A.; Kanraj, D.; Dev, V.G.; Santhanam, M. Performance evaluation of sugarcane bagasse ash blended cement in concrete. *Cem. Concr. Compos.* **2015**, *59*, 77–88. [CrossRef]
6. Teixeira, S.R.; De Souza, A.E.; de Almeida Santos, G.T.; Vilche Pena, A.F.; Miguel, A.G. Sugarcane bagasse ash as a potential quartz replacement in red ceramic. *J. Am. Ceram. Soc.* **2008**, *91*, 1883–1887. [CrossRef]
7. Pacheco-Torgal, F.; Labrincha, J.; Leonelli, C.; Palomo, A.; Chindaprasit, P. *Handbook of Alkali-Activated Cements, Mortars and Concretes*; Elsevier: Amsterdam, The Netherlands, 2014.
8. Van Jaarsveld, J.; Van Deventer, J.; Lorenzen, L. Factors affecting the immobilization of metals in geopolymerized flyash. *Metall. Mater. Trans. B* **1998**, *29*, 283–291. [CrossRef]
9. Saafi, M.; Andrew, K.; Tang, P.L.; McGhon, D.; Taylor, S.; Rahman, M.; Yang, S.; Zhou, X. Multifunctional properties of carbon nanotube/fly ash geopolymeric nanocomposites. *Constr. Build. Mater.* **2013**, *49*, 46–55. [CrossRef]
10. Zhang, H.Y.; Kodur, V.; Cao, L.; Qi, S.L. Fiber Reinforced Geopolymers for Fire Resistance Applications. *Procedia Eng.* **2014**, *71*, 153–158. [CrossRef]
11. Ganesan, K.; Rajagopal, K.; Thangavel, K. Evaluation of bagasse ash as supplementary cementitious material. *Cem. Concr. Compos.* **2007**, *29*, 515–524. [CrossRef]
12. Hernández, J.M.; Middendorf, B.; Gehrke, M.; Budelmann, H. Use of wastes of the sugar industry as pozzolana in lime-pozzolana binders: Study of the reaction. *Cem. Concr. Res.* **1998**, *28*, 1525–1536. [CrossRef]
13. Singh, N.; Singh, V.; Rai, S. Hydration of bagasse ash-blended portland cement. *Cem. Concr. Res.* **2000**, *30*, 1485–1488. [CrossRef]
14. Chusilp, N.; Jaturapitakkul, C.; Kiattikomol, K. Effects of LOI of ground bagasse ash on the compressive strength and sulfate resistance of mortars. *Constr. Build. Mater.* **2009**, *23*, 3523–3531. [CrossRef]
15. Chusilp, N.; Jaturapitakkul, C.; Kiattikomol, K. Utilization of bagasse ash as a pozzolanic material in concrete. *Constr. Build. Mater.* **2009**, *23*, 3352–3358. [CrossRef]
16. Schmitt, R. *Electromagnetics Explained: A Handbook for Wireless/RF, EMC, and High-Speed Electronics*; Newnes: London, UK, 2002.
17. Hanjitsuwan, S.; Hunpratub, S.; Thongbai, P.; Maensiri, S.; Sata, V.; Chindaprasirt, P. Effects of NaOH concentrations on physical and electrical properties of high calcium fly ash geopolymer paste. *Cem. Concr. Compos.* **2014**, *45*, 9–14. [CrossRef]
18. Topark-Ngarm, P.; Chindaprasirt, P.; Sata, V. Setting time, strength, and bond of high-calcium fly ash geopolymer concrete. *J. Mater. Civ. Eng.* **2015**, *27*, 04014198. [CrossRef]
19. Nuruddin, M.F.; Malkawi, A.B.; Fauzi, A.; Mohammed, B.S.; Almattarneh, H.M. Geopolymer concrete for structural use: Recent findings and limitations. In Proceedings of the IOP Conference Series: Materials Science and Engineering, Guangdong, China, 1 June 2016; IOP Publishing: Bristol, UK, 2016; p. 012021.
20. Kantakam, S.; Pimraksa, K.; Ngamjarurojana, A.; Chindaprasirt, P.; Chaipanich, A. Investigation on the dielectric properties of 0–3 lead zirconate titanate-geopolymer composites. *Ferroelectrics* **2013**, *451*, 84–89. [CrossRef]
21. Aradoaei, M.; Pepenar, I. Considerations on the dielectric properties and thermal profile of geopolymeric composites with ferro/ferrimagnetic inserts. In Proceedings of the 2014 International Conference and Exposition on Electrical and Power Engineering (EPE), Iasi, Romania, 16 October 2016; IEEE: Piscataway, NJ, USA, 2016; pp. 891–896.
22. Vlasceanu, I.N.; Gharzouni, A.; Tantot, O.; Lalande, M.; Elissalde, C.; Rossignol, S. Geopolymer as dielectric materials for ultra-wideband antenna applications: Impact of magnetite addition and humidity. *Open Ceram.* **2020**, *2*, 100013. [CrossRef]
23. Jumrat, S.; Chatveera, B.; Rattanadecho, P. Dielectric properties and temperature profile of fly ash-based geopolymer mortar. *Int. Commun.* **2011**, *38*, 242–248. [CrossRef]
24. Hanjitsuwan, S.; Chindaprasirt, P.; Pimraksa, K. Electrical conductivity and dielectric property of fly ash geopolymer pastes. *Int. J. Miner. Metall.* **2011**, *18*, 94–99. [CrossRef]
25. Mebrouki, A.; Belas, N.; Bendani, K.; Bouhamou, N. A Self-Compacting Cement Paste Formulation using Mixture Design. *J. Appl. Sci.* **2009**, *9*, 4127–4136. [CrossRef]
26. Kantro, D.L. Influence of Water-Reducing Admixtures on Properties of Cement Paste—A Miniature Slump Test. *Cem. Concr. Aggreg. CCAGDP* **1980**, *2*, 95–102.
27. Astm, C. *109. Standard Test Method for Compressive Strength of Hydraulic Cement Mortars (Using 2-in. or [50 mm] Cube Specimens), Annual Book of ASTM Standards*; ASTM: West Conshohocken, PA, USA, 2020.
28. Payakaniti, P.; Pinitsoontorn, S.; Thongbai, P.; Amornkitbamrung, V.; Chindaprasirt, P. Electrical conductivity and compressive strength of carbon fiber reinforced fly ash geopolymeric composites. *Constr. Build. Mater.* **2017**, *135*, 164–176. [CrossRef]

29. Rehman, S.K.U.; Imtiaz, L.; Aslam, F.; Khan, M.K.; Haseeb, M.; Javed, M.F.; Alyousef, R.; Alabduljabbar, H. Experimental investigation of NaOH and KOH mixture in SCBA-based geopolymer cement composite. *Materials* **2020**, *13*, 3437. [CrossRef] [PubMed]
30. Ribeiro, D.V.; Morelli, M.R. Effect of calcination temperature on the pozzolanic activity of Brazilian sugar cane bagasse ash (SCBA). *Mater. Res.* **2014**, *17*, 974–981. [CrossRef]
31. Al-Shathr, B.; Shamsa, M.; al-Attar, T. Relationship between amorphous silica in source materials and compressive strength of geopolymer concrete. In Proceedings of the 3rd International Conference on Buildings, Construction and Environmental Engineering, BCEE3-2017, Sharm el-Shiekh, Egypt, 23–25 October 2017; EDP Sciences: Les Ulis, France, 2018; p. 02019.
32. Yao, X.; Zhang, Z.; Zhu, H.; Chen, Y. Geopolymerization process of alkali–metakaolinite characterized by isothermal calorimetry. *Thermochim. Acta* **2009**, *493*, 49–54. [CrossRef]
33. Adam, A.A.; Horianto, X. The effect of temperature and duration of curing on the strength of fly ash based geopolymer mortar. *Procedia Eng.* **2014**, *95*, 410–414. [CrossRef]
34. Duxson, P.; Fernández-Jiménez, A.; Provis, J.L.; Lukey, G.C.; Palomo, A.; van Deventer, J.S. Geopolymer technology: The current state of the art. *J. Mater. Sci.* **2007**, *42*, 2917–2933. [CrossRef]
35. Steveson, M.; Sagoe-Crentsil, K. Relationships between composition, structure and strength of inorganic polymers. *J. Mater. Sci.* **2005**, *40*, 2023–2036. [CrossRef]
36. Taylor, W. Application of infrared spectroscopy to studies of silicate glass structure: Examples from the melilite glasses and the systems Na_2O-SiO_2 and Na_2O-Al_2O_3-SiO_2. *Proc. Indian Natl. Sci.* **1990**, *99*, 99–117. [CrossRef]
37. Voll, D.; Lengauer, C.; Beran, A.; Schneider, H. Infrared band assignment and structural refinement of Al-Si, Al-Ge, and Ga-Ge mullites. *Eur. J. Mineral.* **2001**, *13*, 591–604. [CrossRef]
38. Kiefer, J.; Stärk, A.; Kiefer, A.L.; Glade, H. Infrared spectroscopic analysis of the inorganic deposits from water in domestic and technical heat exchangers. *Energies* **2018**, *11*, 798. [CrossRef]
39. Huang, X.; Liu, L.; Zhao, X.; Tang, C.; Wang, X. Properties of phosphorus-slag-based cementitious pastes for stabilizing lead. *Materials* **2019**, *12*, 3831. [CrossRef] [PubMed]
40. Rodriguez-Blanco, J.D.; Shaw, S.; Benning, L.G. The kinetics and mechanisms of amorphous calcium carbonate (ACC) crystallization to calcite, via vaterite. *Nanoscale* **2011**, *3*, 265–271. [CrossRef] [PubMed]
41. Fang, T.-T.; Liu, C. Evidence of the internal domains for inducing the anomalously high dielectric constant of $CaCu_3Ti_4O_{12}$. *Chem. Mater.* **2005**, *17*, 5167–5171. [CrossRef]
42. Tuichai, W.; Danwittayakul, S.; Maensiri, S.; Thongbai, P. Investigation on temperature stability performance of giant permittivity (In+ Nb) in co-doped TiO_2 ceramic: A crucial aspect for practical electronic applications. *RSC Adv.* **2016**, *6*, 5582–5589. [CrossRef]
43. Kaatze, U.; Hübner, C. Electromagnetic techniques for moisture content determination of materials. *Meas. Sci. Technol.* **2010**, *21*, 082001. [CrossRef]
44. McCarter, W.J.; Chrisp, T.M.; Starrs, G.; Blewett, J. Characterization and monitoring of cement-based systems using intrinsic electrical property measurements. *Cem. Concr. Res.* **2003**, *33*, 197–206. [CrossRef]
45. Cui, X.M.; Liu, L.P.; He, Y.; Chen, J.Y.; Zhou, J. A novel aluminosilicate geopolymer material with low dielectric loss. *Mater. Chem. Phys.* **2011**, *130*, 1–4. [CrossRef]

polymers

MDPI

Article

Suberin Fatty Acid Hydrolysates from Outer Birch Bark for Hydrophobic Coating on Aspen Wood Surface

Anuj Kumar [1,*], Risto Korpinen [1], Veikko Möttönen [2] and Erkki Verkasalo [2]

[1] Natural Resources Institute Finland, Production Systems, Tietotie 2, 02150 Espoo, Finland; risto.korpinen@luke.fi
[2] Natural Resources Institute Finland, Production Systems, Yliopistokatu 6B, 80100 Joensuu, Finland; veikko.mottonen@luke.fi (V.M.); erkki.verkasalo@luke.fi (E.V.)
* Correspondence: anuj.kumar@luke.fi; Tel.: +35-8295-322-088

Abstract: Bark extracts are sustainable sources of biopolymers and have great potential to replace fossil-based polymers in wood coating applications. The present study investigated the applicability of suberin fatty acids hydrolysate (SFA) extracted from the outer bark of silver birch (*Betula pendula* Roth.) for coating of aspen wood (*Populus tremula* L.). The SFA combined with maleic anhydride (MA) and octadecyltrichlorosilane (OTS) as a curing agent was prepared in ethanol and used in surface coating. The water contact angle, surface reflectance spectra, FTIR, and SEM-EDS were used to characterize the physical and chemical properties of the coated wood samples. Further, the long-term stability of the SFA coating was analyzed via artificial aging. The wood surface became hydrophobic, as the contact angle for the water droplet (WCA) was over ~120°, and was stable for all of the prepared combinations of SFA, MA, and OTS.

Keywords: birch bark; wood coating; suberin fatty acids; hydrophobicity; artificial aging

Citation: Kumar, A.; Korpinen, R.; Möttönen, V.; Verkasalo, E. Suberin Fatty Acid Hydrolysates from Outer Birch Bark for Hydrophobic Coating on Aspen Wood Surface. *Polymers* **2022**, *14*, 832. https://doi.org/10.3390/polym14040832

Academic Editors: Sabu Thomas and Maya Jacob John

Received: 26 January 2022
Accepted: 18 February 2022
Published: 21 February 2022

1. Introduction

As a hydrophilic, hygroscopic, porous, and fibrous material, wood is especially vulnerable to water sorption, because water penetrates rapidly into the wood structure causing swelling and eventually a loss of mechanical strength as well as providing conditions for biological degradation [1,2]. Several wood modification methods have been developed to improve the overall chemical and physical properties of the material. Wood modification can be divided into two main categories: active and passive wood modification. Active modification includes chemical, thermal, and enzymatic modifications of the wood cell wall and surface, whereas passive modification consists mainly of impregnation of the pores and lumen space of the cell wall with chemicals [3,4]. Wood surface coating is one of the active chemical modification techniques to improve the service life of wood and wood products, and various synthetic polymeric coatings have been investigated for this purpose [2,5]. However, in recent years, polymers and monomers derived from bio-based materials have attracted enormous interest due to the dwindling of non-renewable feedstock such as petrochemicals [6,7].

The wood processing industry provides large amounts of bark as a side stream, and most of it is incinerated for energy instead of cascading for potential green chemicals [8–10]). The global annual generation of bark is estimated to be between 300 and 400 million m^3 [11]. In birch bark, in particular, the chemical composition of the outer and inner bark differs considerably. The outer bark of silver birch (*Betula pendula* Roth.) is made up of 45% suberin, 40% extractives, 9% lignin, 4% hemicelluloses, and 2% cellulose, on average [12]. The extractives mainly consist of different triterpenoids but only few phenolics—in contrast to inner bark with its very high phenolic contents [13,14]. Birch wood is especially rich in the valuable biochemicals such as the naturally occurring biopolyester suberin and the triterpenoids betulinol and lupeol [12,15]. The content of betulinol and lupeol in outer bark varies between 30% and 35%, and the content of suberin can be up to 40–50% [16]. Birch bark containing both inner and outer bark from a pulp mill has been reported to contain 5.9% suberin [16]. Suberin fatty acids have various functional applications such as a hybrid co-polymer and polyols for polyurethane [17,18], a source of organosolv lignin [19], a thermoset resin from epoxy ω-hydroxyacids and methacrylates [20], a water vapor barrier hybrid film of cellulose [21], and as a hydrophobic coating for a lignocellulosic fiber network [15]. Handiso et al. [22] evaluated the physico-chemical properties of cellulose surface modified with suberin and suberin fatty acids, and the outcomes from this strongly support the use of SFA in cellulose modification for better hydrophobic and barrier properties.

In this study, the suberin fatty acids hydrolysate (SFA) of outer birch bark was investigated as a potential hydrophobic coating application on aspen wood surface. Two crosslinking agents, maleic anhydride (MA) and octadecyltrichlorosilane (OTS), were used in several combinations. The coated wood samples were evaluated by Fourier transform infrared spectroscopy (FTIR), surface reflectance spectra, and scanning electron microscopy with energy-dispersive X-ray analysis (SEM-EDX). In addition, the water contact angle was measured using the sessile drop method. Further, the effectiveness of the hydrophobic properties of wood samples was evaluated by exposing them to artificial aging.

2. Materials and Methods

2.1. Materials

The outer bark of silver birch (*Betula pendula* Roth.) was manually removed from freshly cut stems (tree diameter: 200–300 mm), followed by air-drying at room temperature. The outer bark was then ground using a cutting mill Fritsch PULVERISETTE Type 15.903 (Fritsch GmbH, Germany) with a sieve cassette having 4 × 4 mm^2 square openings. The ground outer bark was then freeze-dried and stored in an airtight polyethylene bag.

European aspen wood (*Populus tremula* L.) was procured from a local sawmill in southern Finland and prepared as specimens free from knots and cracks with straight grains and normal growth ring widths. The wood samples were prepared to uniform dimensions (100 × 45 × 5 mm) and oven-dried at 105 °C for 24 h prior to coating.

Isopropyl alcohol 99.8% (Merck KGaA, Darmstadt, Germany), sodium hydroxide 99.0% (Merck KGaA, Darmstadt, Germany), ethanol 94.0% (Altia Oyj, Rajamäki, Finland), and sulfuric acid 95% (VWR International S.A.S., Briare, France) were used in the experiments. Then, 2 molar sulfuric acid was prepared by pouring 112 mL of concentrated sulfuric acid into 600 mL of maxima ultrapure water. The final volume was adjusted with ultra-pure water to 1000 mL after the solution was cooled to room temperature.

2.2. Suberin Fatty Acids Extraction Process

The dry solid content of the ground and dried outer bark was 97.8%. The hydrolysis and extraction were carried out according to a protocol described by Korpinen et al. (2019) [15]. In the process, 150 g o.d. of outer bark was used with a liquid-to-bark ratio of 10 mL/g. Ethanol was used as the solvent with a solvent-to-water ratio of 9:1 (v/v). On o.d. bark, 20% NaOH was applied, and the bark was boiled in the solution for 3 h. The betulinol fraction and suberin fatty acids were then separated. The extractions were repeated three times. The different fractions were characterized separately and mixed before the wood surface treatment.

2.3. Wood Samples and Coating Solutions Preparation

The wood samples without any coating were selected as a control (**C**). Figure 1 shows a schematic of the extraction and coating process of SFA on wood surface. The following SFA-based coating formulations were prepared and applied on the wood surface:

a. SFA and maleic anhydride (MA) (treatment name, **SFA–MA**): Fixed amounts (70:30) of SFA and maleic anhydride as a hardener were mixed and boiled at 45 °C in ethanol. A manual brush coating was used to prepare a uniform two-layer coating of SFA–MA on wood and dried for 24 h at 105 °C;

b. Treatment of coated samples with octadecyltrichlorosilane (OTS) (treatment name, **SFA–MA + OTS**): SFA–MA-coated samples dipped into a solution (1/100) (v/v) of OTS and n-hexane for 30 min and dried for 24 h at 105 °C;
c. **OTS** treatment of sample: Wood samples dipped in a solution (1/100) (v/v) of OTS and n-hexane for 30 min and dried at 24 h at 105 °C;
d. **SFA–OTS**: In 100 mL ethanol, 5 g SFA and 500 μL OTS were mixed and boiled at 120 °C (10 min) and cooled down to room temperature and applied manually to the wood surface (two coatings).

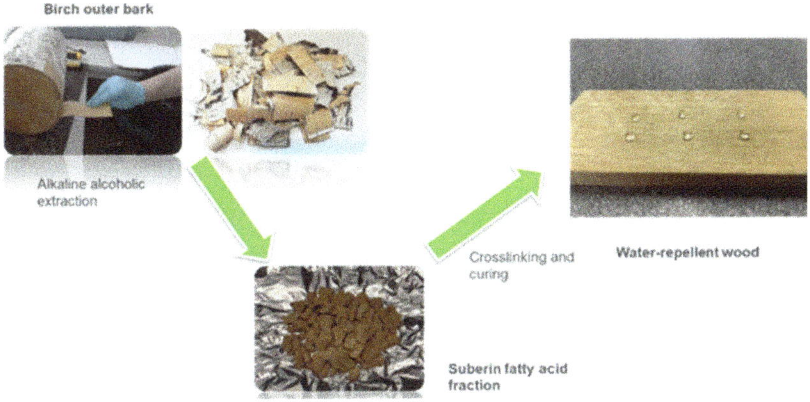

Figure 1. Schematic of SFA extraction and hydrophobic coating on wood.

2.4. GC and GC-MS Analysis

Suberin fatty acids were derivatized using a mixture of N,O-bis(trimethylsilyl)trifluoroacetamide (BSTFA):chlorotrimethylsilane (TMCS):pyridine (120:20:20) and heated for 50 min at 70 °C. The silylated samples were quantified by gas chromatography flame ionization detection (GC-FID), and the peak identities were confirmed by gas chromatography mass spectrometry (GC-MS) as described by Korpinen et al. (2019) [15].

2.5. SEM and EDS Analysis

The surfaces of coated wood samples and uncoated reference samples were evaluated using scanning electron microscopy (SEM, JEOL JSM-5500LV, Osaka, Japan). The specimens were coated with a thin evaporated layer of carbon to improve conductivity prior to the analysis. The prepared specimens were mounted 12 mm from lenses. The chamber pressure $< 1.0 \times 10^{-5}$ (high vacuum) and acceleration voltages of 15 and 10 kV were used to measure the specimen's surface. Further, energy-dispersive X-ray spectroscopy (EDS) was applied with ESPRIT v.2.1.2.17832 coupled with SEM.

2.6. Color Measurement

The surface reflectance spectra of coated wood samples and uncoated reference samples were measured at intervals of 8 mm in diameter for each specimen in the visible light wavelength range of 360–740 nm using a Konica Minolta CM-2600d spectrophotometer (Konica Minolta Inc., Tokyo, Japan). Spectral data were converted to CIEL*a*b* color coordinates using a 2° standard observer and a D65 light source for lightness (L*), redness (a*), and yellowness (b*), according to the CIEL*a*b* color space (ISO 11664-4:2008). For each sample group, the mean and standard deviations of the color coordinates were calculated.

One-way ANOVA was performed using the Origin (Lab 9.1, MA, USA) statistical data analysis software to calculate the samples' mean difference at $p < 0.05$ levels for the color differences in samples before and after artificial aging.

2.7. Water Contact Angle Measurement

The wettability of the coated wood samples and uncoated reference samples was evaluated based on the water contact angle (WCA) using the sessile drop method. The contact angles were measured by means of computer-aided analysis (OCA 50 instrument coupled with SCA 20 CA software, DataPhysics Instruments, Berlin, Germany) of elliptical shapes of liquid drops as observed in an optical goniometer and recorded by a digital camera installed in axial extension of the lens and statistics of 5 droplets per specimen. Drops of approximately 4–6 μL, the volumes being calculated from images of the drops, were applied by means of a dispenser.

2.8. ATR-FTIR Measurement

Fourier transform infrared spectrophotometer (Shimadzu Cooperation, Kyoto, Japan, IRPrestige-21/IRAffinity-1/FTIR-8000 series) coupled with IRsolution software were used to gain, control, and process the data. Semi-thin layers were cut from wood specimens with a razor blade and then dried at 60 °C for two hours. The prepared specimens were scanned while using an attenuated total reflection (ATR) setup in the absorbance range of 400–4000 cm^{-1} with a scanning rate of 2 cm^{-1} and 50 scans per run.

2.9. Artificial Aging

The artificial aging of coated wood samples and uncoated reference samples were performed to compare and understand the long-term performance of prepared coating formulations. The aged samples were characterized to see the changes due to the fact of aging. The artificial aging was carried out according to ISO 4892-2:2013 using an Atlas Xenotest 440 weather testing chamber (Ametek Inc., Baton Rouge, LA, USA) for 500 h. Xenon light exposure at 60 Wm^{-2} in the range of 300–400 nm at 65 °C was used (black standard).

3. Results and Discussion

3.1. Analysis of the SFA Hydrolysates

Two integral and characteristic peaks of a long aliphatic chain in suberin appeared at 2919 and 2851 cm^{-1} as shown in Figure 2. A high-intensity band vibration at 1738 cm^{-1} of the carbonyl groups of typical esters was present in suberin. The symmetric and asymmetric bonds associated with vinyl groups, such as C–O and C–H, were present at 1245, 1164, and 722 cm^{-1}. The characteristic epoxy band was present at 845 cm^{-1} of hydroxyl-fatty acids, i.e., 9,10-epoxy-18-hydroxy-18:0 acid [23,24], which accounted for ~10% of the dry birch outer bark [25]. The bands of saturated aliphatic chains of vinyl (C–H) in-plane bending were present at 1463 cm^{-1} [22].

The chemical composition of the suberin fatty acids in the hydrolysate can be seen in Table 1. The monomer composition was similar to the results obtained by Pinto et al. (2009) [10]. The main difference was found in the amount of 9,10,18-trihydroxy-18:0 acid; the trihydroxy acid was formed during the hydrolysis in ethanol instead of isopropanol. The chlorinated substances were most probably artefacts from the silylation of the samples for the chemical analysis. The small amounts of triterpenoids, such as betulinol and betulinic acid, were derived from the first filtration process, where the betulinol fraction was separated [9]. The filtration process was not fully complete, but further separation techniques are required if fully pure fractions are needed.

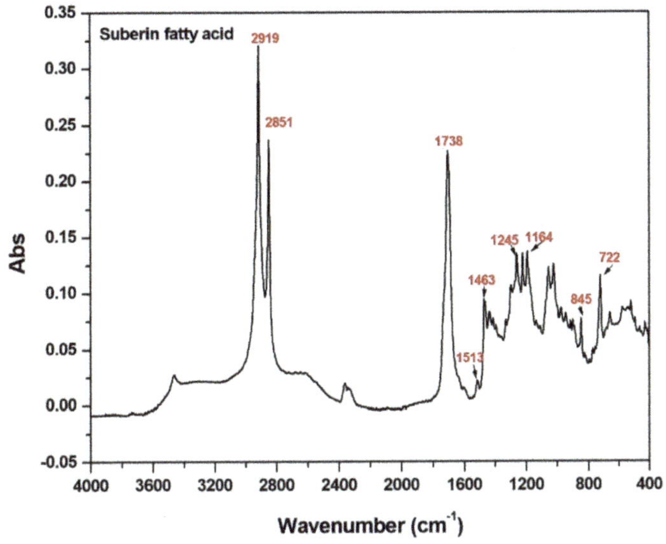

Figure 2. FTIR ATR spectrum corresponding to the SFA hydrolysate.

Table 1. The composition of suberin fatty acid (SFA) hydrolysate from birch outer bark estimated using GC-MS.

Compound	MEAN mg g^{-1} (o.d.)	SD
16:0 Fatty acid (palmitic acid)	0.83	0.06
Ferulic acid	5.48	1.42
17:0 fatty acid	0.12	0.02
18:2 Fatty acid (linoleic acid)	0.97	0.03
18:1 Fatty acid (oleic acid)	0.37	0.00
18:0 Fatty acid (stearic acid)	0.09	0.01
16-Hydroxy-16:0 acid	1.38	0.11
20:0 Fatty acid (arachidic acid)	0.19	0.03
1,16-Dioic-16:0 acid	3.36	0.25
18-Hydroxy-(9)18:1 acid	41.55	4.36
9,16- and 10,16-Dihydroxy-16:0 acids	13.70	0.22
18-Hydroxy-18:0 acid	1.31	0.17
1,18-Dioic-(9)18:1 acid	16.64	0.42
22:0 Fatty acid	0.00	0.00
1,18-Dioic-18:0 acid	6.93	0.53
9,18-Dihydroxy-(9)18:1 acid	5.22	0.02
9,10-Epoxy-18-hydroxy-18:0 acid	138.71	28.18
20-Hydroxy-20:1 acid	5.07	0.48
Dihydroxyoctadecanoic acid	3.73	0.13
20-Hydroxy-20:0 acid	8.60	1.52

Table 1. *Cont.*

Compound	MEAN mg g^{-1} (o.d.)	SD
1,20-Dioic-20:1 acid	2.89	0.04
24:0 Fatty acid	0.22	0.05
1,20-Dioic-20:0 acid	7.44	1.17
9,10,18-Trihydroxy-18:0 acid	130.53	40.72
(9)10-Chloro-10(9),18-dihydroxy-18:0 acid	22.99	11.72
22-Hydroxy-22:1 acid	1.21	0.08
22-Hydroxy-22:0 acid	11.29	3.17
1,22-Dioic-22:0 acid	13.59	7.10
24-Hydroxy-24:0 acid	0.00	0.00
Lupenone (lup-20(29)-en-3-one)	0.00	0.00
Sitosterol	0.66	0.08
Olean-12-ene-3,28-diol	0.94	0.18
Lupeol	0.24	0.04
Betulinol	20.49	2.22
Betulinic acid	18.17	2.49
Monogynol A (lupane-3b,20-diol)	0.00	0.00
Lupaen-3b,20,28-triol	0.00	0.00
Total identified (mg g^{-1})	484.91	25.07
Total eluted (mg g^{-1})	573.24	21.71

3.2. Analysis of SFA Hydrolysates-Based Coatings

3.2.1. ATR-FTIR

The infrared spectra of uncoated reference aspen wood (Sample C) showed absorption bands at 3345 cm^{-1} of the O–H group stretching vibration from the intermolecular and intramolecular hydrogen bonds [26] in Figure 3. The C–H stretching absorption between 2850 and 2930 cm^{-1} in the methyl and methylene groups [1,27] and the fingerprint region between 1800 and 900 cm^{-1} showed the absorption bands of functional groups present such as C–O at 1732 and C–C at 1505 cm^{-1} [28]. Several other important vibration modes present in lignin and carbohydrates were also observed at 1157, 1107, and 1052 cm^{-1}, etc. [28]. The OTS-coated sample showed characteristics peaks at ~2850 and ~2920 cm^{-1} for terminal methyl groups with different intensities and Si–O–C at 1192 cm^{-1}. The OTS treatment usually deposited the hydrophobic self-assembled monolayers via a hydrolysis and condensation process [1]. In a wood polymer system, the OTS is hydrolyzed, most likely by free –OH groups, while layered n-alkylsiloxane (PODS) gels are formed and polymerized into OTS layers on the wood surfaces [29].

In this study, two crosslinking agents were used to crosslink the hydroxy groups present in the SFA hydroxylates. At first, MA was used as a crosslinking agent, and the SFA underwent a maleation reaction due the reaction with MA and maleated SFA having two-way interactions with hydroxy groups; it reacted with epoxides first and the remaining hydroxy groups later [15]. After SFA-based coating with different combinations of crosslinkers, the FTIR spectra showed the formation and changes in the characteristics of the functional groups of the wood surface as shown in Figure 3. The stretching peak at 1030 cm^{-1} associated to cellulose C6–O6H C–O stretching in lignin shifted to a lower absorbance intensity, and a major shift appeared in the SFA–OTS combination. Similar to this, several other functional groups (i.e., 1165 cm^{-1} (C1–O–C4′ antisymmetric stretching in cellulose and hemicellulose); 1462 cm^{-1} (lignin C–H bending); 1736 cm^{-1} (C=O stretching);

aliphatic chain (CH3 and CH3 at 2919 and 2851 cm^{-1}) were more visible in the SFA–OTS coating system. The peak at 1248 cm^{-1} associated with C–O stretching in carboxylic acid shifted to a lower absorbance intensity after all the different coatings, possibly due to the bonding with reactive moieties of SFA and crosslinking agents. Another important characteristic peak belonged to SFA's ester group at 1736 cm^{-1}, and hemicellulose (C=O) significantly shifted to a higher peak absorbance for the SFA–OTS coating system, while it did not change for other combinations of coatings. Importantly, the OH stretching at ~3350 cm^{-1} also changed after SFA-based coating systems.

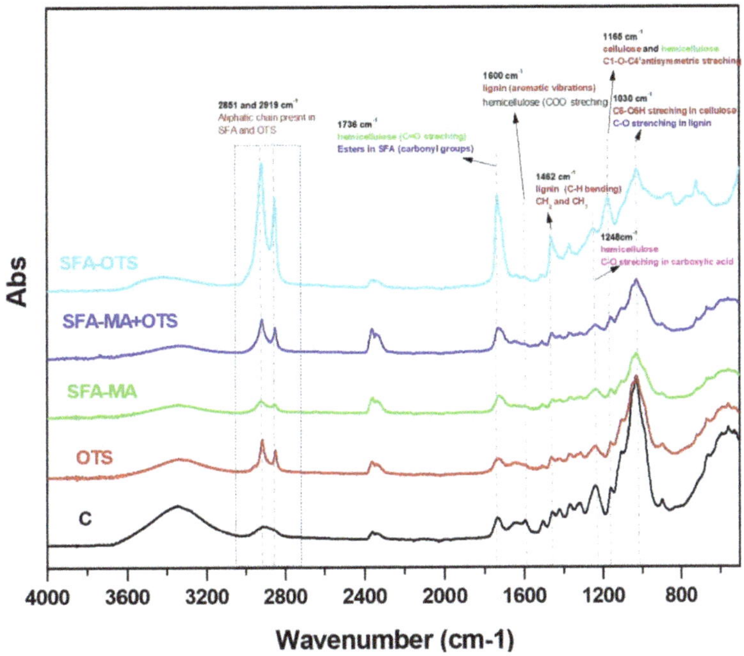

Figure 3. ATR-FTIR spectra of aspen wood coated with SBF hydrolysate-based coatings and uncoated reference wood (sample C).

3.2.2. SEM-EDS

The SEM images of the longitudinal section of the samples are presented in Figure 4. The uncoated reference sample C showed a defibrillated longitudinal surface, because we used rough wood samples with no surface finishing to obtain a better penetration of the coating and clearly visible open voids. The EDS scanning mainly showed the presence of the chemical compounds of carbon (62.24 atom%) and oxygen (37.76 atom%). The SEM micrograph of sample SFA–MA looked similar to the reference sample C; however, EDS scanning revealed a change in the chemical compositions of the C atom% that increased from 62.24% to 65.75% and of the O atom% that decreased from 37.76% to 34.35%. The changes were mainly due to the interaction of SFA–MA and free hydroxyl groups on the wood surface [15].

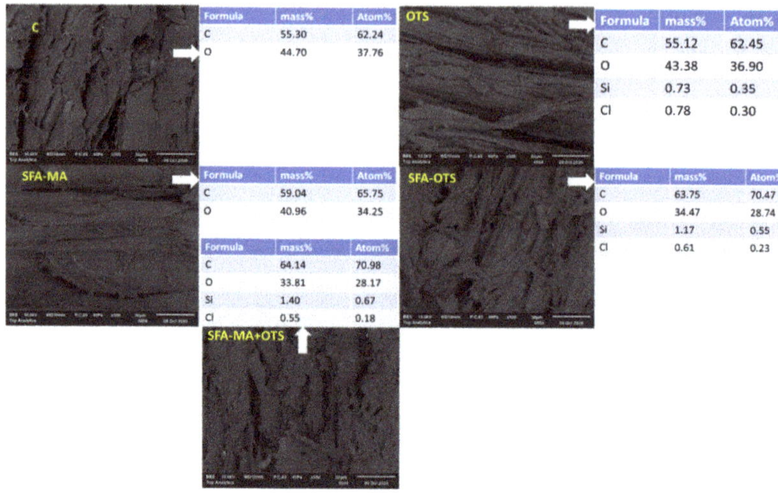

Figure 4. SEM-EDS analysis of aspen wood coated with hydrolysate-based coatings and an uncoated reference wood (sample C). The SEM micrographs were captured at a scale of 100 μm. The EDS scanning was performed at the same scale and surface.

The OTS coating seemed to reduce the surface roughness of specimens due to the deposition of self-assembly monolayers [1,29]. EDS scanning confirmed the deposition of silicon (0.35 atom%), residual chlorine (0.30 atom%) as well as increased carbon (62.45 atom%) and reduced oxygen (36.90 atom%). It was also confirmed by the ATR-FTIR results via the formation of S–O–C bonding at 1192 cm^{-1} that the OTS hydrolyzed first and reacted with free –OH groups (see also [1]). The SEM-EDS of the sample SFA–MA + OTS provided a different prospective on the coated wood surface, as the wood was first coated with SFA–MA and the second layer of OTS was formed after curing on the surface. The SEM image showed a uniform surface, and most of the roughness and voids were filled due to the coatings. EDS scanning confirmed significant changes in the atom% of carbon and oxygen, and silicon and chlorine also appeared due the additional OTS coating. Similar changes were observed in SEM-EDS scanning for the sample coated with SFA–OTS composition.

3.2.3. Water Contact Angle (WCA)

The mean water contact angle of the uncoated reference (sample C) was 60 ± 10°, and it was reduced to 17° after 7 s from when the water droplet was absorbed onto the wood (Figure 5). The WCA of the OTS sample was 140 ± 5°, and it remained stable even after 60 s (139 ± 5°). This change occurred due to the deposition of hydrophobic layers of an alkyl group and silicon groups [1]. The SFA–MA coating also increased the wood's hydrophobicity due to the deposition of long-chain fatty acids. The WCAs of the coated specimens were 126 ± 5°, reducing slightly to 124 ± 2° after 60 s. A similar behavior was observed when cellulosic paper was coated with an SFA–MA formulation [15,22]. The combination of SFA–MA and OTS coating improved the WCA slightly in comparison with SFA–MA only, as it is clearly shown in Figure 4. The SFA–OTS combination provided the highest WCA value of all the protocols tested, being 142 ± 5° and remaining constant for 60 s.

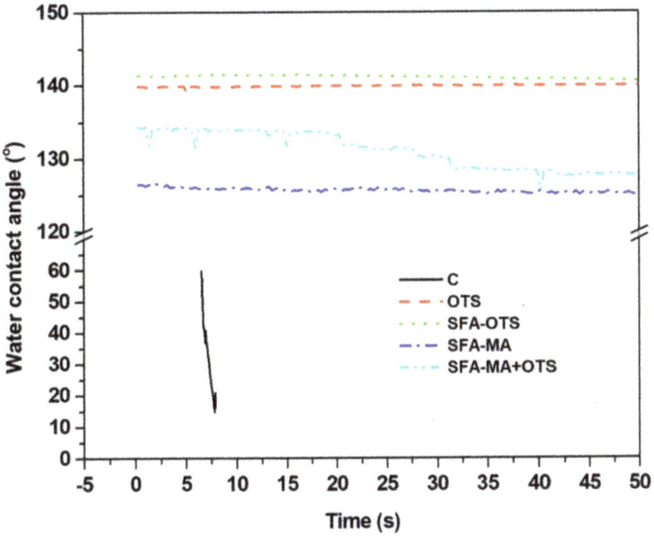

Figure 5. Water contact angle (WCA) of aspen wood coated with SBF hydrolysate-based coatings and an uncoated reference wood (sample C) according to time after applying water droplets on the wood's surface.

3.2.4. Color

In the color measurement, the uncoated reference (sample C) averaged 89 ± 2 for lightness, 2 ± 1 for redness, and 21 ± 2 for yellowness (Figure 6). After OTS coating, the color of the wood's surface did not change significantly. Instead, significant changes were observed after SFA–OTS coating, the lightness reducing to 67 ± 3, redness increasing to 7 ± 2, and yellowness increasing to 27 ± 3. The SFA hydrolysates were the main reason behind the color changes. Similar changes were observed in other coating combinations where SFA was a part of the coating. The largest color changes were observed for the SFA–MA-coated samples. Interestingly, the second coating of OTS on the sample SFA–MA + OTS had the largest impact on the color of the wood's surface. The SFA-based coatings changed the color of the wood surface, making it more reddish and yellowish.

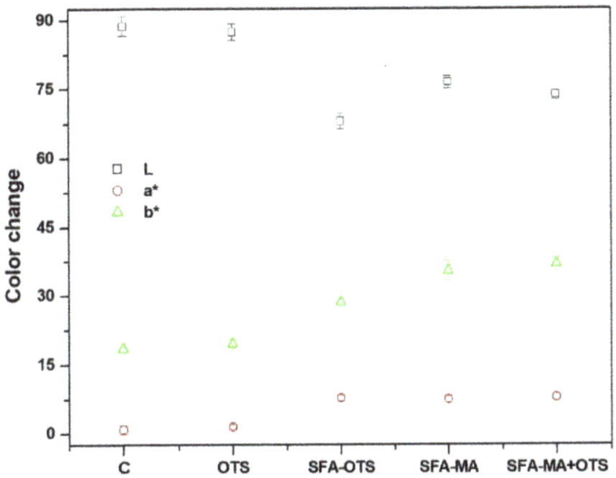

Figure 6. Color of aspen wood coated with SBF hydrolysate-based coatings and an uncoated reference wood (sample C) using the Lab coordinate system, representing lightness (L*), redness (a*), and yellowness (b*).

3.3. Artificial Aging

3.3.1. ATR-FTIR

Comparison of FTIR spectra of unexposed and aged wood surfaces with different coatings are shown in Figure 7. The significant changes in the lignin aromatic structure were observed due to the exposure to xenon light in the uncoated reference (sample C). The two observed changes appeared at 1596 (aromatic skeletal vibration plus C=O stretch) and at 1510 cm^{-1} (aromatic skeletal vibration). The intensity of all these bands and peaks almost diminished due to the exposure to UV light [30]. The vibration peak at 1268 cm^{-1} of syringyl ring vibration and the C–O stretch of lignin and xylan also disappeared due to the fact of aging under xenon light. On the other hand, the change in the polysaccharides' absorption bands at 1157 (C–O–C vibration in cellulose and hemicellulose), 1107, and 1052 cm^{-1} in comparison to the unexposed sample [28].

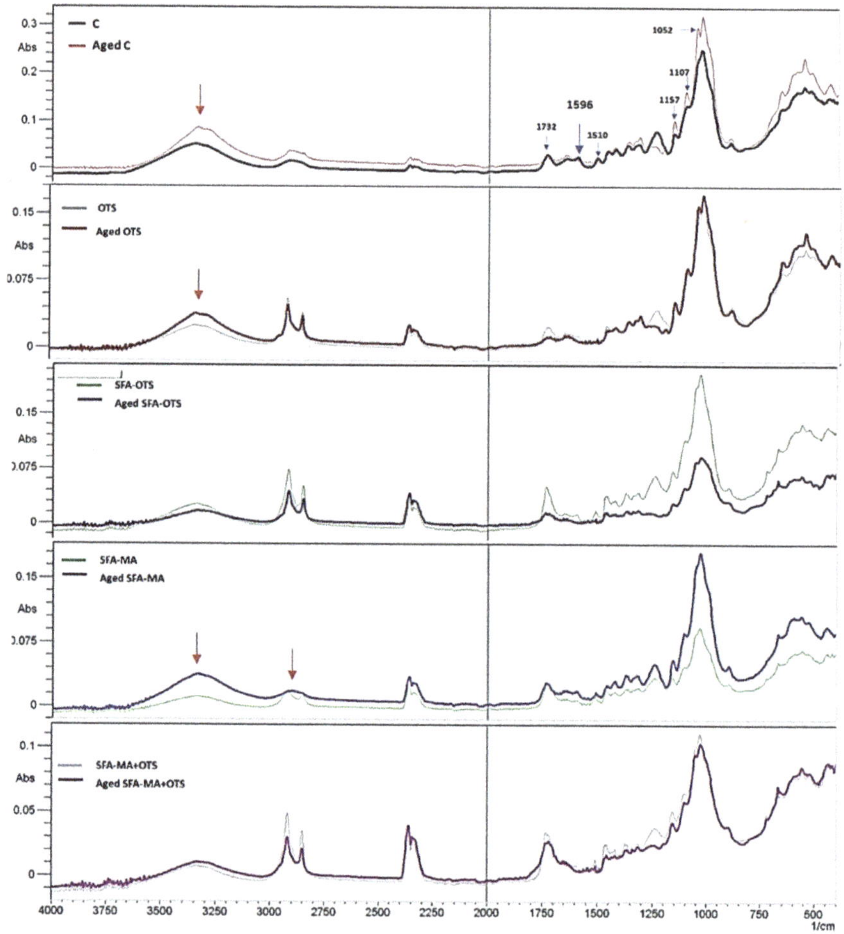

Figure 7. ATR-FTIR spectra of aspen wood coated with SBF hydrolysate-based coatings and artificially aged (500 h) samples.

In the case of the coated samples, significant changes after aging were observed in the SFA–MA samples. The peaks associated to C–H stretching absorption between 2850 and 2930 cm^{-1} in the methyl and methylene groups disappeared, and the intensity vibration peak at 1268 cm^{-1} increased significantly, the results being associated with the presence of a syringyl ring. The peak associated to the carbonyl group at approximately 3400 cm^{-1} showed an increased intensity due to the fact of aging for the uncoated refence (sample C), OTS samples, and SFA–MA samples. The samples being coated with combinations, including SFA and OTS, showed high stability in terms of surface chemistry alteration due to the fact of aging.

3.3.2. Water Contact Angle

Usually, most of the protection coating on wood surfaces deteriorates with time due to the continuous change in external environmental conditions; therefore, the water contact angle was measured after the artificial aging of wood surfaces (Figure 8). The WCA of all samples decreased after 500 h of aging. The uncoated reference (sample C) became highly hydrophilic due to the degradation of the aromatic lignin, the WCA being ~47°, and the water droplet being immediately absorbed into the wood. However, all the coating formulations showed a better hydrophobicity with a WCA of over 115° after 500 h of aging. The OTS and its combination with SFA showed a stability towards the aging and all such samples showed a WCA over 120°. Maybe the presence of silico, oxygen, and carbon in the OTS coating provided resistance towards the artificial aging [1].

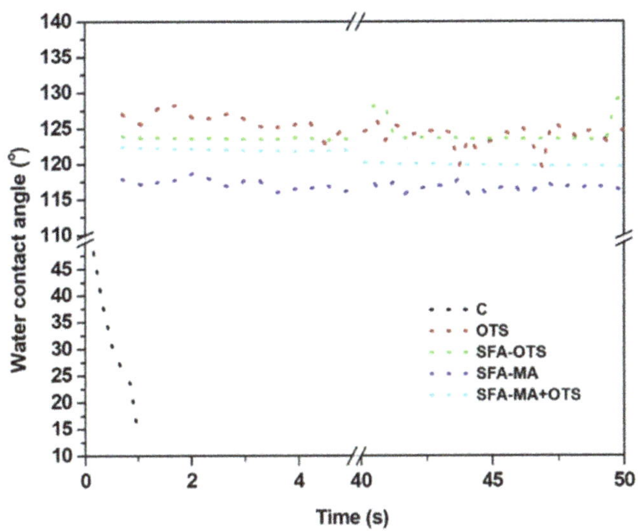

Figure 8. Water contact angle (WCA) of artificially aged (500 h) aspen wood coated with SBF hydrolysate-based coatings and an uncoated reference wood (sample C) according to time after applying water droplets on wood surface.

3.3.3. Color

As can be seen in Figures 9 and 10, the color of the wood surfaces changed much due to the aging. The uncoated reference (sample C) became whiter due to the aging, as yellowness decreased with time. A similar behavior was observed in the OTS samples. In the SFA-based samples, the lightness increased with the aging time; in contrast, especially yellowness but also redness decreased.

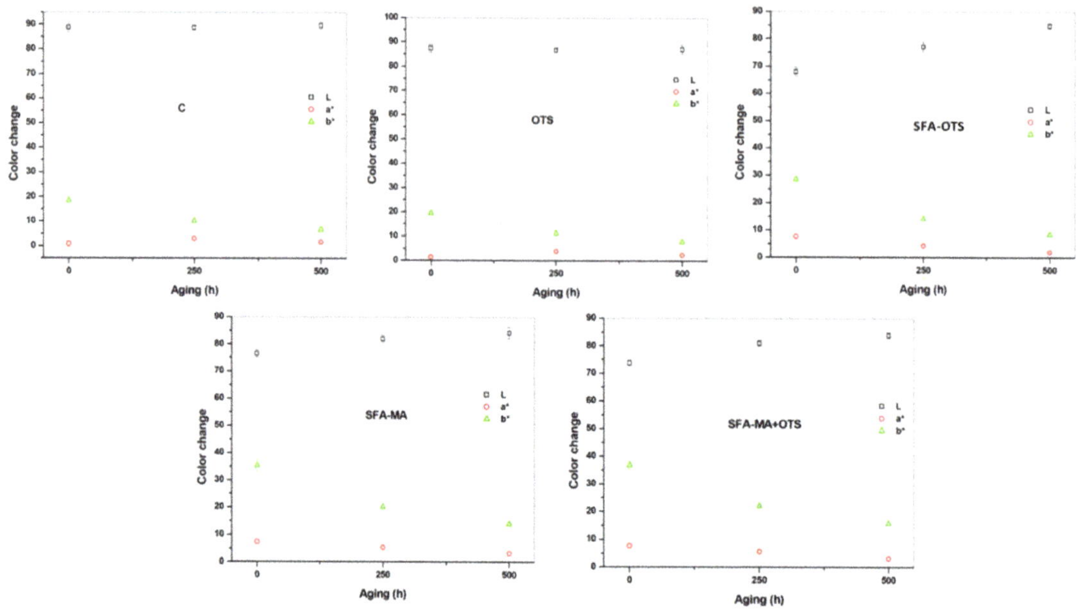

Figure 9. Color of aspen wood coated with SFA hydrolysate-based coatings and an uncoated reference wood (sample C) after artificial aging for 0, 250, and 500 h by xenon exposure using the Lab coordinate system, representing lightness (L*), redness (a*), and yellowness (b*).

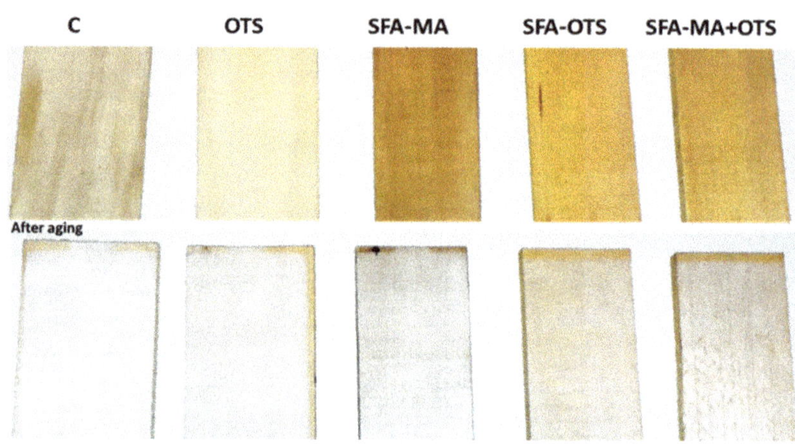

Figure 10. Examples of aspen wood coated with SBF hydrolysates-based coatings and an uncoated reference wood (sample C) before and after artificial aging (500 h) by xenon light exposure.

4. Conclusions

In this study, the suberin fatty acid hydrolysates (SBF) extracted from outer birch bark with a majority composition of 9,10-epoxy-18-hydroxy-18:0 acid was investigated as bio-based polymeric materials for wood coating applications. SFA alone was not able to form any bonding with the wood surface; thus, two types of crosslinking agents, MA and OTS, were used. Different combinations of SFA with OTS and MA were used for wood coating, i.e., OTS alone, SFA–OTS, SFA–MA, and SFA–MA + OTS. The physico-chemical analysis revealed that all SFA-based coating combinations significantly improved the wood surface's hydrophobicity. The use of OTS as a crosslinking agent for SFA was investigated for the very first time in this study. Even a layer of OTS on the SFA–MA coat contributed to the hydrophobicity of the coating. The color of the wood surface changed notably with the different coatings. Further, the long-term stability of coated wood samples was evaluated using artificial aging up to 500 h in a xenon weather testing chamber, and the aged samples were evaluated for physico-chemical properties. The results showed that all of the coating combinations were stable. The color changes were most notable in the SFA-based coatings: the lightness increased and, especially, the yellowness, but also the redness decreased. Yellowness decreased also in the OTS-coated and uncoated woods, making the wood lighter.

In terms of crosslinker impact on an SFA-coating formulation, MA is mostly used in bio-based polyester crosslinking and is relatively successful in various application. OTS was used for the very first time as a crosslinker for SFA, and it performed better than MA. However, the combination of both MA and OTS performed better in comparison to all combinations. The extraction process of suberin fatty acids from birch bark could affect the quality of bio-based polyesters. Thus, important considerations need to be taken into account during the extraction process in the future course of research work. In conclusion, suberin fatty acid hydrolysates from birch bark could be potential bio-based, renewable substitutes for fossil-based polymeric materials to be used in wood coating.

Author Contributions: Writing first draft: A.K.; Experiments: A.K., R.K., V.M.; Writing, editing and reviewing: A.K., R.K., V.M., E.V., Funding: R.K., Management: A.K., R.K. All authors have read and agreed to the published version of the manuscript.

Funding: Natural Resources Institute Finland for the project "Green Chemistry from Forest: Innovative Processes and Products and Biopolyester-Based Hydrophobic and Antimicrobial Coating of Lignocellulosic Materials (PolyCoat)".

Informed Consent Statement: Not applicable.

Data Availability Statement: Not applicable.

Acknowledgments: The authors would like to acknowledge the strategic funding from the Natural Resources Institute Finland for the project "Green Chemistry from Forest: Innovative Processes and Products and Biopolyester-Based Hydrophobic and Antimicrobial Coating of Lignocellulosic Materials".

Conflicts of Interest: The authors declare no conflict of interest.

References

1. Kumar, A.; Richter, J.; Tywoniak, J.; Hajek, P.; Adamopoulos, S.; Šegedin, U.; Petrič, M. Surface modification of Norway spruce wood by octadecyltrichlorosilane (OTS) nanosol by dipping and water vapour diffusion properties of the OTS-modified wood. *Holzforschung* **2017**, *72*, 45–56. [CrossRef]
2. Petrič, M. Surface Modification of Wood. *Rev. Adhes. Adhes.* **2013**, *1*, 216–247. [CrossRef]
3. Hill, C.A. *Wood Modification: Chemical, Thermal and Other Processes*; John Wiley & Sons: Hoboken, NJ, USA, 2007; Volume 5.
4. Kumar, A.; Jyske, T.; Petrič, M. Delignified Wood from Understanding the Hierarchically Aligned Cellulosic Structures to Creating Novel Functional Materials: A Review. *Adv. Sustain. Syst.* **2021**, *5*, 2000251. [CrossRef]
5. Petrič, M.; Oven, P. Determination of wettability of wood and its significance in wood science and technology: A critical review. *Rev. Adhes. Adhes.* **2015**, *3*, 121–187. [CrossRef]
6. Rajput, S.D.; Mahulikar, P.P.; Gite, V.V. Biobased dimer fatty acid containing two pack polyurethane for wood finished coatings. *Prog. Org. Coat.* **2014**, *77*, 38–46. [CrossRef]
7. Kumar, A.; Petrič, M.; Kričej, B.; Žigon, J.; Tywoniak, J.; Hajek, P.; Škapin, A.S.; Pavlič, M. Liquefied-Wood-Based Polyurethane–Nanosilica Hybrid Coatings and Hydrophobization by Self-Assembled Monolayers of Orthotrichlorosilane (OTS). *ACS Sustain. Chem. Eng.* **2015**, *3*, 2533–2541. [CrossRef]
8. Gandini, A.; Neto, C.; Silvestre, A.J.D. Suberin: A promising renewable resource for novel macromolecular materials. *Prog. Polym. Sci.* **2006**, *31*, 878–892. [CrossRef]
9. Holmbom, B. Extraction and Utilisation of Non-Structural Wood and Bark Components. In *Biorefining of Forest Resources*; Paperi ja Puu Oy: Espoo, Finland, 2011.
10. A Sustainable Bioeconomy for Europe—Strengthening the Connection between Economy, Society and the Environment 2018. Available online: https://ec.europa.eu/research/bioeconomy/pdf/ec_bioeconomy_strategy_2018 (accessed on 20 October 2021).
11. Pásztory, Z.; Mohácsiné, I.R.; Gorbacheva, G.; Börcsök, Z. The utilization of tree bark. *BioResources* **2016**, *11*, 7859–7888. [CrossRef]

12. Pinto, P.C.R.O.; Sousa, A.F.; Silvestre, A.J.D.; Neto, C.P.; Gandini, A.; Eckerman, C.; Holmbom, B. Quercus suber and Betula pendula outer barks as renewable sources of oleochemicals: A comparative study. *Ind. Crops Prod.* **2009**, *29*, 126–132. [CrossRef]
13. Kähkönen, M.P.; Hopia, A.I.; Vuorela, H.J.; Rauha, J.-P.; Pihlaja, K.; Kujala, T.S.; Heinonen, M. Antioxidant Activity of Plant Extracts Containing Phenolic Compounds. *J. Agric. Food Chem.* **1999**, *47*, 3954–3962. [CrossRef]
14. Roitto, M.; Siwale, W.; Tanner, J.; Ilvesniemi, H.; Julkunen-Tiitto, R.; Verkasalo, E. *Characterization of Extractives in Tree Biomass and By-Products of Plywood and Saw Mills from Finnish Birch in Different Climatic Regions for Value-Added Chemical Products*; FPInnovations: Montreal, QC, Canada, 2015; pp. 174–181.
15. Korpinen, R.I.; Kilpeläinen, P.; Sarjala, T.; Nurmi, M.; Saloranta, P.; Holmbom, T.; Koivula, H.; Mikkonen, K.S.; Willför, S.; Saranpää, P.T. The Hydrophobicity of Lignocellulosic Fiber Network Can Be Enhanced with Suberin Fatty Acids. *Molecules* **2019**, *24*, 4391. [CrossRef]
16. Ferreira, R.; Garcia, H.; Sousa, A.F.; Freire, C.S.R.; Silvestre, A.J.D.; Rebelo, L.P.N.; Pereira, C.S. Isolation of suberin from birch outer bark and cork using ionic liquids: A new source of macromonomers. *Ind. Crops Prod.* **2013**, *44*, 520–527. [CrossRef]
17. Cordeiro, N.; Belgacem, M.N.; Gandini, A.; Neto, C. Urethanes and polyurethanes from suberin 2: Synthesis and characterization. *Ind. Crops Prod.* **1999**, *10*, 1–10. [CrossRef]
18. Rizikovs, J.; Godina, D.; Makars, R.; Paze, A.; Abolins, A.; Fridrihsone, A.; Meile, K.; Kirpluks, M. Suberinic Acids as a Potential Feedstock for Polyol Synthesis: Separation and Characterization. *Polymers* **2021**, *13*, 4380. [CrossRef]
19. Grzybek, J.; Sepperer, T.; Petutschnigg, A.; Schnabel, T. Organosolv Lignin from European Tree Bark: Influence of Bark Pretreatment. *Materials* **2021**, *14*, 7774. [CrossRef]
20. Torron, S.; Semlitsch, S.; Martinelle, M.; Johansson, M. Polymer Thermosets from Multifunctional Polyester Resins Based on Renewable Monomers. *Macromol. Chem. Phys.* **2014**, *215*, 2198–2206. [CrossRef]
21. Heinämäki, J.; Halenius, A.; Paavo, M.; Alakurtti, S.; Pitkänen, P.; Pirttimaa, M.; Kirsimäe, K.; Kogermann, K.; Yliruusi, J. Suberin fatty acids isolated from outer birch bark improve moisture barrier properties of cellulose ether films intended for tablet coatings. *Int. J. Pharm.* **2015**, *489*, 91–99. [CrossRef]
22. Handiso, B.; Valle-Delgado, J.J.; Johansson, L.-S.; Hughes, M. The physicochemical properties of cellulose surfaces modified with (depolymerised) suberin and suberin fatty acid. *Ind. Crops Prod.* **2021**, *159*, 113070. [CrossRef]
23. Rüdiger, A.; Hendil-Forssell, P.; Hedfors, C.; Martinelle, M.; Trey, S.; Johansson, M. Chemoenzymatic Route to Renewable Thermosets Based on a Suberin Monomer. *J. Renew. Mater.* **2013**, *1*, 124–140. [CrossRef]
24. Li, C.; Fan, H.; Aziz, T.; Bittencourt, C.; Wu, L.; Wang, D.-Y.; Dubois, P. Biobased epoxy resin with low electrical permissivity and flame retardancy: From environmental friendly high-throughput synthesis to properties. *ACS Sustain. Chem. Eng.* **2018**, *6*, 8856–8867. [CrossRef]
25. Ekman, R. The Suberin Monomers and Triterpenoids from the Outer Bark of Betula verrucosa Ehrh. *Holzforschung* **1983**, *37*, 205–211. [CrossRef]
26. Wei, X.; Tao, D.; Sheng, X.; Zhang, M. Modification of Poplar Wood Using Polyhexahydrotriazine and Its Effect on Hygroscopicity. *J. Wood Chem. Technol.* **2018**, *38*, 214–223. [CrossRef]
27. Pandey, K. A study of chemical structure of soft and hardwood and wood polymers by FTIR spectroscopy. *J. Appl. Polym. Sci.* **1999**, *71*, 1969–1975. [CrossRef]
28. Pandey, K.K.; Pitman, A.J. Weathering characteristics of modified rubberwood (Hevea brasiliensis). *J. Appl. Polym. Sci.* **2002**, *85*, 622–631. [CrossRef]
29. Kumar, A.; Staněk, K.; Ryparová, P.; Hajek, P.; Tywoniak, J. Hydrophobic treatment of wood fibrous thermal insulator by octadecyltrichlorosilane and its influence on hygric properties and resistance against moulds. *Compos. Part B Eng.* **2016**, *106*, 285–293. [CrossRef]
30. Pandey, K.K.; Vuorinen, T. Comparative study of photodegradation of wood by a UV laser and a xenon light source. *Polym. Degrad. Stab.* **2008**, *93*, 2138–2146. [CrossRef]

polymers

MDPI

Article

Development and Characterization of Plantain (*Musa paradisiaca*) Flour-Based Biopolymer Films Reinforced with Plantain Fibers

Ramiro Venegas [1], Andres Torres [1], Ana M. Rueda [1], Maria A. Morales [1], Mary J. Arias [2] and Alicia Porras [1,*]

[1] Grupo de Diseño de Productos y Procesos (GDPP), Department of Chemical and Food Engineering, Universidad de los Andes, CR 1 18a 12, Bogota 111711, Colombia; ra.venegas@uniandes.edu.co (R.V.); af.torres@uniandes.edu.co (A.T.); am.rueda13@uniandes.edu.co (A.M.R.); ma.morales12@uniandes.edu.co (M.A.M.)

[2] Chemical Engineering Program, School of Engineering, Universidad Tecnológica de Bolívar, Parque Industrial y Tecnológico Carlos Vélez Pombo km 1 Vía Turbaco, Cartagena 130012, Colombia; mariast@utb.edu.co

* Correspondence: n-porras@uniandes.edu.co; Tel.: +57-1-339-4999 (ext. 1775)

Abstract: Agroindustrial wastes are a cheap and abundant source of natural fibers and macromolecules that can be used in the manufacturing of biocomposites. This study presents the development and thermo-mechanical characterization of a bio-composite film (TPF/PF), made of thermoplastic banana flour (TPF) matrix and plantain fibers (PF). Fabricated materials were characterized by physical analysis, chemical composition, Fourier-transformed spectroscopy (FTIR), thermal analysis (TGA), mechanical analysis, and scanning electronic microscopy (SEM). The physical analysis showed that TPF and PF have a low density and high affinity to water resulting in a lightweight, renewable, and biodegradable TPF/PF composite. The chemical composition and spectra analysis of the fiber showed that PF is a potential candidate for reinforcing composites due to its high α-cellulose and low lignin content. The thermal analysis determined that TPF degrades at a lower temperature than PF, therefore the matrix sets the processing temperature for TPF/PF composite films. The mechanical test showed an improvement in the tensile properties of the composite in comparison to neat TPF. Tensile strength and Young's modulus were improved by 345% and 1196%, respectively, when PF fibers was used. Good bonding and mechanical interlocking of PF to the TPF were identified by SEM. Therefore, potential biocomposites can be developed using natural fibers and thermoplastic starches obtained from plantain agroindustrial wastes.

Keywords: bio-based; plantain; natural fibers; agroindustrial waste; starch biopolymer

Citation: Venegas, R.; Torres, A.; Rueda, A.M.; Morales, M.A.; Arias, M.J.; Porras, A. Development and Characterization of Plantain (*Musa paradisiaca*) Flour-Based Biopolymer Films Reinforced with Plantain Fibers. *Polymers* **2022**, *14*, 748. https://doi.org/10.3390/polym14040748

Academic Editors: Maya Jacob John and Sabu Thomas

Received: 28 January 2022
Accepted: 11 February 2022
Published: 15 February 2022

1. Introduction

Plastics are an essential part of modern economies and have been extensively used in a variety of fields, such as food packaging, electronics, aerospace, and more. However, the disposal of solid wastes in which plastics are included has proven to be a challenging endeavor. The issue of solid waste disposal has turned into an urgent environmental problem worldwide, especially petrol-derived plastics wastes. Therefore, governments and researchers worldwide have intensified their efforts to develop novel biocomposites in their quest to replace conventional plastic manufacturing with biocomposites [1].

Biocomposites are defined as biocompatible and/or eco-friendly composites, which can consist of organic polymers such as polysaccharides and proteins [2]. Such materials are important research topics in modern science and technology because they can help industries achieve a more carbon-neutral production and replace the use of non-renewable resources, like oil and gas, with renewable resources such as biomass [3]. Studies in the field of biocomposites have focused mainly on fibers such as hemp [4], flax [5], sisal [6], and bamboo [7]. Nevertheless, researchers have kept identifying new sources of potential natural fiber reinforcement that can improve the mechanical and physical properties of biocomposites, such as *Abelmoschus esculentus* [8], *Grewia tilifolia* [9], *Sterculia urens* [10], and *Prosopis juliflora* [11]. Following this trend, fibers extracted from the *Musa paradisiaca* (plantain) tree are highlighted as a viable composite reinforcement due to its favorable physical and mechanical characteristics [12]. Chamorro et al. [13] successfully tested the possibility of incorporating *Musa paradisiaca* fibers into thermosetting polymeric resins such as polyester and epoxy resins.

Exploring different composite matrices, organic biodegradable macromolecules, like proteins and polysaccharides, are becoming ever more attractive to researchers, due to their ability to form bio-based polymers. Even though macromolecules are found abundantly in most organic solid wastes in landfills and other trash disposal facilities, they continue to be left to decompose or even incinerate instead of using these wastes' high potential as a source of bio-based polymers for the manufacturing industry [14]. Among these macromolecules, starch has been implemented in several studies [15–18] as the main component in the fabrication of biodegradable films and coating, indicating the high interest of researchers in implementing starch-based biopolymers, also known as thermoplastic starch, for product manufacturing. Its biodegradability, high commercial availability, and easy processing make starch very attractive for polymer production. Due to starch being mainly present in the edible part of the fruit, recent studies began to evaluate the possibility of using fruit-peel waste flour as a source of starch. Such flours contain not only starch but also proteins, lipids, fibers, and other polysaccharides [16,19,20]. The interest in using starch-based biopolymers, in combination with natural fiber reinforcements, lies in the possibility of creating added value products by using agroindustrial wastes.

The banana fruit and its varieties, such as plantain, are a renewable source from which both starch and natural fibers can be extracted. Unfortunately, it is used solely as a food crop, meaning that most of the residues of its life cycle (peels and pseudostem) have no current use in the industry [21], making finding new applications a research challenge. In particular, plantain as a source of starch for bio-based polymer production has been studied in the last decade. F.M. Pelissari et al. [19] compared the physical, thermal, mechanical, and morphological properties of both plantain starch-based and plantain flour-based polymer films, highlighting the potential of flour-based polymer films as a greener and cheaper alternative to the use of pure starch for manufacturing, as well as championing the tenets of a circular economy.

Even though the use of natural fibers as composite reinforcement and the development of bio-based matrices has been evaluated, the integration of natural fibers and starch bio-based matrices has not been deeply explored in literature. Therefore, the production of biocomposites combining plantain flour bio-based polymeric matrices and plantain natural fibers could combine both fields in the manufacturing of added-value products. This new approach in the development of biocomposites would assess the possibility of implementing most of the wastes of a single crop's production (plantain fibers) and consumption (plantain peels) stages.

Hence, this work presents the development and characterization of a bio-composite film of plantain flour thermoplastic starch (TPF) reinforced with plantain tree fibers (PF). The chemical, thermal, mechanical, and morphological characterization of TPF, untreated PF, and the composite material (TPF/PF) are shown through chemical composition analysis, Fourier-transformed infrared analysis (FTIR), thermos-gravimetric analysis (TGA), tensile tests, and scanning electronic microscopy (SEM).

2. Materials and Methods

2.1. Materials

Unripe plantain bananas of the variety "Harton" were acquired during the 2021 harvest in local markets in Bogota, Colombia, and used to obtain the plantain flour. This flour is extracted to produce TPF, in combination with industrial-grade glycerin (87% *w*/*w*) provided by Roda Químicos S.A.S. Plantain fiber bundles with a diameter between

94–189 μm were provided by farmers from San Agustín, Huila, Colombia, to implement the fiber as reinforcement. This fiber was previously cut and combed manually by the providers.

2.2. Chemical Composition of Plantain Fibers

A chemical composition analysis of the plantain fiber was carried out following TAPPI methods. Ash, lignin, cellulose, and extractive contents were determined. A minimum of 3 samples were used in each test.

2.2.1. Ash Content

This test was performed following the TAPPI T211 Standard Method [22]. For this procedure, a 1 g dry sample was exposed to a three-step drying cycle to avoid flames during the experiment. First, the sample was heated at 105 °C for 1 h, then heated at 250 °C for another hour, and finally at 575 °C for 4 h. This procedure guarantees good carbonization and the calcination of the sample. For this test, a VIBRA HT224R analytical scale and a Thermolyne 62700 furnace were used. The ash content was calculated as the recovered ash mass (m_{ash}) divided by the initial sample mass (m_{sample}), as explained in Equation (1).

$$\text{Ash content } (\%) = \frac{m_{ash}}{m_{sample}} \times 100 \tag{1}$$

2.2.2. Lignin Content

This test was performed following the TAPPI T222 Standard Method [23]. A 1 g dry sample was treated with 72% (w/w) sulfuric acid for 2 h. The sample was then macerated and stirred using a glass rod to dissolve extractive contents entirely during this process. Next, water was added to the solution until a 3% sulfuric acid mixture was obtained. Afterward, the solution was heated for 4 h to maintain constant volume to get lignin to coagulate. Lignin content was filtered from the solution via glass vacuum filtration, washed with water at 90 °C, dried in a Memmert UNE 200 forced convection oven at 105 °C, cooled, and finally weighed. The lignin content was calculated as the ratio of recovered lignin mass (m_{lig}) to the initial sample mass (m_{sample}), as seen in Equation (2).

$$\text{Lignin content } (\%) = \frac{m_{lig}}{m_{sample}} \times 100 \tag{2}$$

2.2.3. Cellulose Content

This test was performed following the TAPPI T203 Standard Method [24]. A 1.5 g sample was extracted using 17.5% sodium hydroxide solution at 25 °C. Alpha (α), beta (β), and gamma (γ) cellulose were determined via oxidation with potassium dichromate; therefore a two-step oxidation was carried out to determine the content of each cellulose type correctly. In the first oxidation, beta and gamma cellulose content was separated from α-cellulose content because α-cellulose is insoluble in sodium hydroxide. α-cellulose content was then determined as the difference between the initial specimen (100%) and the undissolved fraction, as seen in Equation (3).

$$\alpha\text{-cellulose content } (\%) = 100 - \frac{6.85 \times (V_2 - V_1) \times N \times 20}{A \times W} \tag{3}$$

V_1 corresponds to the sample titration volume; V_2 corresponds to the blank titration volumen; N is the normality of the titrate (ferrous amonium); A is the volume of the sample used, and W is the dry weigth of the fiber sample. Subsequently, γ-cellulose was recovered in the second oxidation of the solution, leaving behind the β-cellulose precipitated content. γ-cellulose content was determined using Equation (4).

$$\gamma\text{-cellulose content } (\%) = \frac{6.85 \times (V_4 - V_3) \times N \times 20}{25 \times W} \tag{4}$$

where V_3 corresponds to the sample titration volume after filtering the precipitated β-cellulose, and V_4 corresponds to the blank titration volume. Finally, β-cellulose content was determined by the difference between the first and second oxidations, see Equation (5).

$$\beta\text{-cellulose content } (\%) = 100 - (\alpha\text{-cellulose} + \beta\text{-cellulose}) \quad (5)$$

2.2.4. Extractive Content

This test was performed following the TAPPI T204 Standard Method [25]. A 4 g sample was oven-dried at 105 °C and milled to obtain a 0.4 mm granule size. Using an ethanol-benzene mixture (1:2, *v*/*v*), the sample was boiled at 75 ± °C in a Soxhlet apparatus for 5 h. Afterward, the solvent mixture was evaporated, and the residue was weighed. Extractive content was calculated as the ratio of the extract dry weight (m_{ext}) to the initial sample mass (m_{sample}), as seen in Equation (6).

$$\text{Extractive content } (\%) = \frac{m_{ext}}{m_{sample}} \times 100 \quad (6)$$

2.3. Film and Composite Preparation

Plantain flour was extracted from plantain peels as a preliminary step to prepare the TPF films. Peels were cut into small pieces, and the endocarp was separated from the cellulose-rich outer peel. Then, the starch-rich endocarp was dried for 90 min at 105 °C and subsequently ground using a Pulversisette 19 Universal Cutting Mill to obtain a fine flour powder. Finally, the flour was sieved using a 100 μm sieve, achieving homogeneous particle sizes and removing unwanted cellulose residue.

Films were prepared by casting method [15,16,19], a process in which a film-forming suspension (FFS) is applied over a surface and dried until a film is obtained. The FFS was obtained by homogenizing an aqueous solution of 4% *w*/*w* flour employing mechanical stirring at a constant temperature between 70 °C and 80 °C for 30 min to dissolve the starch content of the flour adequately. Industrial-grade glycerin was added at 2% *w*/*w* to the solution and processed at the same conditions for 15 more minutes. The produced suspension was poured into a mold to obtain a constant thickness film. The suspension was dried in a Memmert UNE 200 forced convection oven at 50 °C for 1.5 h to achieve a fast initial plasticization. Finally, the molds were left to dry approximately for 72 h at 20 °C and 50% relative humidity. The final product is a film, which acts as the matrix for the composite.

The composite films were obtained using the casting method described previously. PF was placed and fixed in a unidirectional orientation on the mold. Approximately 60 individual fibers were placed at a separation distance of 1 mm to permeate the fibers thoroughly. FFS was added to the previously conditioned mold with the fibers in place and left to dry for approximately 72 h at 20 °C and 50% relative humidity. The resulting film contained 6.54% *w*/*w* fiber content.

2.4. Physical Characterization

The thickness of each specimen was measured using an Ono Sokki GS-332 Linear Sensor Gauge. To obtain an average thickness, five different measures, each 30.5 mm apart, were taken along the length of the specimen. The density of TPF and TPF/PF was determined as the ratio between the mass and volume. The moisture content of each specimen was determined by a Precisa XM 60 Thermobalance and then averaged, following the ASTM D4442 standard procedure [26]. A minimum of 10 samples of each material were used for this test.

2.5. Fourier-Transformed Infrared Spectroscopy (FTIR)

An FTIR test was carried out to analyze the functional groups present in the TPF, PF, and TPF/PF. For this test, all specimens were ground and mixed with KBr in a ratio of 100:1. The mixture was later homogenized and pressed at 9 MPa for 30 s to obtain a pellet.

The wavelength range analyzed was between 4000 cm^{-1} to 400 cm^{-1}. A Thermo Nicolet 380 FT-IR was used for these tests.

2.6. Thermo-Gravimetric Analysis (TGA)

Thermo-gravimetric analysis (TGA) tests for the TPF, PF, and TPF/PF, were carried out using a TA Instruments Q600 analyzer and following the ASTM E1131 [27] standard procedure. Samples were heated from room temperature (25 °C) to 600 °C at a rate of 10 °C/min in a nitrogen atmosphere.

2.7. Mechanical Characterization

Tensile properties of the TPF and TPF/PF were determined by an Instrom 3367 Universal Testing System following the D882-10 ASTM standard [28]. Rectangular specimens, each 25 mm wide and 152.4 mm long, were tested. The specimens were cut utilizing die-cutting. The initial crosshead speed and grip separation were set at 12.5 mm/min and 125 mm, respectively.

Additionally, tensile properties of PF were determined by an Instrom 3367 Universal Testing System according to the C1557 ASTM standard [29]. A 50 mm long gauge length with a 10 mm/min crosshead speed where used in this test.

A minimum of 10 specimens for each material were used for each test.

2.8. Scanning Electron Microscopy (SEM)

PF and the fractured surfaces of TPF/PF samples were observed by FE-MEB LYRA 3 TESCAN (SEM). Before the analysis, specimens were coated with a thin gold layer, increasing sample conductivity, using a Desk® IV apparatus.

2.9. Statistical Analysis

An ANOVA one-way test was carried out to determine if the physical and mechanical properties of TPF and TPF/PF presented a significant difference. A p-value lower than 0.05 (confidence level of 95%) was considered statistically significant [30]. To perform the statistical analysis, Minitab 18 Statistical Software was used.

3. Results

3.1. Chemical Composition of Plantain Fibers

Natural fibers are composed mainly of cellulose and lignin with lesser amounts of other components such as ash and extractives. The composition of a natural fiber directly affects its physical and mechanical properties [31]; therefore, a chemical composition analysis of PF was carried out. As seen in Table 1, the main component is cellulose (81.05%), from which 69.09%, 0.36%, and 11.40% correspond to alpha (α), beta (β), and gamma (γ) cellulose, respectively. The main cellulose component in plantain fiber, α-cellulose, is an advantage for composites, because it has been reported to have high strength and stiffness per weight, allowing the formation of long fibrous cells [32]. Further, β-cellulose corresponds to degraded cellulose, and γ-cellulose is composed mainly of hemicellulose. In comparison to other natural fibers (see Table 2), PF shows a high cellulose content, making it a competitive reinforcement choice because it correlates to better mechanical performance [33,34]. In summary, PF's high α-cellulose content makes it a suitable reinforcement for composite applications.

Table 1. Chemical composition of PF.

Component	Percent (%)
Ash	3.50 ± 0.32
Extractives	2.28 ± 0.86
Lignin	13.17 ± 1.25
Cellulose [a]	81.05
α-cellulose	69.09 ± 0.36
β-cellulose	0.40 ± 0.25
γ-cellulose	11.40 ± 0.75

[a] Calculated from the difference of total constituents. Values are given as mean ± standard deviation.

Similarly, as shown in Table 2, PF has a low/intermediate lignin content (13.17%), correlating to better fiber–matrix interface and increasing their compatibility [35]. This results in a more reliable adhesion to polymeric matrices when employed in composite manufacturing. Furthermore, the ash content (3.5%) and extractives content (2.28%) (see Table 1) are in the range of other natural fibers reported by other authors, 0.6–8% [36] and 0.8–14.3% [37], respectively. In conclusion, the chemical composition analysis displays that plantain fibers are a potential reinforcement alternative for the manufacturing of composite materials.

Table 2. Chemical composition of plantain fibers compared to other cellulose-based natural fibers [32,34,38,39].

Fiber	Lignin (%)	Cellulose (%)
Jute	12–13	61–71
Hemp	3	68
Sisal	8–10	65–78
Flax	3	71–81
Piassava	48.4	31.6
Bamboo	21–31	26–60
Caraua	7.5	81
Oil palm	29	40–50
Coir	40–45	32–43
Bagasse	25.3	52
Plantain (Present study)	13.2	81.1

3.2. Physical Characterization

In composite engineering, the physical characteristics of the matrix and the reinforcement dictate the final properties of the composite; therefore, a physical characterization (moisture content and density) of TPF matrix, PF fiber, and TPF/PF composite was carried out. As seen in Table 3, the moisture content of TPF is 10.42%, this value is in the range of other plantain thermoplastic starch films found in literature, 6.50–15.24% [17,19,40]. This can be an advantage for composite applications because moisture creates a plasticizer effect, increasing the flexibility of the bio-based polymer [41]. Additionally, PF displays a low moisture content (8.39%) when compared to other widely used natural fibers, such as jute (12.00%), sisal (11.00%), coir (10.00%), and hemp (9.00%) [38]. When TPF and PF are implemented into a composite material, the result was a TPF/PF composite with an average moisture content of 9.90%. TPF/PF composite appears to have a high affinity to water molecules, due to the presence of hydrophilic components found in TPF and PF, such as starch, cellulose, and, in lower amounts, proteins. According to the ANOVA test, there was no significant difference in moisture content between TPF and TPF/PF (p-value = 0.051). In general, this hydrophilic composite behavior promotes its faster biodegradation at the end of its life cycle, which is an advantage when designing short life-cycle products.

Table 3. Average density and moisture content of the TPF, PF, and TPF/PF.

Property	TPF	PF	TPF/PF
Moisture (%)	10.42 ± 0.01	8.39 ± 0.72	9.90 ± 0.01
Density (g/cm^3)	1.14 ± 0.04	0.83 ± 0.05	1.10 ± 0.06

Values are given as mean ± standard deviation.

Regarding density, TPF's value is 1.14 g/cm^3 (see Table 3) which is on par with other plantain starch and flour bio-based polymers [19], and lower than that of synthetic polymer resins such as epoxy (1.2 g/cm^3 [42]), phenolic (1.4 g/cm^3 [43]), and cyanate ester (1.2 g/cm^3). In the case of the fiber, it has a density of 0.83 g/cm^3, a value lower than other widespread natural fibers like hemp (1.48 g/cm^3), flax (1.5 g/cm^3), sisal (1.5 g/cm^3), and coir (1.2 g/cm^3) [38]. Finally, the composite displays a density of 1.10 g/cm^3, a good alternative to be used for lightweight products applications. Nevertheless, the ANOVA test determined that no significant difference in the densities of TPF and TPF/PF was present (p-value = 0.070). In general, the physical properties of TPF and PF allowed the manufacturing of a lightweight composite material.

3.3. Fourier-Transformed Infrared Spectroscopy (FTIR)

Figure 1 shows results from the FTIR spectra analysis for TPF, PF, and TPF/PF. Such analysis was conducted to have a better comprehension of the functional groups present in each material. TPF presents an energy absorption band around 3370 cm^{-1}. These can be attributed to stretching –OH groups caused by the formation of hydrogen bonds [44]. Energy absorption bands around 2920 cm^{-1} are evidence of the presence of C–H and CH_2 groups, most probably coming from amylose and amylopectin content [45], both are polysaccharides found in plantain flour. The peak around 1690 cm^{-1} corresponds to amide I, a functional group present mainly in proteins, in which the stretching of C=O groups occurs [20]. Other amide groups such as amide III are also present, correlating to the bands around 1323 cm^{-1} [46]. These amide groups are found mainly in organic proteins, meaning that plantain peel flour does contain protein material. Additional energy absorption bands around 1152 cm^{-1}, 1077 cm^{-1}, and 1021 cm^{-1} were associated with stretching the C–C, C–O, and C–O–H bonds of starch [47,48]. Finally, an energy absorption band was also observed around 960 cm^{-1}, indicating the presence of glycosidic bonds in starch due to amylopectin α-16 bonds [45]. Even though different plantain varieties were used, the FTIR is in accordance with a similar analysis carried out by F.M Pelissari et al. [19] on plantain starch and flour polymeric films, as well as other starch-based films from other sources such as corn [49] and cassava [50]. In general, the main functional groups detected in TPF correspond to polysaccharides and glycosidic bonds which allow the formation of the polymeric matrix.

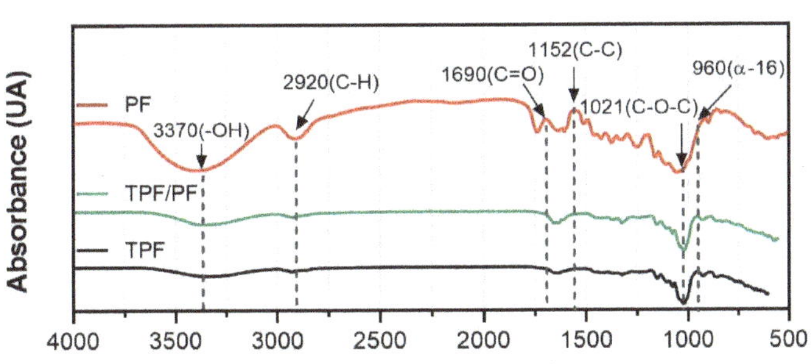

Figure 1. Fourier-transformed spectroscopy of the TPF, PF, and TPF/PF.

The previous chemical characterization revealed that the main components of PF were cellulose and lignin, substances also detected by the FTIR analysis. PF presents an energy absorption band around 3370 cm^{-1}, confirming the presence of α-cellulose due to the stretching –OH groups caused by the formation of hydrogen bonds [44]. The peak around 2920 cm^{-1} correlates to the symmetric and asymmetric stretching of C–H and CH_2 groups, also related to α-cellulose [32,51]. Bands around 1152 cm^{-1}, 1077 cm^{-1}, and 1021 cm^{-1}, indicating the presence of asymmetrical stretching of C–C and C–O–C bonds also found in cellulose. Additionally, the peak around 1638 cm^{-1} corresponds to C=C bonds found primarily in lignin [51,52]. Finally, TPF/PF combines the behavior observed in both TPF and PF, the materials used as constituents of the composite.

3.4. *Thermo-Gravimetric Analysis (TGA)*

The thermal stability of materials can determine processing conditions in composite applications; therefore, a thermo-gravimetric analysis was carried out for the TPF matrix, PF fiber, and TPF/PF composite. As seen in Figure 2, the matrix displays an initial phase, up to 120 °C, which is attributed to moisture evaporation [53–55]. A second phase takes place between 130 °C and 220 °C, due to the evaporation of a glycerol-rich phase. Finally, between 230 °C and 350 °C, a weight-loss step correlating to the degradation of the starch-rich phase is observed [56–58]. A residual mass percentage of around 17% was found at 600 °C. This thermal behavior agrees with other starch bio-based polymeric films reported by other authors in literature such as cassava starch films [52,55], sugar palm starch films [57], and corn starch films [58]. In summary, the thermo-gravimetric analysis determined that the degradation of TPF starts around 130 °C; therefore, it should be processed at temperatures below that.

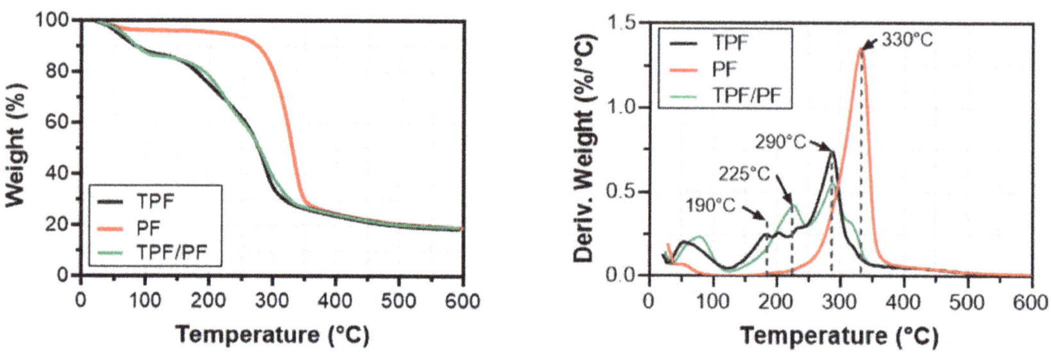

(a) (b)

Figure 2. TGA (**a**) and DTGA (**b**) for TPF, PF, and TPF/PF.

Organic material in PF started to degrade at 240 °C, reaching a peak degradation temperature at 330 °C. Nevertheless, the degradation process of PF fibers extended well into the 500 °C range. This is attributed to the presence of lignocellulosic and inorganic substances found in the fiber [59]. These results show a similar thermal profile to that of other natural fibers such as okra (peak degradation at 335 °C), curaua (peak degradation at 350 °C), and jute (peak degradation at 283 °C), as well as other plantain thermal degradation analyses of other authors [60,61]. A 19% residual mass percentage was found at 600 °C. In brief, TGA determined that PF should be processed at temperatures below 240 °C to avoid fiber degradation.

TPF/PF shows thermal stability that combines its constituents' thermal behavior. It exhibits an initial phase, up to 120 °C, representing the loss of moisture. This is followed by a second phase, between 130 °C and 220 °C, showing the loss of the glycerol-rich phase of the matrix and the early degradation of PF. Finally, the third phase from 230 °C onwards shows the degradation of both the starch-rich phase of the matrix (between 230 °C and 350 °C) and PF's degradation process (between 240 °C up to 500 °C). The peak temperatures of the degradation of the starch-phase (290 °C) and PF (330 °C) can be found in the third phase. A residual mass percentage of 17% was found at 600 °C. In conclusion, the TPF/PF composite should not be exposed to temperatures above 130 °C to prevent the degradation of the matrix.

3.5. Mechanical Characterization

As seen in Figure 3, all three materials show an initial linear region corresponding to an elastic behavior, followed by a plastic region characterized by a deviation from linearity until the final rupture of the material. Something to highlight in the TPF/PF curve is the steps present near the rupture of the sample. This behavior is attributed to the unidirectional configuration of the TPF/PF composite, in which fibers tend to break progressively until a total sample rupture is reached.

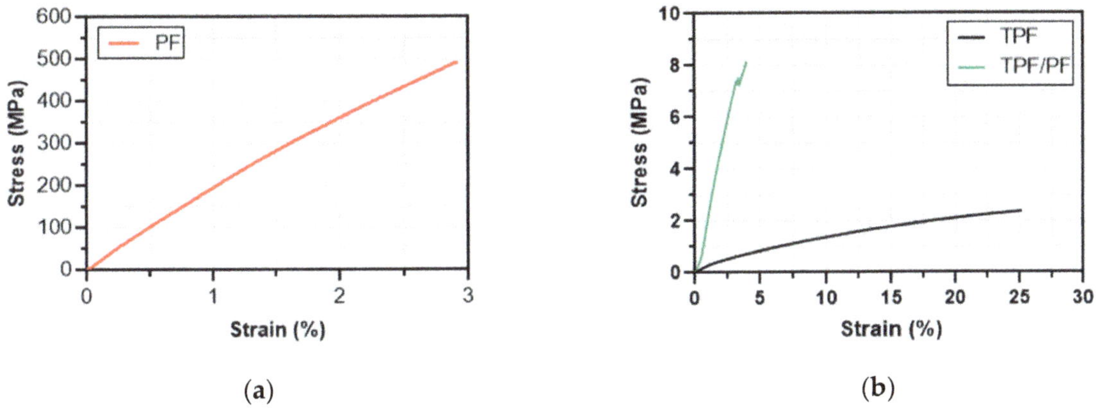

Figure 3. Stress vs. strain curves of PF (**a**), and TPF and TPF/PF (**b**).

TPF films averaged a tensile strength and Young's modulus of 2.36 MPa and 23.52 MPa, respectively (see Table 4). These results are comparable and, in some cases, higher than other starch and flour bio-based polymeric films such as batata starch and rice starch bio-based films (see Table 5). On the other hand, PF averaged a tensile strength and Young's modulus of 440.24 MPa and 15.67 GPa, respectively. The fibers displayed a mechanical performance on par with other widely used natural fibers such as jute, sisal, and bagasse (see Table 6), meaning that plantain fibers could be used as reinforcement for a wide variety of polymeric matrices. Once PF fibers were incorporated into TPF films, the resulting TPF/PF composite averaged a tensile strength and Young's modulus of 8.16 MPa and 281.34 MPa, respectively. This represents a 345.76% increase in tensile strength and an 1196% increase in Young's modulus when compared to TPF films. In addition, specific properties of TPF/PF also increased by 518.84% for specific tensile strength and 1793.51% for specific modulus, when compared to TPF films (see Table 4). In conclusion, the TPF/PF composite films displayed a good mechanical performance. Its attractive properties show potential to be used, for example, as single-use flexible plastic packaging, where biodegradability, non-toxicity, and intermediate tensile properties are needed to achieve a circular-economy product.

Table 4. Average mechanical properties of TPF, PF, and TPF/PF.

Property	TPF	PF	TPF/PF
Tensile strength (MPa)	2.36 ± 0.21	440.30 ± 146.95	8.16 ± 1.66
Elongation (%)	24.53 ± 2.82	2.94 ± 0.77	4.09 ± 0.53
Young's modulus (MPa)	23.62 ± 3.10	15,668.81 ± 5534.94	281.34 ± 38.58
Specific strength (MPa/g cm^3)	2.07 ± 0.21	530.12 ± 146.95	10.74 ± 1.66
Specific modulus (MPa/g cm^3)	20.64 ± 2.82	18,878.08 ± 5534.94	370.18 ± 38.58

Values are given as mean ± standard deviation.

Table 5. Average mechanical properties of various starch-based polymeric films [32,62–66].

Source	Tensile Strength (MPa)	Young's Modulus (MPa)	Elongation at Break (%)
Corn starch	1.23 ± 0.07	2.31 ± 0.06	53.13 ± 0.81
Batata starch	3.51 ± 0.16	25.09 ± 2.66	70.74 ± 5.67
Rice starch	3.70 ± 0.92	9.33 ± 0.47	39.50 ± 0.92
Banana flour	1.00 ± 0.10	2.70 ± 0.70	49 ± 2.00
Banana/beet flour	1.14 ± 0.03	3.3 ± 0.10	53 ± 1.00
Chemically modified starch	12.8	1169	8.3
Commercial starch	7.90 ± 0.07	73.10 ± 5.78	33.40 ± 0.41
TPF (Present study)	2.36 ± 0.21	23.62 ± 3.10	24.53 ± 2.82

Values are given as mean ± standard deviation.

Table 6. Average mechanical properties of various natural fibers commonly used as composite reinforcements [32,35,67–70].

Fiber	Tensile Strength (MPa)	Young's Modulus (GPa)	Elongation at Break (%)
Jute	187–773	13–26.5	1.5–3.1
Hemp	580–1110	70	1.6–4.5
Sisal	507–885	9.4–22	1.9–3
Flax	343–1035	27.6	1.2–3
Date	58–203	2–7.5	5–10
Bagasse	20–290	N/A	1–3
PF (Present study)	211–599	10–21	2.2–3.8

3.6. Scanning Electron Microscopy (SEM)

SEM micrographs of PF fiber and tensile fractured TPF/PF composite were carried out to analyze their morphology and identify failure mechanisms in composite films. Figure 4a–c shows an overview of both longitudinal and cross-section of the fiber. The fiber was seen to have longitudinal ridges, impurities, and a waxy epidermis covering its surface. Moreover, the cross-section view shows that plantain fibers are organized as a bundle of round and hollow fibrils bonded together, a common structure for natural fibers. This binding of fibrils made possible by the effect of lignocellulosic intercellular material explains the low density of natural fibers [52].

With respect to TPF/PF, Figure 4d–f shows that the PF is completely embedded in the matrix, but its inner sections were not filled by resin. Fiber fracture occurred at different heights, due to the different mechanical performances of each fibril. Matrix fracture and fiber pull-out were also identified as failure mechanisms found on the specimens. The pulled fiber was indeed bonded to the polymeric matrix, as evidenced by the fiber surface marked on the matrix. This phenomenon indicates mechanical interlocking, which explains the improvement of the mechanical properties of the composite compared to those of the film. Nevertheless, fibers can also lack bonding with the polymeric matrix, resulting in gaps between them. This lack of fiber–matrix bonding is attributed to the waxy epidermis and impurities previously identified on the fiber's surface [71]. In order to remove lignin waxes and impurities found on the surface, surface treatments such as mercerization and alkalization are some of the most common and compelling possibilities. These types of treatments have been shown to improve fiber/matrix adhesion and surface roughness, resulting in better mechanical interlocking [72–74]. In brief, the PF bonded successfully to the TPF films, but a better matrix/fiber interfacial interaction could be achieved by treating the fiber's surface, resulting in better mechanical properties in TPF/PF.

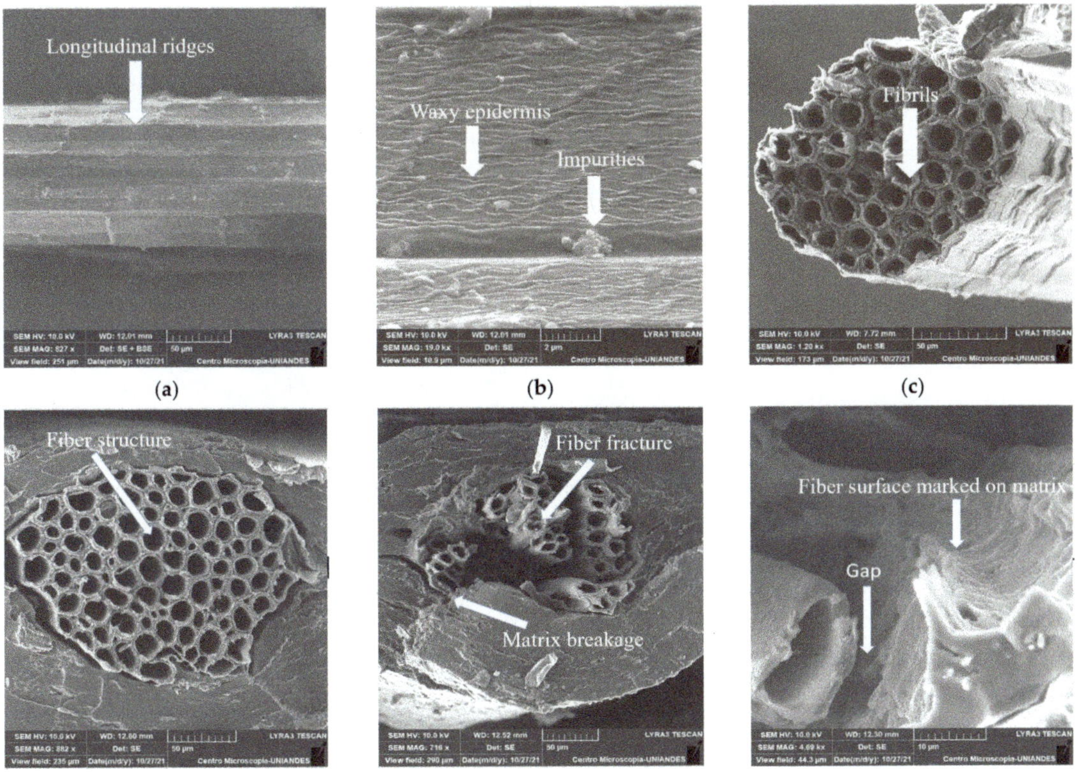

Figure 4. SEM images of PF at (**a**) ×827, (**b**) ×19,000, (**c**) ×1200 and TPF/PF (**d**) ×882, (**e**) ×716, (**f**) ×4690 longitudinal and cross sections.

4. Conclusions

In this study, a biocomposite from agroindustrial plantain waste with attractive mechanical properties and eco-friendly character was developed, contributing to the industrial conversion towards a circular economy.

The overall physical characterization shows that both the plantain flour-based biopolymer film (TPF) and plantain fibers (PF) are low-density and low-moisture materials when compared to synthetic polymer resins and other natural fibers, respectively. As a result, the TPF/PF composite has the advantage of being a lightweight, renewable and biodegradable material.

The chemical composition and spectral analysis of PF showed a high amount of α-cellulose and low amounts of lignin. These characteristics make PF a good candidate for reinforcing composite materials. In addition, the FTIR spectra of the TPF matrix and TPF/PF composite revealed the presence of OH and CH_2 groups and amide I and III groups, which are found primarily in polysaccharides and proteins, respectively.

With regards to thermal degradation, it has been determined that, for TPF matrices, it starts at 130 °C, while PF fibers start degrading at 240 °C. The TPF/PF composite displayed the behavior of its constituent materials; therefore its processing temperature is set to below 130 °C to avoid unwanted degradation of the material.

The mechanical properties of both TPF and PF proved to be comparable to other starch bio-based resins and natural fibers, respectively. When combined into TPF/PF, composite films displayed a superior performance in comparison to TPF films, showing plantain fibers as effective reinforcement. SEM analysis showed a good bonding between TPF and PF; and fiber fracture, pull-out, and matrix breakage were determined to be the main failure mechanisms in the TPF/PF composite. The rough surface of the fibers created mechanical interlocking with the matrix, resulting in enhanced mechanical characteristics. Nevertheless, a waxy layer was identified on the fiber's surface, reducing its adhesion to the matrix and causing gaps between the fiber and matrix. Chemical treatment of the fibers would improve this issue.

Author Contributions: Conceptualization, R.V.; methodology, R.V.; formal analysis, R.V., A.T., A.M.R., M.A.M., M.J.A., and A.P.; investigation, R.V.; resources, A.P.; data curation, R.V., A.T., A.M.R.; writing—original draft preparation, R.V.; writing—review and editing, R.V., M.J.A., M.A.M., and A.P.; supervision, M.J.A., M.A.M., and A.P. All authors have read and agreed to the published version of the manuscript.

Funding: This research was funded by the Chemical and Food Engineering Department and the Vice Presidency for Research & Creation publication fund program at the Universidad de los Andes.

Institutional Review Board Statement: Not applicable.

Informed Consent Statement: Not applicable.

Data Availability Statement: The data are available from the corresponding author upon request.

Acknowledgments: The authors wish to express gratitude to Universidad de Los Andes, in particular to the Chemical Engineering and Mechanical Engineering lab teams, for making this research possible with their help.

Conflicts of Interest: The authors declare no conflict of interest.

References

1. Mohanty, A.K.; Misra, M.; Drzal, L.T. Sustainable Bio-Composites from Renewable Resources: Opportunities and Challenges in the Green Materials World. *J. Polym. Environ.* **2002**, *10*, 19–26. [CrossRef]
2. Haraguchi, K. Biocomposites. In *Encyclopedia of Polymeric Nanomaterials*; Kobayashi, S., Müllen, K., Eds.; Springer: Berlin/Heidelberg, Germany, 2015; pp. 124–130, ISBN 978-3-642-29648-2.
3. Masutani, K.; Kimura, Y. Biobased Polymers. In *Encyclopedia of Polymeric Nanomaterials*; Kobayashi, S., Müllen, K., Eds.; Springer: Berlin/Heidelberg, Germany, 2015; pp. 118–124, ISBN 978-3-642-29648-2.
4. Madsen, B.; Hoffmeyer, P.; Lilholt, H. Hemp Yarn Reinforced Composites—II. Tensile Properties. *Compos. Part A Appl. Sci. Manuf.* **2007**, *38*, 2204–2215. [CrossRef]
5. Goutianos, S.; Peijs, T.; Nyström, B.; Skrifvars, M. Development of Flax Fibre Based Textile Reinforcements for Composite Applications. *Appl. Compos. Mater.* **2006**, *13*, 199–215. [CrossRef]
6. Oksman, K.; Wallström, L.; Berglund, L.; Toledo Filho, R. Morphology and Mechanical Properties of Unidirectional Sisal—Epoxy Composites. *J. Appl. Polym. Sci.* **2002**, *84*, 2358–2365. [CrossRef]
7. Abdul Khalil, H.P.S.; Bhat, I.U.H.; Jawaid, M.; Zaidon, A.; Hermawan, D.; Hadi, Y.S. Bamboo Fibre Reinforced Biocomposites: A Review. *Mater. Des.* **2012**, *42*, 353–368. [CrossRef]
8. De Rosa, I.M.; Kenny, J.M.; Puglia, D.; Santulli, C.; Sarasini, F. Morphological, Thermal and Mechanical Characterization of Okra (*Abelmoschus esculentus*) Fibres as Potential Reinforcement in Polymer Composites. *Compos. Sci. Technol.* **2010**, *70*, 116–122. [CrossRef]
9. Jayaramudu, J.; Guduri, B.R.; Varada Rajulu, A. Characterization of New Natural Cellulosic Fabric *Grewia tilifolia*. *Carbohydr. Polym.* **2010**, *79*, 847–851. [CrossRef]
10. Jayaramudu, J.; Reddy, S.M.; Varaprasad, K.; Sadiku, R.; Sinha Ray, S.; Rajulu, A.V. Preparation and Properties of Biodegradable Films from Sterculia Urens Short Fiber/Cellulose Green Composites. *Carbohydr. Polym.* **2013**, *93*, 622–627. [CrossRef]
11. Saravanakumar, S.S.; Kumaravel, A.; Nagarajan, T.; Sudhakar, P.; Baskaran, R. Characterization of a Novel Natural Cellulosic Fiber from *Prosopis juliflora* Bark. *Carbohydr. Polym.* **2013**, *92*, 1928–1933. [CrossRef]

12. Adeniyi, A.; Ighalo, J.O.; Onifade, D. Banana and Plantain Fiber-Reinforced Polymer Composites. *J. Polym. Eng.* **2019**, *39*, 597–611. [CrossRef]
13. Cadena Ch, E.M.; Vélez R, J.M.; Santa, J.F.; Otálvaro, G.V. Natural Fibers from Plantain Pseudostem (*Musa paradisiaca*) for Use in Fiber-Reinforced Composites. *J. Nat. Fibers* **2017**, *14*, 678–690. [CrossRef]
14. Zarate, L.; Romero, A.; Bengoechea, C.; Partal, P.; Guerrero, A. Thermo-Mechanical and Hydrophilic Properties of Polysaccharide/Gluten-Based Bioplastics. *Carbohydr. Polym.* **2014**, *112*, 24–31. [CrossRef] [PubMed]
15. Gutiérrez, T.J.; Guzmán, R.; Medina Jaramillo, C.; Famá, L. Effect of Beet Flour on Films Made from Biological Macromolecules: Native and Modified Plantain Flour. *Int. J. Biol. Macromol.* **2016**, *82*, 395–403. [CrossRef]
16. Pelissari, F.M.; Andrade-Mahecha, M.M.; Sobral, P.J.d.A.; Menegalli, F.C. Optimization of Process Conditions for the Production of Films Based on the Flour from Plantain Bananas (*Musa paradisiaca*). *LWT Food Sci. Technol.* **2013**, *52*, 1–11. [CrossRef]
17. Żołek-Tryznowska, Z.; Holica, J. Starch Films as an Environmentally Friendly Packaging Material: Printing Performance. *J. Clean. Prod.* **2020**, *276*, 124265. [CrossRef]
18. Diyana, Z.N.; Jumaidin, R.; Selamat, M.Z.; Ghazali, I.; Julmohammad, N.; Huda, N.; Ilyas, R.A. Physical Properties of Thermoplastic Starch Derived from Natural Resources and Its Blends: A Review. *Polymers* **2021**, *13*, 1396. [CrossRef]
19. Pelissari, F.M.; Andrade-Mahecha, M.M.; Sobral, P.J.d.A.; Menegalli, F.C. Comparative Study on the Properties of Flour and Starch Films of Plantain Bananas (*Musa paradisiaca*). *Food Hydrocoll.* **2013**, *30*, 681–690. [CrossRef]
20. Andrade-Mahecha, M.; Tapia-Blácido, D.R.; Menegalli, F.C. Development and Optimization of Biodegradable Films Based on Achira Flour. *Carbohydr. Polym.* **2012**, *88*, 449–458. [CrossRef]
21. Gañán, P.; Zuluaga, R.; Restrepo Osorio, A.; Labidi, J.; Mondragon, I. Plantain Fibre Bundles Isolated from Colombian Agro-Industrial Residues. *Bioresour. Technol.* **2008**, *99*, 486–491. [CrossRef]
22. TAPPI. *Ash in Wood, Pulp, Paper and Paperboard: Combustion at 525 °C (Five-Year Review of T 211 Om-02)*; TAPPI: Peachtree, GA, USA, 2007.
23. TAPPI. *Acid-Insoluble Lignin in Wood and Pulp (Reaffirmation of T 222 Om-02)*; TAPPI: Peachtree, GA, USA, 2006.
24. TAPPI. *Alpha-, Beta- and Gamma-Cellulose in Pulp*; TAPPI: Peachtree, GA, USA, 1999.
25. TAPPI. *Solvent Extractives of Wood and Pulp (Proposed Revision of T 204 Cm-97)*; TAPPI: Peachtree, GA, USA, 2007.
26. ASTM 4442. *Standard Test Methods for Direct Moisture Content Measurement of Wood and Wood-Based Materials*; ASTM: West Conshohocken, PA, USA, 2020.
27. ASTM E1131-08. *Standard Test Method for Compositional Analysis by Thermogravimetry*; ASTM: West Conshohocken, PA, USA, 2014.
28. ASTM D882-10. *Standard Test Method for Tensile Properties of Thin Plastic Sheeting*; ASTM: West Conshohocken, PA, USA, 2012.
29. ASTM C1557. *Standard Test Method for Tensile Strength and Young's Modulus of Fibers*; ASTM: West Conshohocken, PA, USA, 2017.
30. Montgomery, D.C. *Design and Analysis of Experiments*, 2nd ed.; Wiley: New York, NY, USA, 1984.
31. Komuraiah, A.; Kumar, N.S.; Prasad, B.D. Chemical Composition of Natural Fibers and Its Influence on Their Mechanical Properties. *Mech. Compos. Mater.* **2014**, *50*, 359–376. [CrossRef]
32. Porras, A.; Maranon, A.; Ashcroft, I.A. Characterization of a Novel Natural Cellulose Fabric from Manicaria Saccifera Palm as Possible Reinforcement of Composite Materials. *Compos. B Eng.* **2015**, *74*, 66–73. [CrossRef]
33. Bogoeva-Gaceva, G.; Avella, M.; Malinconico, M.; Bužarovska, A.; Grozdanov, A.; Gentile, G.; Errico, M. Natural Fiber Eco-Composites. *Polym. Compos.* **2007**, *28*, 98–107. [CrossRef]
34. Williams, G.; Wool, R. Composites from Natural Fibers and Soy Oil Resins. *Appl. Compos. Mater.* **2000**, *7*, 421–432. [CrossRef]
35. Mohanty, A.; Misra, M.; Drzal, L. Surface Modifications of Natural Fibers and Performance of the Resulting Biocomposites: An Overview. *Compos. Interfaces* **2001**, *8*, 313–343. [CrossRef]
36. Chokshi, S.; Parmar, V.; Gohil, P.; Chaudhary, V. Chemical Composition and Mechanical Properties of Natural Fibers. *J. Nat. Fibers* **2020**, 1–12. [CrossRef]
37. Poletto, M.; Ornaghi, H.L.; Zattera, A.J. Native Cellulose: Structure, Characterization and Thermal Properties. *Materials* **2014**, *7*, 6105–6119. [CrossRef]
38. Faruk, O.; Bledzki, A.K.; Fink, H.-P.; Sain, M. Biocomposites Reinforced with Natural Fibers: 2000–2010. *Prog. Polym. Sci.* **2012**, *37*, 1552–1596. [CrossRef]
39. Bledzki, A.K.; Reihmane, S.; Gassan, J. Properties and Modification Methods for Vegetable Fibers for Natural Fiber Composites. *J. Appl. Polym. Sci.* **1996**, *59*, 1329–1336. [CrossRef]
40. García-Ramón, J.; Carmona-García, R.; Valera-Zaragoza, M.; Aparicio-Saguilán, A.; Bello-Pérez, L.; Aguirre-Cruz, A.; Alvarez-Ramirez, J. Morphological, Barrier, and Mechanical Properties of Banana Starch Films Reinforced with Cellulose Nanoparticles from Plantain Rachis. *Int. J. Biol. Macromol.* **2021**, *187*, 35–42. [CrossRef]
41. Godbillot, L.; Dole, P.; Joly, C.; Rogé, B.; Mathlouthi, M. Analysis of Water Binding in Starch Plasticized Films. *Food Chem.* **2006**, *96*, 380–386. [CrossRef]
42. Bhatia, S.; Angra, S.; Khan, S. Mechanical and Wear Properties of Epoxy Matrix Composite Reinforced with Varying Ratios of Solid Glass Microspheres. *J. Phys. Conf. Ser.* **2019**, *1240*, 012080. [CrossRef]
43. Brydson, J.A. 23—Phenolic Resins. In *Plastics Materials (Seventh Edition)*; Brydson, J.A., Ed.; Butterworth-Heinemann: Oxford, UK, 1999; pp. 635–667. ISBN 978-0-7506-4132-6.
44. Silverstein, R.; Webster, F.; Kiemle, D. *Identificação Espectrofotométrica de Compostos Orgânicos*, 7th ed.; LTC: Rio de Janeiro, Brazil, 2007; ISBN 978-85-216-1521-7.

45. Kizil, R.; Irudayaraj, J.; Seetharaman, K. Characterization of Irradiated Starches by Using FT-Raman and FTIR Spectroscopy. *J. Agric. Food Chem.* **2002**, *50*, 3912–3918. [CrossRef] [PubMed]
46. Singh, B.R.; Vickers, T. Infrared Analysis of Peptides and Proteins. *Appl. Spectrosc.* **2001**, *55*, 108.
47. Van Soest, J.J.G.; Tournois, H.; de Wit, D.; Vliegenthart, J.F.G. Short-Range Structure in (Partially) Crystalline Potato Starch Determined with Attenuated Total Reflectance Fourier-Transform IR Spectroscopy. *Carbohydr. Res.* **1995**, *279*, 201–214. [CrossRef]
48. Huang, C.; Jeng, R.; Sain, M.; Saville, B.; Hubbes, M. Production, Characterization, and Mechanical Properties of Starch Modified by Ophiostoma Spp. *BioResources* **2006**, *1*, 257–269. [CrossRef]
49. Estrada-Monje, A.; Alonso-Romero, S.; Zitzumbo-Guzmán, R.; Estrada-Moreno, I.A.; Zaragoza-Contreras, E.A. Thermoplastic Starch-Based Blends with Improved Thermal and Thermomechanical Properties. *Polymers* **2021**, *13*, 4263. [CrossRef]
50. Jiménez-Regalado, E.J.; Caicedo, C.; Fonseca-García, A.; Rivera-Vallejo, C.C.; Aguirre-Loredo, R.Y. Preparation and Physicochemical Properties of Modified Corn Starch–Chitosan Biodegradable Films. *Polymers* **2021**, *13*, 4431. [CrossRef]
51. Dan-asabe, B.; Yaro, S.A.; Yawas, D.; Aku, S.; Samotu, I.; Abubakar Zaria, U.; Obada, D. Mechanical, Spectroscopic and Micro-Structural Characterization of Banana Particulate Reinforced PVC Composite as Piping Material. *Tribol. Ind.* **2016**, *38*, 255–266.
52. Rodríguez, L.J.; Cardona, C.A.; Orrego, C.E. Water Uptake, Chemical Characterization, and Tensile Behavior of Modified Banana–Plantain Fiber and Their Polyester Composites. *Polym. Compos.* **2015**, *37*, 2960–2973. [CrossRef]
53. Ayala Valencia, G.; Agudelo, A.; Zapata, R. Effect of Glycerol on the Electrical Properties and Phase Behavior of Cassava Starch Biopolymers. *DYNA* **2012**, *79*, 138–147.
54. Nuwamanya, E.; Baguma, Y.; Wembabazi, E.; Rubaihayo, P. A Comparative Study of the Physicochemical Properties of Starches from Root, Tuber and Cereal Crops. *Afr. J. Biotechnol.* **2011**, *10*, 12018–12030. [CrossRef]
55. Marques, P.T.; Lima, A.M.F.; Bianco, G.; Laurindo, J.B.; Borsali, R.; Le Meins, J.-F.; Soldi, V. Thermal Properties and Stability of Cassava Starch Films Cross-Linked with Tetraethylene Glycol Diacrylate. *Polym. Degrad. Stab.* **2006**, *91*, 726–732. [CrossRef]
56. Liu, X.; Yu, L.; Liu, H.; Chen, L.; Li, L. In Situ Thermal Decomposition of Starch with Constant Moisture. *Polym. Degrad. Stab.* **2008**, *93*, 260–262. [CrossRef]
57. Sanyang, M.L.; Sapuan, S.M.; Jawaid, M.; Ishak, M.R.; Sahari, J. Effect of Plasticizer Type and Concentration on Tensile, Thermal and Barrier Properties of Biodegradable Films Based on Sugar Palm (*Arenga pinnata*) Starch. *Polymers* **2015**, *7*, 1106–1124. [CrossRef]
58. Shi, R.; Zhang, Z.; Liu, Q.; Han, Y.; Zhang, L.; Chen, D.; Tian, W. Characterization of Citric Acid/Glycerol Co-Plasticized Thermoplastic Starch Prepared by Melt Blending. *Carbohydr. Polym.* **2007**, *69*, 748–755. [CrossRef]
59. Jimenez, A.; Betancourt, S.; Gañán, P.; Cruz, L. Degradación Térmica de Fibras Naturales Procedentes de La Calceta de Plátano. (Estudio Cinético). *Rev. Latinoam. Metal. Mater.* **2009**, *29*, 215–219.
60. Kusić, D.; Božič, U.; Monzon, M.; Paz Hernández, R.; Bordón, P. Thermal and Mechanical Characterization of Banana Fiber Reinforced Composites for Its Application in Injection Molding. *Materials* **2020**, *13*, 3581. [CrossRef]
61. Benítez, A.N.; Monzón, M.D.; Angulo, I.; Ortega, Z.; Hernández, P.M.; Marrero, M.D. Treatment of Banana Fiber for Use in the Reinforcement of Polymeric Matrices. *Measurement* **2013**, *46*, 1065–1073. [CrossRef]
62. Song, H.; Choi, I.; Lee, J.-S.; Chung, M.-N.; Yoon, C.S.; Han, J. Comparative Study on Physicochemical Properties of Starch Films Prepared from Five Sweet Potato (*Ipomoea batatas*) Cultivars. *Int. J. Biol. Macromol.* **2021**, *189*, 758–767. [CrossRef]
63. Alqahtani, N.; Alnemr, T.; Ali, S. Development of Low-Cost Biodegradable Films from Corn Starch and Date Palm Pits (*Phoenix dactylifera*). *Food Biosci.* **2021**, *42*, 101199. [CrossRef]
64. Hasan, M.; Gopakumar, D.A.; Olaiya, N.G.; Zarlaida, F.; Alfian, A.; Aprinasari, C.; Alfatah, T.; Rizal, S.; Abdul Khalil, H.P.S. Evaluation of the Thermomechanical Properties and Biodegradation of Brown Rice Starch-Based Chitosan Biodegradable Composite Films. *Int. J. Biol. Macromol.* **2020**, *156*, 896–905. [CrossRef] [PubMed]
65. Ali, A.; Xie, F.; Yu, L.; Liu, H.; Meng, L.; Khalid, S.; Chen, L. Preparation and Characterization of Starch-Based Composite Films Reinfoced by Polysaccharide-Based Crystals. *Compos. B Eng.* **2018**, *133*, 122–128. [CrossRef]
66. Bodirlau, R.; Teaca, C.-A.; Spiridon, I. Influence of Natural Fillers on the Properties of Starch-Based Biocomposite Films. *Compos. B Eng.* **2013**, *44*, 575–583. [CrossRef]
67. Djafari Petroudy, S.R. 3—Physical and Mechanical Properties of Natural Fibers. In *Advanced High Strength Natural Fibre Composites in Construction*; Fan, M., Fu, F., Eds.; Woodhead Publishing: Sawston, UK, 2017; pp. 59–83, ISBN 978-0-08-100411-1.
68. Ochi, S. Development of High Strength Biodegradable Composites Using Manila Hemp Fiber and Starch-Based Biodegradable Resin. *Compos. A Appl. Sci. Manuf.* **2006**, *37*, 1879–1883. [CrossRef]
69. Mahmud, S.; Hasan, K.M.; Jahid, M.; Mohiuddin, K.; Zhang, R.; Zhu, J. Comprehensive Review on Plant Fiber-Reinforced Polymeric Biocomposites. *J. Mater. Sci.* **2021**, *56*, 7231–7264. [CrossRef]
70. Alawar, A.; Hamed, A.M.; Al-Kaabi, K. Characterization of Treated Date Palm Tree Fiber as Composite Reinforcement. *Compos. B Eng.* **2009**, *40*, 601–606. [CrossRef]
71. Mohanty, A.K.; Wibowo, A.; Misra, M.; Drzal, L.T. Effect of Process Engineering on the Performance of Natural Fiber Reinforced Cellulose Acetate Biocomposites. *Compos. A Appl. Sci. Manuf.* **2004**, *35*, 363–370. [CrossRef]
72. Kabir, M.M.; Wang, H.; Lau, K.T.; Cardona, F. Chemical Treatments on Plant-Based Natural Fibre Reinforced Polymer Composites: An Overview. *Compos. Part B Eng.* **2012**, *43*, 2883–2892. [CrossRef]

73. Yan, L.; Yuan, X. Improving the Mechanical Properties of Natural Fibre Fabric Reinforced Epoxy Composites by Alkali Treatment. *J. Reinf. Plast. Compos.* **2012**, *31*, 425–437. [CrossRef]
74. Yan, L. Effect of Alkali Treatment on Vibration Characteristics and Mechanical Properties of Natural Fabric Reinforced Composites. *J. Reinf. Plast. Compos.* **2012**, *31*, 887–896. [CrossRef]

polymers

MDPI

Article

Investigating the Effects of Tobacco Lignin on Polypropylene

Irfan Tahir, John Rapinac, Abdulaziz Abutunis and Venkata Gireesh Menta *

Mechanical & Industrial Engineering, University of Minnesota Duluth, Duluth, MN 55812, USA; tahir037@d.umn.edu (I.T.); rapin002@d.umn.edu (J.R.); abutu001@d.umn.edu (A.A.)
* Correspondence: vkmenta@d.umn.edu; Tel.: +1-218-726-6925

Abstract: The utilization of eco-friendly materials, such as lignin, for higher value product applications became increasingly important as environmental concerns due to global warming increased. Melt blending is one of the easy ways to increase the usage of lignin in commercial applications. However, the degradation of the final product performance and increase in the production time and costs are of major concern. In the current work, the effects of blending lignin, extracted from tobacco plants, with polypropylene (PP) on the injection molding parameters, physical, thermal and mechanical properties are investigated. Blends of lignin (5, 15 and 30% by wt.) with PP were prepared using a Filabot single screw extruder. Results show that tensile strength decreases by 3.2%, 9.9% and 5.4% at 5 wt. %, 15 wt. %, and 30 wt. % of lignin addition, respectively. The tensile stiffness was almost unaffected by the addition of up to 15% lignin, but a 23% increase was observed at 30 wt. % loading. When compared to lignin processed via expensive processes, such as acetylation, tobacco lignin showed superior performance. The DSC results show unaffected crystallization and melting temperatures but a decrease in enthalpies and percentage of crystallinity. The SEM and optical micrographs of the coupon cross-sections show that the extrusion process has achieved a uniform distribution of lignin particles in the PP. Thermogravimetric analysis results show that tobacco lignin accelerates the onset decomposition temperature but does not influence the decomposition peak temperature. The increase in lignin content did not have a significant influence on the injection molding parameters, implying no additional processing costs for adding lignin to the PP. Overall, the performance of the tobacco lignin is comparable, if not better, than that of processed lignin reported in the literature.

Keywords: lignin; polypropylene; tobacco lignin; blends; characterization; biopolymers

check for updates

Citation: Tahir, I.; Rapinac, J.; Abutunis, A.; Menta, V.G. Investigating the Effects of Tobacco Lignin on Polypropylene. *Polymers* **2022**, *14*, 706. https://doi.org/10.3390/polym14040706

Academic Editors: Maya Jacob John, Sabu Thomas and Cédric Delattre

Received: 22 December 2021
Accepted: 9 February 2022
Published: 12 February 2022

1. Introduction

With an increase in the usage of plastics in applications ranging from household goods to 3D printed prototype parts, their impact on the environment is a growing concern. Using biopolymers, such as lignin, to substitute petroleum-based plastics is a widely accepted approach towards environmental sustainability. Derived from the Latin word lignum, meaning wood, lignin is an organic polymer found in the support tissues of plants, which is essential in the formation of plant cell walls [1]. After cellulose, it is the second most abundant renewable carbon source on Earth [2]. Annually, around 50 million tons of lignin are produced worldwide, as a by-product of mostly paper pulping industries, out of which only 2% is utilized commercially [3]. Recently, lignin has stirred interest among researchers, as it has shown to be one of the most promising candidates for blending with thermoplastic polymers [4]. Melt blending with petroleum-based plastics is one of the most convenient and inexpensive ways of using lignin. However, almost no work has been reported on lignin extracted from tobacco stalks, which generally goes into waste. As tobacco lignin can be obtained at a low cost and is fully biodegradable, replacing even part of petroleum-based plastics with lignin poses economic and environmental allure, especially if the physical properties and processing parameters of the blended products do not vary significantly. One such plastic is polypropylene (PP), which is used in various consumer products, such as household appliances, laboratory equipment, packaging materials and piping systems. It is estimated that the annual demand for PP will continue to grow from USD 76.00 billion in 2021 to USD 108.57 billion in 2028, at a compound annual growth rate (CAGR) of 5.2% [5].

Blending lignin with polymers has been studied for multiple decades now [6–8]. Kharade et al. reported the influence of lignin addition to PP on the mechanical properties [4]. The authors found that the tensile strength reduced with the increase in lignin content, while the impact properties remained constant and melt viscosity increased. Pucciariello et al. reported a significant reduction in strain but observed improvement in Young's modulus [9]. Peng et al. studied the weathering and thermal properties of PP blended with wood flour, cellulose and lignin [10]. The authors found that the addition of lignin resulted in less of a reduction in flexural strength and modulus, compared to wood flour and cellulose. Alexy et al. studied the processing stability, mechanical properties and thermal properties of PP/lignin blends [11]. The authors found lignin to be a good processing stabilizer and an initiator for thermal degradation at higher concentrations.

Though lignin is a good candidate for biopolymer production, the high polarity and hydrophobicity of lignin makes it difficult to simply blend with other polymers and results in degrading the performance of the host polymer [6–11]. To improve the miscibility, procedures such as grafting acetyl or ethyl groups on hydroxyls, and usage of plasticizers is typical. Maldhure et al. reported significant improvements in mechanical properties by the addition of modified lignin to PP (up to 25 wt. %) compared to unmodified lignin [12]. Similar improvements in properties have been consistently reported by multiple authors, when secondary processes, such as esterification, alkalization, and co-grafting, are used [13–15]. However, these secondary processes are difficult to scale and increase the cost of lignin, making them unattractive to commercial adaptation. The objective of the current work is to investigate the usage of tobacco lignin as low-cost fillers in petroleum-based plastics, maximizing the percentage of renewable resources, while minimizing the processing modifications and loss of mechanical properties in the host polymer. Hence, lignin extracted from tobacco using a one-step process, without any secondary modifications or plasticizers, was identified and investigated.

Another major issue limiting the adaptability of lignin in commercial applications can be attributed to the wide variance in lignin quality. The recovery methods used are one of the key causes of the variability and poor performance of lignin. Commercially available lignin is typically obtained using procedures where cellulose and hemicellulose are the primary products and lignin is a byproduct. As a result, the lignin quality receives less attention. Processes aimed at generating high-quality lignin have the potential to alleviate a number of the difficulties raised. The lignin used in the present work is obtained using one such process, developed by Attis Innovations Inc., where the lignin is the primary product and the quality of lignin is the main focus.

Tobacco has been one of the most important industries in at least three states (North Carolina, Georgia and Kentucky) in the USA and accounts for over 185,000 jobs. It is needless to list the harmful effects of tobacco products on human health. The current work aims to use the byproducts of harmful tobacco production to reduce the usage of petroleum-based polymers. Also, the effects of lignin, extracted from tobacco, on plastics have not been reported earlier. In the present study, lignin has been added to commercial PP at concentrations of 5 wt. %, 15 wt. % and 30 wt. %. Physical, mechanical and thermal properties of the blends were studied. Additionally, the effect of lignin addition on the processing parameters of PP was also investigated.

2. Materials and Methods

PP from M. Holland Inc. (Northbrook, IL, USA) in pellet form was used. The density of PP pellets was 0.90 g/cm^3 and melt flow index (MFI) was 12 g/10 min. Lignin extracted from tobacco in the form of powder without any secondary modifications was obtained from Attis Innovations Inc., Milton, GA, USA. Attis Innovations Inc. claims to use a proprietary one-step butonal-organosolv processing technology to carefully extract and purify lignin, with consistent and high quality. No other additives were added to the mixture. Both materials were vacuum dried at 60 °C for 9 h to remove any moisture content. Materials were stored in a controlled environment.

PP pellets and tobacco lignin powder (5, 15, 30 wt. %) were weighed separately and placed in a container for thorough manual mixing for 5 min. The mixtures were then compounded at 180 °C in a Filabot EX2 single screw extruder (Filabot Inc., Barre, VT, USA). It has an extrusion rate of over 0.91 kg per hour. Extrusion parameters are as follows:

Nozzle diameter: 2.85 mm
Pellet size: 3.18 mm
L/D ratio: 12
Compression ratio: 2:5:1
Drive force: 9.6 Nm
Feed screw and drive: 35 RPM

The obtained filaments were then pelletized and re-extruded a second time to improve the dispersion of lignin particles in PP. The filaments were then pelletized into injection moldable pellets and stored in vacuum sealed containers.

A Morgan Press G-125T injection molding machine (Morgan Industries Inc., Long Beach, CA, USA) was used for producing tensile test coupons following the ASTM D638 Type IV standard. A three-piece injection mold tool was designed and manufactured from Aluminum 6061 (see Figure 1) as lignin is typically known for its adhesion which can adversely affect the mold. Process parameters that resulted in parts without any defects such as short shots, warpage or flash were identified for neat PP and each lignin blend via trial and error process. Any required changes in processing parameters were recorded and the variations with change in lignin loadings were studied. Test coupons are shown in Figure 2.

(a) (b)

Figure 1. Three-piece injection mold designed for lignin blends: (**a**) showing the three separate pieces; (**b**) assembled mold with the sprue.

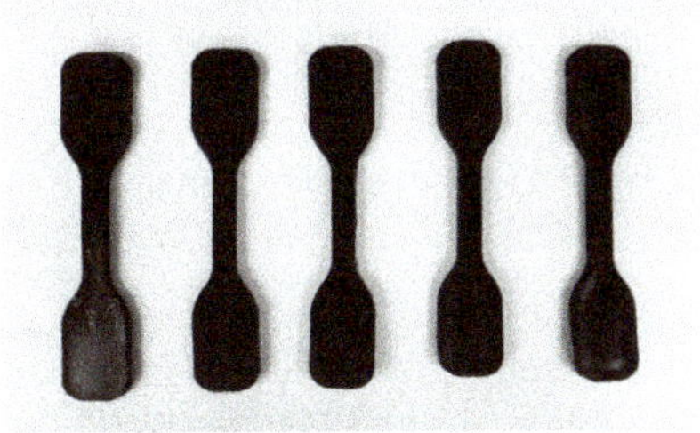

Figure 2. 30 wt. % PP-lignin ASTM D638 Type IV tensile test coupons.

Thermal characterization of the blends was performed using a TA Instruments Discovery DSC 250 (TA Instruments, New Castle, DE, USA) in accordance with ASTM D3418-15. All the samples of about 5 mg were heated at a rate of 10 °C/min from 25 °C to 180 °C, held at constant temperature for 5 min, rapidly quenched to 25 °C, and then reheated to 180 °C (the second heating scan). Thermogravimetric analysis (TGA) was carried out for all PP/lignin blends to analyze thermal characteristics and long-term heat degradation of blend material using TA Instrument Discovery TGA 550 (TA Instruments, New Castle, DE, USA). The samples were subjected to a heating rate of 10 °C/min in the heating range of 40–800 °C. Once the mass loss plateau was established, the environment in the furnace was switched from inert to reactive oxygen environment.

The density of the test coupons was measured using a Mettler Toledo precision balance following the ASTM D792 standard. Hardness tests were performed according to the ASTM D785-08 standard using a PTC Instrument shore D hardness tester. Tensile tests were performed on the test coupons using an Instron 5585H universal testing machine in accordance with ASTM D638 at a crosshead speed of 5 mm/min. Roller clamps were used to hold the coupons. Ultimate tensile strength (UTS) and Young's modulus (E) were calculated for each test sample. At least five samples were tested at each blend percentile and the average values were reported with standard deviation for all the tests.

The cross-sections of the test specimens were analyzed using a Jeol JSM-6590LV Scanning Electron Microscope (SEM) from JEOL USA Inc., Peabody, MA, USA equipped with INCA X-act EDS detector for elemental analysis. Samples were bonded to aluminum stubs and their surfaces were observed in a low vacuum environment (30 Pa) using the accelerating voltage of 20.0 kV. Leica DVM6 optical microscope (Leica Microsystems Inc., Buffalo Grove, IL, USA) was also used to measure lignin particle size distribution to gain a better understanding of the blend morphology and its relation to the mechanical properties of the blend samples.

3. Results

3.1. Processing Stability of the Blends

Table 1 shows the injection molding parameters used for producing PP and PP-lignin blends. Other than increasing the injection pressure from 3500 psi to 4500 psi, no further changes were needed for producing the blend samples. Test coupons were fabricated easily, without any challenges, even at 30 wt. % of tobacco lignin loading and without any additives. Hence, adding lignin to PP does not require any major modifications to the current manufacturing process and does not incur additional major processing/utility costs.

Table 1. Injection molding parameters.

Blend Composition	Barrel Temp (°F)	Nozzle Temp (°F)	Clamping Pressure (psi)	Plate Temp (°F)	Injection Pressure (psi)	Injection/ Dwell Time (s)
PP	420	450	7000	150	3500	5
5% Lignin	420	450	7000	150	4500	5
15% Lignin	420	450	7000	150	4500	5
30% Lignin	420	450	7000	150	4500	5

3.2. Differential Scanning Calorimetry (DSC)

Figure 3 shows the DSC thermograms of PP, pure lignin and blends. Figure 4 shows the first and second heating curves of lignin blends at 30 wt. %. DSC is an acceptable method for measuring the glass transition temperature (T_g) of lignin but it is often difficult to detect the T_g of lignin, due to the complexity of its chemical structure [16]. Typically, the T_g of lignin ranges from 90 °C to 180 °C. In this case, the T_g of lignin could not be identified. The crystallization temperature (T_c), melt temperature (T_m), exothermic enthalpy of crystallization (H_c), and endothermic enthalpy of melting (H_m) of the blends are reported in Table 2. As seen in the table, the peak melting and crystallization temperatures almost remained constant with the addition of lignin content. However, the exothermic and endothermic enthalpies gradually decreased with the increase in lignin content. The exothermic enthalpy dropped by 9%, 17%, and 50% at 5 wt. %, 15 wt. %, and 30 wt. % lignin addition, respectively. The endothermic enthalpy diminished by 11%, 22%, and 54%, at 5 wt. %, 15 wt. %, and 30 wt. % lignin addition, respectively. Typically, a decrease in enthalpies implies fewer bond formations and can significantly influence the behavior of the polymer. To further understand, the percentage of crystallinity in lignin blends was calculated by dividing the crystallization enthalpy by the heat of fusion from a perfect crystal of PP, as shown in the equation below:

$$\text{Percent crystallinity equation}: \; X_c = \frac{\Delta H_m}{\Delta H^{\circ}{}_m} \times 100\%$$

where X_c is the percentage of crystallinity, ΔH_m is the measured heat of fusion from the crystallinity peak, and $\Delta H^{\circ}{}_m$ is the heat of fusion from a perfect crystal. The heat of fusion from a perfect crystal of PP is known to be 207 (J/g) [2]. The average percentage of crystallinity of PP and the lignin blends are shown in Table 2. The results show that the percentage of crystallinity gradually decreased with an increase in lignin content, which should typically translate into a significant reduction in the mechanical performance.

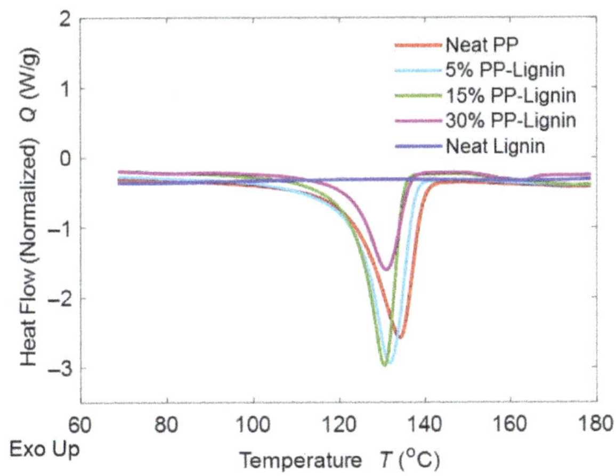

Figure 3. DSC Traces of PP/Lignin Blends.

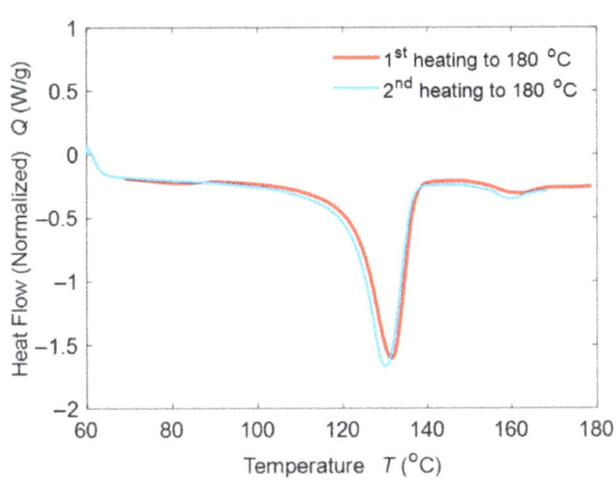

Figure 4. 1st and 2nd heating curves of 30 wt. % PP-Lignin blends.

Table 2. DSC Test Results.

Blend Composition	T_c (°C)	Exothermic Enthalpy (J/g)	T_m (°C)	Endothermic Enthalpy (J/g)	Crystallinity (%)
Neat PP	115.67 ± 0.15	174.00 ± 0.69	132.07 ± 2.35	182.23 ± 3.52	84.06 ± 0.33
5% Lignin	115.50 ± 0.35	158.70 ± 0.57	133.69 ± 2.00	162.83 ± 4.04	76.67 ± 2.45
15% Lignin	115.37 ± 0.57	144.57 ± 0.73	131.73 ± 1.46	142.71 ± 5.50	69.84 ± 3.40
30% Lignin	114.83 ± 0.93	87.50 ± 0.74	132.10 ± 2.16	84.07 ± 1.88	42.27 ± 3.60

3.3. Thermogravimetric Analysis (TGA)

Figures 5 and 6 show the TGA and DTGA curves of the lignin blends. A typical one-stage decomposition was observed for all the samples, except for the 30 wt. % PP-lignin blend [17]. Table 3 shows the decomposition onset temperature, end temperature, peak temperature and residual mass at 600 °C. The decomposition onset temperature is defined as the temperature at which 5 wt. % loss occurs. The temperature at which maximum weight loss is observed is the peak temperature of the derivative curve. From Figure 5, it can be seen that the degradation temperature of PP starts at 388.72 °C and continues to degrade until 451.58 °C. With the increase in the lignin content, the degradation temperature seems to drop to 344.12 °C at 5 wt. %, 331.27 °C at 15 wt. % and 224.61 °C at 30 wt. % lignin addition. A steady decrease in the decomposition onset temperature was observed with the increase in the lignin content in the PP. With only 5 wt. % addition of lignin, an 11% drop in the onset temperature was observed. Increasing the lignin content to 30 wt. % resulted in a 42% decrease in the onset temperature. Lignin typically has been considered to improve the thermal stability of the host polymers. However, tobacco lignin seems to have a destabilizing effect on the thermal degradation of blends, especially at higher concentrations [12].

3.4. Density

Figure 7 shows and compares the density of the blends. The density remains unchanged with the addition of lignin. With a significant decrease in the percentage of crystallinity from DSC, the density is expected to drop. However, the higher density of lignin appears to be compensating for the reduction in density due to lower crystallization. Similar behavior was observed with tensile performance, where only a slight reduction was observed, despite the significant decrease in the percentage of crystallinity.

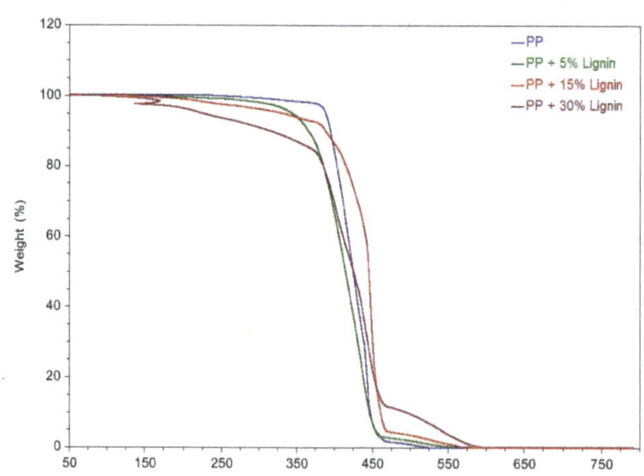

Figure 5. TGA curves for different blends of PP-lignin.

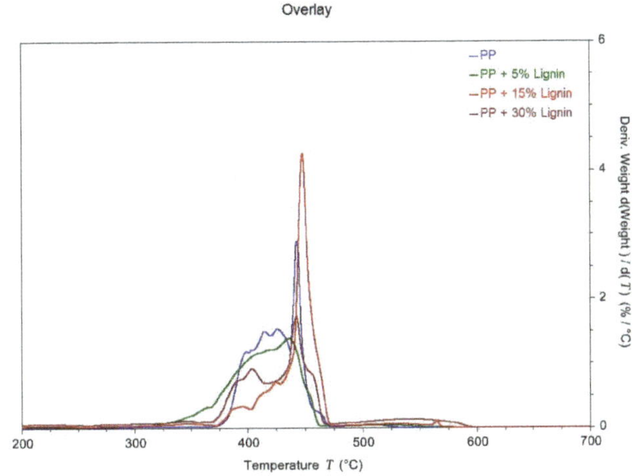

Figure 6. DTGA curves of different blends of PP-Lignin.

Table 3. Temperatures and Char Residue Characteristics of PP-Lignin Composties.

Blend Composition	T_{onset} (°C)	T_{End} (°C)	Decomposition Peak Temperature (°C)	Residual Mass at 600 °C (%)
Neat PP	388.72 ± 1.75	451.58 ± 0.97	444.01± 0.68	0.03
5% Lignin	344.12 ± 0.98	453.12 ± 1.45	440.35 ± 0.75	0.07
10% Lignin	331.27 ± 1.27	457.53 ± 2.14	448.78 ± 1.35	0.07
30% Lignin	224.61 ± 1.86	480.34 ± 1.78	441.80 ± 1.87	0.22

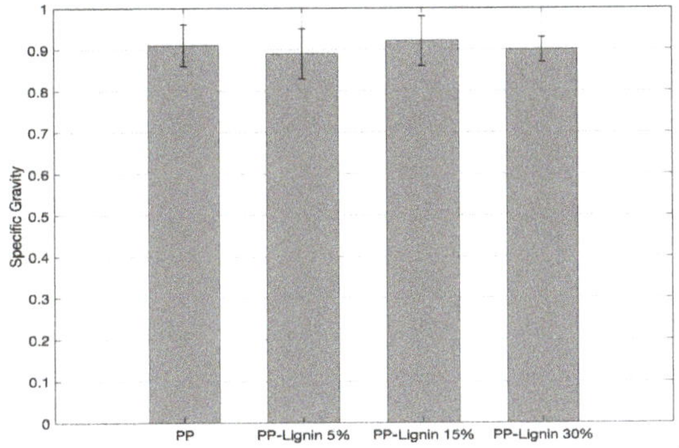

Figure 7. Density of Lignin Blends.

3.5. Shore D Hardness

The Shore D hardness numbers for different blend specimens are presented in Figure 8. In all cases, the addition of lignin increased the hardness of composites. At 5 wt. % and 15 wt. % lignin loading, a slight increase of 3.7% and 4.1% was observed, respectively. However, the increase is within the error range and can statisticaly be considered as unchanged. However, at 30 wt. % lignin addition, an 18.82% increase in hardness value was recorded. These results are in close agreement with the tensile properties.

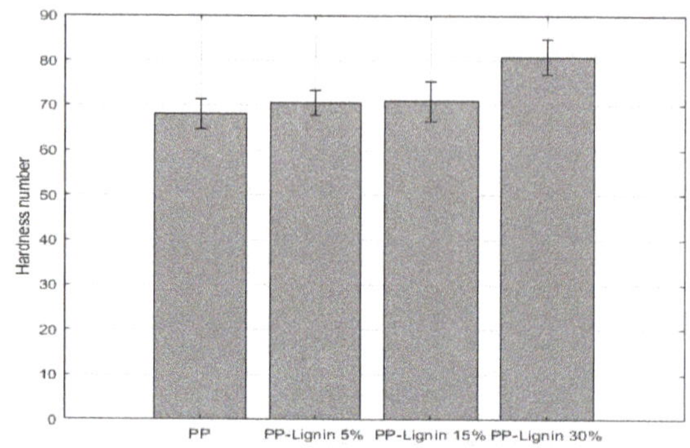

Figure 8. Shore D hardness values of lignin blends.

3.6. Tensile Properties

The tensile properties of PP and lignin blends are shown in Table 4. Figure 9 compares the change in UTS as lignin wt. % is increased in PP-lignin blends. With an increase in 5 wt. %, 15 wt. %, and 30 wt. % of lignin, a drop of 3.2%, 9.9% and 5.4% in tensile strength was observed, respectively. Statistically, the values are within the standard deviation, and it can be stated that tensile strength seems to be unaffected by the addition of lignin. While a similar trend was observed with Young's modulus, an exception in behavior was observed at 30% wt. lignin concentration, with a 20% increase in stiffness. The 30% wt. lignin blend samples also exhibited a brittle failure, unlike other samples. The variation in stiffness properties is very similar to ShoreD hardness values.

Table 4. Tensile properties of PP-lignin blend.

Blend Composition	Ultimate Tensile Strength (MPa)	Young's Modulus (GPa)
Neat PP	21.65 ± 0.310	0.21 ± 0.02
5% Lignin	20.96 ± 0.570	0.21 ± 0.10
10% Lignin	19.50 ± 0.540	0.21 ± 0.03
30% Lignin	20.47 ± 1.25	0.24 ± 0.02

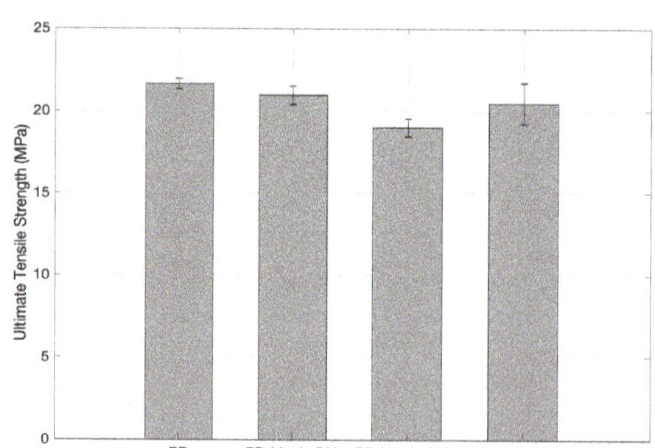

Figure 9. Ultimate tensile strength comparison (UTS).

Table 5 compares the tensile test results obtained in the current work with those from the literature. Kharade and Kale [4], who used dry lignin powder extracted from paper mill waste, found a drop of 59.3% when lignin concentration was increased to 30%. Toriz et al. [6], who used a purified form of kraft lignin, reported a drop of 37.4%. In the present work, a drop of only 5.4% was observed at 30 wt. % lignin loading to PP. Maldhure et al. [11], who used isolated kraft lignin from a paper mill, reported a 27.8% decrease in UTS for 15 wt. %, which is 15.6% higher than the one found in the present work. Overall, the tensile performance of tobacco lignin blends seems to be far superior compared to the ones reported in the literature. With the decrease in crystallinity, the tensile performance is supposed to deteriorate. Yet, a different behavior was observed. To investigate the driving forces behind this behavior, SEM and optical microscopy examinations were conducted.

Table 5. Comparison of Ultimate Tensile Strength of PP-lignin with literature.

PP/Lignin	UTS % Decrease: Kharade and Kale [4]	UTS % Decrease: Toriz et al. [1] [6]	UTS % Decrease: Maldhure et al. [11]	UTS % Decrease: Present Work
95/5	19.5%	8.4%	16.2%	3.1%
85/15	28.1%	18.9%	27.8%	12.2%
70/30	59.3%	37.4%	-	5.4%

[1] Calculated using interpolation as source work increased lignin in increments of 10.

3.7. Scanning Electron Microscopy (SEM) and Optical Microscopy

SEM micrographs of 5 wt. %, 15 wt. % and 30 wt. % blends are shown in Figure 10. At 5 wt. %, lignin seems to be well dispersed with particles barely being visible at 100× magnification. At 15 wt. %, a relatively larger sized lignin particles were observed. At 30 wt. %, more and larger aggregates can be seen. The strong polarity of lignin typically causes agglomeration, especially at high loading concentrations, and the mechanical characteristics decrease as the degree of agglomeration increases. Smaller lignin particles imply that lignin was successfully dispersed in PP [18,19].

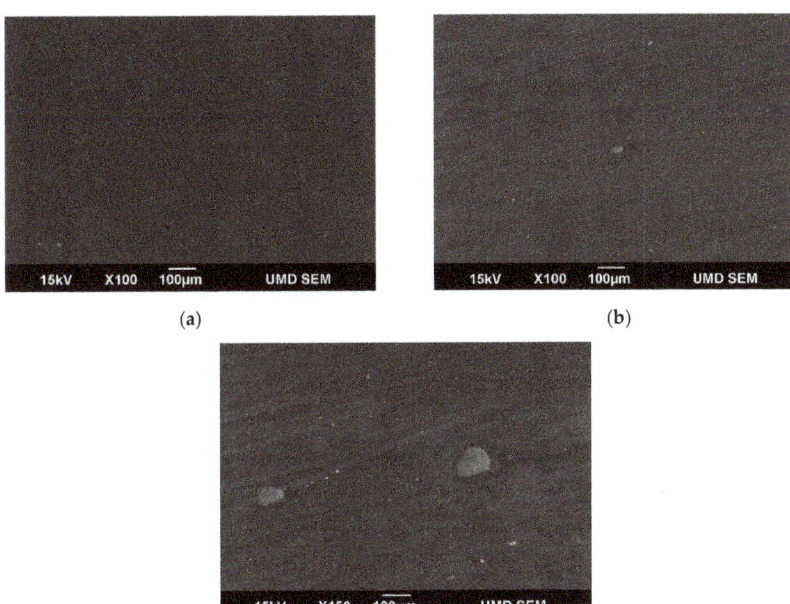

Figure 10. SEM Micrographs of (**a**) 5%, (**b**) 15% and (**c**) 30% PP-Lignin blends.

Since dispersion plays an important role in the mechanical performance of the blends, optical microscopic images of fracture surfaces were taken, and Image J software was used to measure the size and density of lignin particles in the PP. A histogram with probability density curves is displayed in Figure 11. Particle sizes mainly ranged from 25–100 μm with a peak at 50 μm. Earlier works showed that the particle distribution of acetylated lignin was around 55 μm [11]. This implies a slightly better particle size distribution of tobacco lignin was achieved, without any secondary modifications to the lignin. The smaller lignin particle size and uniform dispersion can be attributed to an increase in mechanical performance of the blends, despite the decrease in crystallinity.

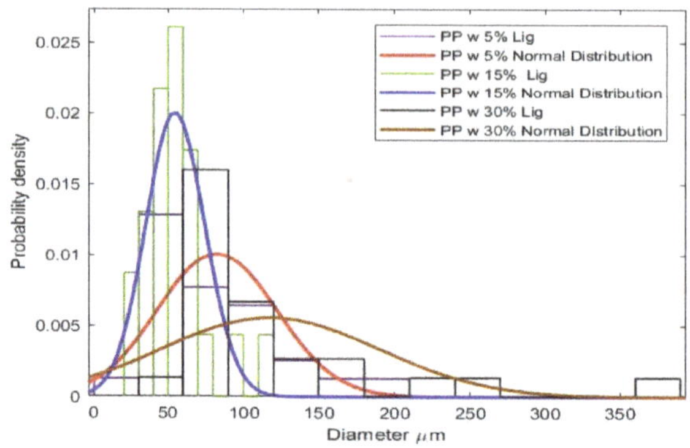

Figure 11. Probability density and histogram of tobacco lignin grain sizes.

4. Conclusions

Physical and mechanical properties of PP/unmodified tobacco lignin (5%, 15% and 30% by wt.), prepared via melt mixing, were studied. The addition of lignin did not have an influence on the tensile strength, but with a 30 wt. % concentration of lignin, Young's modulus increased by 23%. An increase in lignin did not influence the density and specific gravity of the blends. SEM images and optical micrographs of the coupon cross-sections showed that the extrusion process had achieved a uniform distribution of lignin particles. The addition of lignin did not require any modifications of injection molding parameters, except for a slight increase in the pressure, from 3500 psi to 4000 psi at 0 and 5 wt. %, implying that the addition of lignin will not increase any processing costs or delays. The DSC results showed that peak crystallization and melting temperatures did not change with the increase in lignin content, but the percentage of crystallinity decreased significantly. However, the unaffected mechanical properties can be attributed to better and uniform dispersion of lignin particles, along with a good adhesion between lignin and PP. The better mechanical performance also conveys that lignin particles might be acting as reinforcing material in the PP. However, further studies are warranted to understand the behavior better. TGA results showed that tobacco lignin acted as a thermal destabilizer at high temperatures. Comparing mechanical properties of unmodified tobacco lignin blends with those obtained from expensive acetylated lignin materials shows that the current material has superior properties. In conclusion, the addition of tobacco lignin up to 30 wt. % does not negatively affect the PP mechanical properties. Tobacco lignin shows great promise in improving some of the mechanical properties, without requiring any additional expensive processing steps.

Author Contributions: Conceptualization, V.G.M.; methodology, I.T. and J.R., software, A.A., I.T., J.R. and V.G.M.; formal analysis, V.G.M.; investigation, V.G.M.; resources, V.G.M.; data curation, I.T. and J.R.; writing—original draft preparation, V.G.M., I.T. and A.A.; writing—review and editing, V.G.M. and A.A.; visualization, V.G.M.; supervision, V.G.M.; project administration, V.G.M.; funding acquisition, V.G.M. All authors have read and agreed to the published version of the manuscript.

Funding: This research received no external funding.

Institutional Review Board Statement: Not applicable.

Informed Consent Statement: Not applicable.

Data Availability Statement: Not applicable.

Acknowledgments: We want to thank the Natural Resources Research Institute (NRRI) for supplying the novel tobacco lignin materials. We also want to thank Alexey Kacharov, Sergiy Yemet, and Eric Singsaas from NRRI for their support.

Conflicts of Interest: The authors declare no conflict of interest.

References

1. Haghdan, S.; Renneckar, S.; Smith, G. Sources of Lignin, no. 1. In *Lignin in Polymer Composites*; William Andrew Publishing: Norwich, NY, USA, 2016; Volume 10, pp. 1–11.
2. Shrotri, A.; Kobayashi, H.; Fukuoka, A. Catalytic conversion of structural carbohydrates and lignin to chemicals. *Adv. Catal.* **2017**, *60*, 59–123.
3. Naegele, H.; Pfitzer, J.; Ziegler, L.; Inone-Kauffmann, E.; Eisenreich, N. Applications of Lignin Materials and Their Composites. In *Lignin in Polymer Composites*; William Andrew Publishing: Norwich, NY, USA, 2016; pp. 233–244.
4. Kharade, A.; Kale, D. Lignin-filled polyolefins. *J. Appl. Polym. Sci.* **1999**, *72*, 1321–1326. [CrossRef]
5. Fortune Business Insights. *Polypropylene Market Size, Share & COVID-19 Impact Analysis*; Market Research Report; Fortune Business Insights: Pune, India, 2021; pp. 1–200. Available online: https://www.fortunebusinessinsights.com/industry-reports/polypropylene-pp-market-101583 (accessed on 22 December 2021).
6. Kun, D.; Pukánszky, B. Polymer/lignin blends: Interactions, properties, applications. *Eur. Polym. J.* **2017**, *93*, 618–641. [CrossRef]
7. Wang, J.; Manley, R.S.J.; Feldman, D. Synthetic polymer-lignin copolymers and blends. *Prog. Polym. Sci.* **1992**, *17*, 611–646. [CrossRef]
8. Thakur, V.K.; Thakur, M.K.; Raghavan, P.; Kessler, M.R. Progress in green polymer composites from lignin for multifunctional applications: A review. *ACS Sustain. Chem. Eng.* **2014**, *2*, 1072–1092. [CrossRef]
9. Rodrigues, P.C.; Muraro, M.; Garcia, C.M.; Souza, G.P.; Abbate, M.; Schreiner, W.H.; Gomes, M.A. Polyaniline/lignin blends: Thermal analysis and XPS. *Eur. Polym. J.* **2001**, *37*, 2217–2223. [CrossRef]
10. Peng, Y.; Liu, R.; Cao, J. Characterization of surface chemistry and crystallization behavior of polypropylene composites reinforced with wood flour, cellulose, and lignin during accelerated weathering. *Appl. Surf. Sci.* **2015**, *332*, 253–259. [CrossRef]
11. Alexy, P.; Košíková, B.; Podstránska, G. The effect of blending lignin with polyethylene and polypropylene on physical properties. *Polymer* **2000**, *41*, 4901–4908. [CrossRef]
12. Maldhure, A.V.; Ekhe, J.; Deenadayalan, E. Mechanical properties of polypropylene blended with esterified and alkylated lignin. *J. Appl. Polym. Sci.* **2012**, *125*, 1701–1712. [CrossRef]
13. Holsti-Miettinen, R.; Seppälä, J.; Ikkala, O. Effects of compatibilizers on the properties of polyamide/polypropylene blends. *Polym. Eng. Sci.* **1992**, *32*, 868–877. [CrossRef]
14. Dehne, L.; Vila, C.; Saake, B.; Schwarz, K.U. Esterification of Kraft lignin as a method to improve structural and mechanical properties of lignin-polyethylene blends. *J. Appl. Polym. Sci.* **2017**, *134*, 1–8. [CrossRef]
15. Casenave, S.; Aït-Kadi, A.; Riedl, B. Mechanical behaviour of highly filled lignin/polyethylene composites made by catalytic grafting. *Can. J. Chem. Eng.* **1996**, *74*, 308–315. [CrossRef]
16. Gordobil, O.; Robles, E.; Egüés, I.; Labidi, J. Lignin-ester derivatives as novel thermoplastic materials. *RSC Adv.* **2016**, *6*, 86909–86917. [CrossRef]
17. Abdelkhalik, A.; Abdel-Hakim, A.; Makhlouf, G.; El-Gamal, A. Effect of iron poly (acrylic acid-co-acrylamide) and melamine polyphosphate on the flammability properties of linear low-density polyethylene. *J. Therm. Anal. Calorim.* **2019**, *138*, 1021–1031. [CrossRef]
18. Nosova, N.; Roiter, Y.; Samaryk, V.; Varvarenko, S.; Stetsyshyn, Y.; Minko, S. Polypropylene surface peroxidation with heterofunctional polyperoxides. *Macromol. Symp.* **2004**, *210*, 339–348. [CrossRef]
19. Nitz, H.; Semke, H.; Mülhaupt, R. Influence of Lignin Type on the Mechanical Properties of Lignin Based Compounds. *Macromol. Mater. Eng.* **2001**, *286*, 737–743. [CrossRef]

polymers

MDPI

Article

Esterification of Cellulose with Long Fatty Acid Chain through Mechanochemical Method

Jacqueline Lease [1], Tessei Kawano [1] and Yoshito Andou [1,2,*]

[1] Department of Biological Functions Engineering, Graduate School of Life Science and Systems Engineering, Kyushu Institute of Technology, 2-4 Hibikino, Wakamatsu-ku, Kitakyushu, Fukuoka 808-0196, Japan; lease.jacqueline708@mail.kyutech.jp (J.L.); kawano.tessei758@mail.kyutech.jp (T.K.)

[2] Collaborative Research Centre for Green Materials on Environmental Technology, Kyushu Institute of Technology, 2-4 Hibikino, Wakamatsu-ku, Kitakyushu, Fukuoka 808-0196, Japan

* Correspondence: yando@life.kyutech.ac.jp

Abstract: Mechanochemical reaction, a green synthetic esterification route was utilized to prepare long-chain cellulose esters from microcrystalline cellulose. The influence of reaction conditions such as reaction temperature and time were elucidated. Only low dosage of oleic acid, 1-butyl-3-metylimidazolium acetate, and *p*-toluenesulfonyl chloride were required. The success of modification reaction was confirmed by Fourier transforms infrared spectroscopy as a new absorbance peak at 1731 cm^{-1} was observed, which indicated the formation of carbonyl group (C=O). Solid-state nuclear magnetic resonance was also performed to determine the structural property and degree of substitution (DS) of the cellulose oleate. Based on the results, increasing reaction temperature and reaction time promoted the esterification reaction and DS. DS values of cellulose oleates slightly decreased after 12 h reaction time. Besides, X-ray diffraction analysis showed the broadening of the diffraction peaks and thermal stability decreased after esterification. Hence, the findings suggested that grafting of oleic acid's aliphatic chain onto the cellulose backbone lowered the crystallinity and thermal stability.

Keywords: microcrystalline cellulose; mechanochemical esterification; long fatty acid chain; oleic acid; ionic liquid; magnetic mortar and pestle

Citation: Lease, J.; Kawano, T.; Andou, Y. Esterification of Cellulose with Long Fatty Acid Chain through Mechanochemical Method. *Polymers* **2021**, *13*, 4397. https://doi.org/10.3390/polym13244397

Academic Editors: Maya Jacob John and Sabu Thomas

Received: 14 November 2021
Accepted: 13 December 2021
Published: 15 December 2021

1. Introduction

Lignocellulosic biomass is garnering attention among researchers as it is a readily renewable and environmentally benign natural source. Among lignocellulosic materials, cellulose is the most abundant biopolymer in nature composed of anhydroglucose units, which is linked by β-1,4-glucosidic linkages [1]. Cellulose materials offered advantages of biodegradability, biocompatibility, high surface area, low cost, high thermal stability, and high mechanical properties [2,3].

Nevertheless, the absence of thermal transition makes cellulose non-processable and the presence of inherent strong intra- and intermolecular hydrogen bonding through hydroxyl groups often results in pronounced aggregation [4]. The agglomeration led to cellulose being insoluble in water and most of organic solvents. To overcome the aforementioned problems, esterification has been demonstrated as an effective approach to improve the aggregation and thermoplastic behavior of cellulose [5].

Conventionally, chemical esterification of cellulose was carried out via pyridine-acyl chloride or anhydride reactions. These reaction systems produced hydrochloric acid as a byproduct, which resulted in cellulose degradation and is environmentally harmful [6,7]. To date, a number of non-derivative solvent systems, such as *N*,*N*-dimethylacetamide-lithium chloride (DMAc-LiCl) [8], *N*-methylmorpholine *N*-oxide (NMMO) [9], and dimethyl sulfoxide-tetramethylammonium fluoride (DMSO-TBAF) [10] have been found to be efficient in the cellulose dissolution process. However, several drawbacks from the solvent systems include high cost, the required large amount of chemicals, volatility, toxicity, and difficulty in solvent recovery [11,12].

Among potential new solvents, the application of ionic liquids (ILs) as a solvent system has been shown to be a feasible yet effective system. The usage of ILs as suitable solvents produced no toxic or explosive gases and can be recycled for repeated usage [13]. Their negligible vapor pressure, low viscosity, improvement of selectivity and yields, excellent thermal stabilities, as well as excellent dissolution performance for cellulose have made them solvents of choice [14,15]. According to Kakuchi et. al. [16], IL such as 1-ethyl-3-methylimidazolium acetate (EmimOAc) can offer dual functionalities by acting concurrently as a facile solvent for cellulose and as an activating reagent for esterification reaction of cellulose hydroxy groups. In this work, 1-butyl-3-metylimidazolium acetate (BmimOAc) was used as both reaction media and catalyst for the esterification reaction.

Besides ILs, mechanochemical esterification is another green synthetic process used to address the setbacks of environment burden. Mechanochemistry is a chemical reaction induced and sustained by the use of mechanical force (grinding or milling) [17,18]. Ball milling is a well-known technique for benchtop pilot scale study, so it is possible to realize the value of cellulose in industrial sectors due to the efficient chemical transformation generated through mechanical forces [18–20]. Cellulose esterification employing mechanochemical methods has been studied previously, but most of the literature focused on short-chain fatty acid in the presence of solvent during the reaction [5]. Long-chain oleic acid (OA) is predicted to be the best substitute to serve as esterifying agent for cellulose esterification as it does not produce harmful by-products. In addition, oleic acid has excellent flexibility, wider thermal process window, low melting point, easy processing, renewability, and hydrophobicity [5,21]. Herein, mechanochemical esterification with long chain OA in IL is proposed as a facile yet environmentally friendly protocol to synthesize cellulose oleate (CO).

The novel aspect of this work details the synthesis of CO in a green, simple, and scalable way by controlling the reaction conditions. Furthermore, degree of substitution (DS), thermal stability, and the chemical structure of the CO were identified.

2. Materials and Methods

2.1. Materials

Microcrystalline cellulose (MCC) powder with a particle size of 20μm was obtained from Sigma-Aldrich, St. Loius, MO, USA and was dried under vacuum conditions at 60 °C for 6 h before use. A total of ≥95% of 1-butyl-3-metylimidazolium acetate (BmimOAc) was also purchased from Sigma-Aldrich. Oleic acid (OA) was supplied by Wako Pure Chemical Industry, Osaka, Japan. *p*-Toluenesulfonyl chloride (TsCl), which acts as an activating agent, was provided by Tokyo Chemical Industry (Tokyo, Japan) and stored in a desiccator cabinet with silica gels to prevent water absorption. All the reagents were used as received without further purification or treatment.

2.2. Preparation of Cellulose Oleate (CO)

Modification of cellulose with OA in TsCl/BmimOAc system via mechanochemical esterification was depicted (Scheme 1). Before functionalization, MCC powder was dried for 8 h at 70 °C before use to remove any trapped moisture from the atmosphere. A total of 0.5 g of MCC (3 mmol), 5 g of OA (18 mmol), and 3.6 g of BmimOAc (18 mmol) were added and kneaded in a magnet mortar with a constant rotational speed of 150 rpm (100 VAC, 15 W). After 1 h, 3.5 g of TsCl (18 mmol) was added into the mixture to induce the hydroxyl groups of the cellulose backbone into better leaving groups. The mixture was continuously stirred at different reaction temperatures (50, 80, and 100 °C) and reaction time (4, 12, and 24 h). After the mechanochemical esterification process, the product was precipitated in acetone, followed by Soxhlet extraction with methanol to remove unreacted substances for 8 h. The CO was filtered and washed with ethanol twice to completely remove any unreacted oleic acid. Lastly, the modified cellulose was vacuum-dried at 80 °C and stored at ambient conditions.

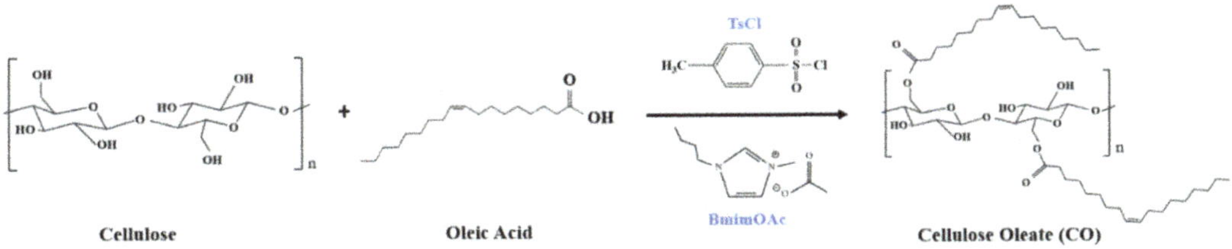

Scheme 1. Schematic representation of the mechanochemical esterification reaction of cellulose in BmimOAc with OA, and TsCl as the activating reagent.

2.3. *Characterizations*

2.3.1. Fourier Transform Infrared (FTIR) Spectroscopy

The chemical structures of as-prepared CO were analyzed by FTIR spectroscopy (Nicolet iS5, Thermo Fisher, Waltham, MA, USA). A total of 2 mg of dried sample was mixed thoroughly with 100 mg potassium bromide (KBr) in a mortar. The mixture was then pressed into a pellet until a translucent film was obtained. Sixteen scans were taken for each run in a wavenumber range of 400–4000 cm^{-1}.

2.3.2. Solid-State Nuclear Magnetic Resonance (ssNMR)

Solid-state 13C NMR spectra of CO was recorded by using ssNMR (JNM-ECA-500 MHz II, JEOL Resonance, Tokyo, Japan) spectrometer. Samples were placed in a zirconia rotor and spun at 6 kHz with 2000 scans. The temperature was set to 298 K and the degree of substitution of cellulose oleates (DS_{NMR}) were calculated from a ratio of the integrals of NMR peaks using the following formula [22]:

$$DS_{NMR} = (n_{cel} \times I_{oleoyl}) / (n_{oleoyl} \times I_{cel}), \quad (1)$$

where, I_{oleoyl} and I_{cel} are the integrations of peaks corresponding to oleoyl carbons and cellulose carbons, respectively. n_{cel} and noleoyl are the number of carbon atoms in cellulose (n_{cel} = 6) and in the corresponding acetyl group (n_{oleoyl} = 18), respectively.

2.3.3. Thermogravimetric Analysis (TGA)

The thermal stability of the unmodified cellulose and modified cellulose were determined using EXSTAR TG/DTA 6200 (SII Nanotechnology Inc., Chiba, Japan) under constant nitrogen flow (100 mL/min). The samples (5–10 mg) were heated from 30–500 °C with a heating rate of 10 °C/min.

2.3.4. Scanning Electron Microscopy (SEM)

Morphologies of the cellulose before and after functionalization were performed using a JCM-6000 SEM (JEOL, Tokyo, Japan). Before the observation, the samples were mounted on an aluminum stub and coated with carbon first using vacuum sputter-coater to improve conductivity and prevent charging of the samples. SEM images were obtained at an acceleration voltage of 15 kV.

2.3.5. X-Ray Diffraction (XRD) Analysis

XRD measurements were characterized on a Rigaku Miniflex II diffractometer using CuKα radiation (λ = 0.154 nm). The samples were exposed to the X-ray beam with the X-ray generator running at 30 kV and 15 mA. Scattered radiation was detected at ambient temperature in the angular region (2θ) of 3° to 70° at a rate of 10°/min and a step size of 0.02. The crystallinity index (CrI) of the cellulose derivatives was calculated by the following equation [23]:

$$\text{CrI } (\%) = [(I_{002} - I_{am})/I_{002} \times 100\%], \quad (2)$$

where, I_{002} is the maximum intensity of the (002) crystal plane reflection of the cellulose and Iam is the minimum intensity between the (002) and (101) peaks.

Crystallite size was calculated by using Scherrer's formula [23]:

$$\text{Crystallite size} = 0.9\ \lambda/(\beta \cos \theta), \quad (3)$$

where, λ is the wavelength of X-ray beam with 0.154 nm, β is the full width at half maximum (FWHM), and θ is the angle of beam reflection.

2.3.6. Differential Scanning Calorimetry (DSC) Analysis

DSC was carried out with an EXSTAR DSC 6220 (SII Nanotechnology Inc., Tokyo, Japan). The sample was first heated from 30 to 150 °C at a scanning rate of 10 °C min^{-1} to provide the same thermal history before measurements. The temperature was maintained at 150 °C for 1 min and then quenched to −50 °C. The second heating scan was conducted from −50 to 200 °C at a scanning rate of 10 °C min^{-1} to examine the glass-transition temperature (Tg). The results of Tg from the Figure S3 were obtained from second heating.

3. Results and Discussion

3.1. Preparation of Cellulose Oleate (CO)

In this work, esterification of microcrystalline cellulose (MCC) with oleic acid (OA) in *p*-Toluenesulfonyl chloride (TsCl) and 1-butyl-3-metylimidazolium acetate (BmimOAc) system was realized by mechanochemical method (Figure 1). When MCC was subjected to the intense milling, the stable crystalline structure and strong inter- and intramolecular hydrogen bonding can be destroyed and the steric effect of OA (long fatty acid chain) was also weakened. This scenario might lead to the increase of the reactivity of hydroxyl groups in the MCC.

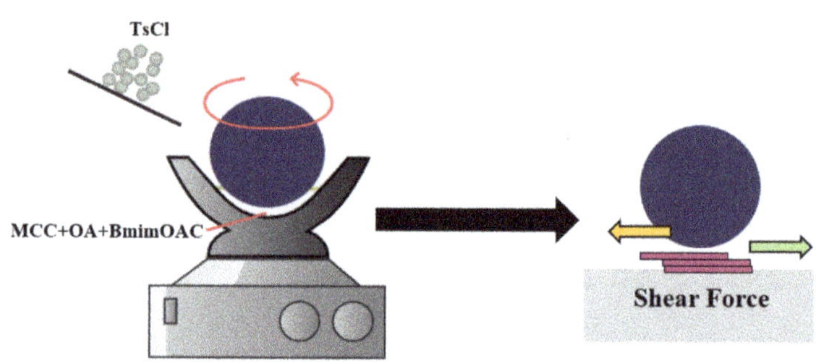

Figure 1. Mechanochemical esterification of MCC by using magnetic agate mortar and pestle.

Here, green mechanochemical reaction could be carried out with low amount of catalysts and solvents due to the physical shearing force if compared to conventional chemical esterification. The binding sites in cellulose induced by intense milling can rapidly react with oleate group, assisting by combining the milling and esterification reaction in the mortar. The chemical structures, thermal stabilities, and degree of substitution of MCC with different reaction temperature and time were analyzed and investigated.

3.2. Chemical Structure of Cellulose Oleates (CO)

The chemical structures of COs were characterized by FTIR and the results were presented in Figure 2. FTIR spectra of the COs showed clear evidence that esterification had occurred from the considerable differences with the unmodified cellulose. After mechanochemical esterification, a new absorbance peak at 1731 cm^{-1} was recorded, which was assigned to the stretching vibration of the carbonyl group (C=O) [24]. From the spectra, the carbonyl peaks increased linearly over reaction temperature.

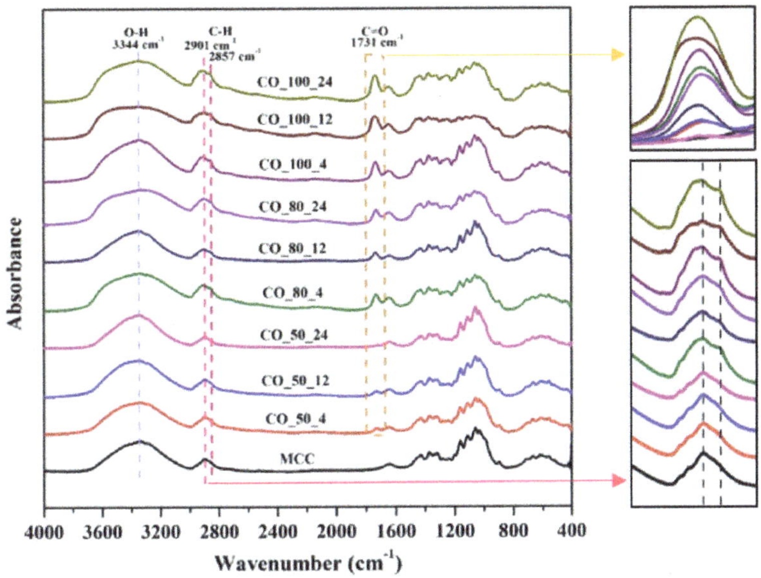

Figure 2. FTIR spectra of cellulose oleates synthesized at different reaction time (4, 12, and 24 h) and reaction temperature (50, 80, and 100 °C).

Besides, it can be clearly observed that the single absorptive band in the unmodified cellulose at 2901 cm^{-1} was changed to double absorptive bands (2901 cm^{-1} and 2857 cm^{-1}) in the modified cellulose, which could be attributed to the introduction of more methylene groups (C–H bonds) from the OA [25]. The characteristic peak of the unsaturated group (H–C=C) of the oleic acid was also observed at 3009 cm^{-1}.

The decrease in intensity of the broad band at about 3344 cm^{-1}, assigned to the cellulose O–H vibration was additional proof of the successful esterification. The decreasing peak intensity in the cellulose hydroxyl region suggested the substitution of hydroxyl group with fatty acid, especially in high reaction temperatures, 80 and 100 °C. The peak changes of O–H vibration were not obvious for the modified cellulose at lower temperature (50 °C), which can be explained by the low rate of fatty acid side chain grafted on the cellulose.

3.3. The Effect of Reaction Conditions on DS of Cellulose Oleate (CO)

Nuclear magnetic resonance (NMR) spectroscopy is divided into liquid-state NMR (lsNMR) and solid-state NMR (ssNMR). After functionalization, CO still cannot be dissolved in organic solvents, but well-dispersed in dimethylformamide (DMF) and tetrahydrofuran (THF) (Figure S1 and Table S1). In the Figure S2, CO also showed better dispersity compared to MCC in DMF and THF after standing for one day. This scenario indicated that the hydroxyl group of cellulose backbone was partially removed and became less polar. In order to obtain reliable results of molecular structures, 13C ssNMR spectroscopy was used for further investigation.

Figure 3a shows the NMR spectra of 100 °C, 24 h cellulose sample. The signals at 105.65, 89.19, 75.24, 72.74, and 65.24 ppm were assigned to the carbon atoms on the anhydroglucose (AGU) unit at C1, C2, C3, C4, C5, and C6, respectively. The appearance of signal at 174.18 ppm was carboxylic carbon (C=O) and the signal at 130 ppm was observed as vinyl group of carbon (C=C). Besides, methyl ($-CH_3$) and ethyl groups ($-CH_2$) at 14 to 35 ppm supported the notion of esterification reaction [26].

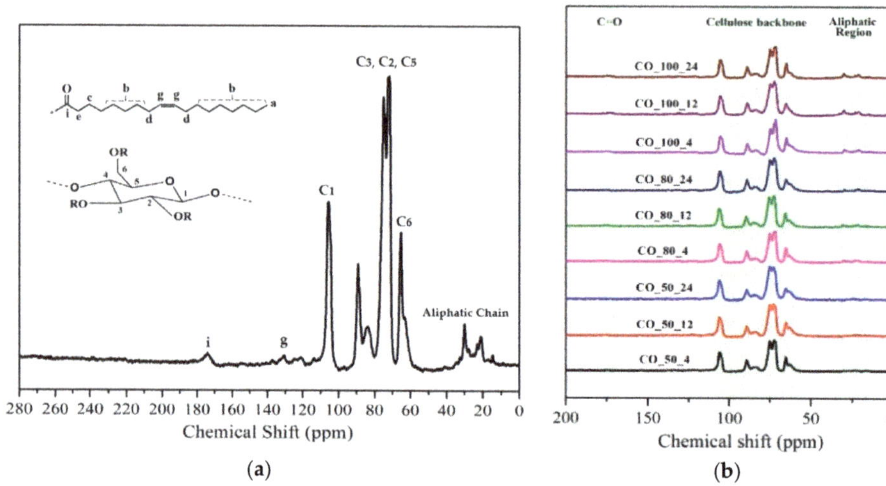

Figure 3. Solid state NMR (ssNMR) spectra of cellulose oleates (**a**) 100 °C and 24 h; (**b**) various reaction temperature and time.

The substitution level of fatty acid is the average number of oleoyl groups per anhydroglucose (AGU) unit, ranging from zero to three. Based on the peak assignments and the corresponding integrals, the DS of the oleoyl group on AGU units were summarized in Table 1. The ratio between the number of carbon atoms in the AGU unit C-2,3,4,5,6/C-1 was fixed at 5:1 and DS were calculated based on the integral of C-1 to oleoyl group ratio [22]. The area under the graph of NMR peaks in relation with standard allows the calculation of DS precisely.

Table 1. Degree of substitution (DS) of cellulose oleates at different reaction conditions.

Samples	Temperature (°C)	Reaction Time (h)	DS
CO_50_4	50	4	0.030
CO_50_12	50	12	0.014
CO_50_24	50	24	0.001
CO_80_4	80	4	0.091
CO_80_12	80	12	0.104
CO_80_24	80	24	0.088
CO_100_4	100	4	0.132
CO_100_12	100	12	0.210
CO_100_24	100	24	0.204

The NMR spectra of modified cellulose under different reaction conditions with their respective signal assignation is presented in Figure 3b. The calculated DS values ranged from 0.001 to 0.210. The findings showed that the modification via mechanochemical method is limited to the surface or the outer layer of the cellulose bundles. Low DS indicated that only partial intramolecular interaction occurred and the ultrastructure of cellulose oleates might not be affected by the reaction [27].

Based on the DS values of COs, increasing the reaction temperature and reaction time were favorable to improve the DS [6,28]. However, the DS slightly decreased after 12 h. This trend could be explained by the possible competition between the esterification reaction and the partial hydrolysis of the ester groups formed as the byproduct of the reaction is water. The presence of moisture in the reaction medium is detrimental to the process [27].

In the case of CO_100_12 and CO_100_24, the DS values of both samples did not show significant changes, which are 0.210 and 0.204. The findings suggested that the hydrolysis process can be prevented at an elevated temperature (100 °C) and the saturation point is reached at 12 h.

3.4. Crystal Structure of Cellulose Oleates (CO)

To gain further insights into the structure changes caused by mechanochemical treatment, X-ray diffraction analysis (XRD) was conducted and the respective patterns of all cellulose samples were shown in Figure 4.

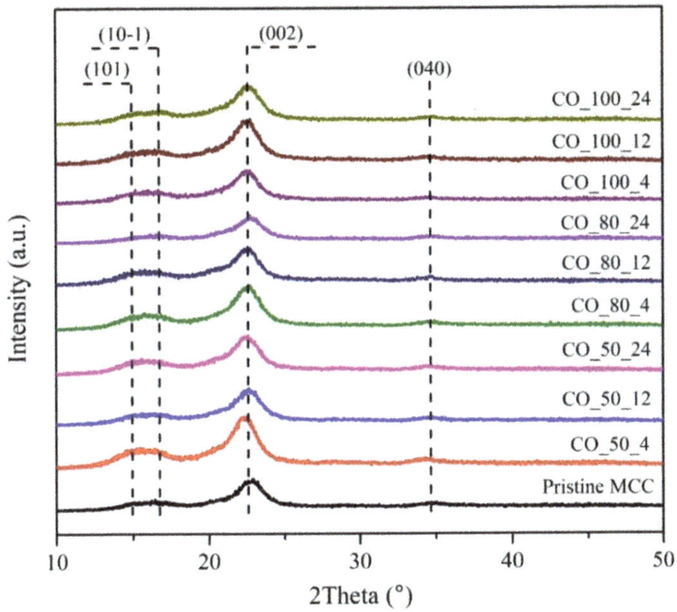

Figure 4. X-ray diffractograms patterns of MCC and modified cellulose (COs).

The pristine cellulose and mechanochemical-treated cellulose showed the main reflection peaks at around 2θ = 14.8°, 16.4°, 22.7°, and 34.7°, which are normally ascribed to the (101), (101), (002), and (040) diffraction planes, respectively. Based on the diffraction peak angles, they were exhibited characteristic peaks as cellulose I. The XRD patterns of modified cellulose remained unchanged in comparison to the microcrystalline cellulose, indicating that the reaction only partially took place without affecting the inner structure of cellulose [29]. The modification is posited to occur at the amorphous regions of the cellulose.

The crystallinity index (CrI), full width at half maximum (FWHM), and crystallite size of each sample were calculated according to their corresponding XRD diffractogram. The results are presented in Table 2.

Table 2. The (002) peak position, maximum intensity, FWHM, crystallinity index, and crystallite size of cellulose and cellulose oleates.

Sample	(002) Peak Position (2θ)	Max Intensity (counts/s)	FWHM (2θ)	CrI (%)	Crystallite Size(nm)
MCC	22.99	141	1.92	86.30	4.41
CO_50_4	22.33	250	1.92	82.17	4.41
CO_50_12	22.83	162	1.92	82.63	4.41
CO_50_24	22.39	186	2.11	79.75	4.01
CO_80_4	22.62	223	2.11	82.97	4.01
CO_80_12	22.72	170	1.92	78.77	4.41
CO_80_24	22.79	121	1.73	84.62	4.89
CO_100_4	22.77	146	1.73	80.00	4.89
CO_100_12	22.83	157	1.92	80.83	4.41
CO_100_24	22.70	163	1.73	79.81	4.89

From Table 2, it was observed that the crystallinity index of pristine MCC is higher than COs based on the Segal calculation. The crystallinity index of COs slightly decreased after esterification due to the attachment of oleoyl side chain [30]. Oleic acid moieties have long alkyl chains, which severed the intra- and intermolecular hydrogen bonds of the cellulose result in a less crystalline structure [31].

The trend of crystallite size of modified MCCs obtained in this study did not support findings from previous studies [23,31]. It was observed previously that the crystallite size and crystallinity is inversely proportional to the reaction time and temperature. The FWHM of the modified cellulose obtained via the mechanochemical method is expected to decrease with the prolong reaction time and temperature. Nonetheless, findings in this study did not abide the observed trends as the crystalline size and crystallinity did not have significant changes. The low DS of cellulose oleate is posited to be one of the reasons that the crystallite size and crystallinity remain almost unchanged, which are in the range of 4.01–4.89 nm and 84.63–78.77%. Further study needs to be carried out to investigate the unexpected behavior.

3.5. *Thermal Stability of Cellulose Oleate (CO)*

In this work, differential scanning calorimetry (DSC) and thermogravimetric analysis (TGA) were used to examine the thermal stability of CO. DSC was performed to investigate the macroscopic manifestation of the macromolecular chains motions by determining glass transition (Tg). Based on Figure S3, CO_50_12 and CO_80_12 had no obvious glass transition, which exhibit the same behavior as pristine MCC.

Theoretically, due to the strong inter- and intramolecular hydrogen bonding, the unmodified MCC has no Tg prior to its decomposition [32]. As the substitution rate of oleoyl increased, the oleoyl side chain weaken the hydrogen bonding interaction. CO_100_12 has relatively higher DS showed one major glass transitions at 116.5 °C, which was attributed to the motion of cellulose backbone and oleoyl side chain, respectively [20].

From the thermogravimetric analysis, the thermal stability of modified cellulose decreased after surface modification. Modified MCCs started to decompose at lower degradation temperature (236–294 °C) than pristine MCC (337 °C).

Thermal stability of modified cellulose was affected by the disruption of hydrogen bonding, which decreased the crystallinity of the cellulose after the substitution with fatty acid [31]. In addition, the intensive mechanical force that promoted decrystallization is expected to be another reason. The small particle size of CO (Figure 5) enabled more exposure to heat due to their larger surface area. Therefore, thermal stability decreased after esterification took place [29].

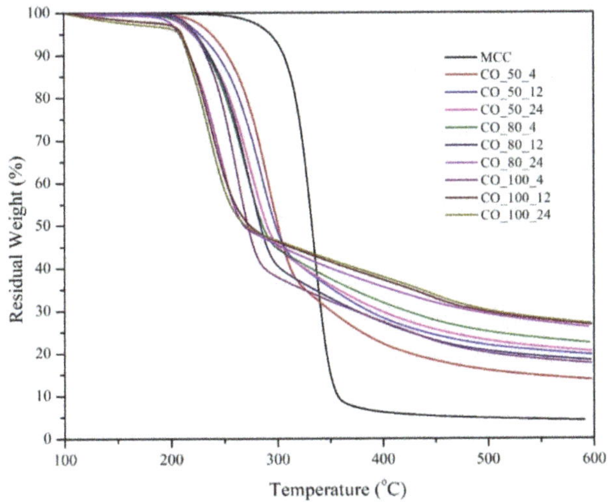

Figure 5. TGA thermograms of unmodified MCC and modified MCC.

In general, thermograms of cellulose esters showed two main degradation steps. However, all of the samples in this study showed one-step degradation curve except CO_100_12 (T_{d2} = 445 °C) and CO_100_24 (T_{d2} = 455 °C). According to Uschanov et al. [33], the absence of the second transition at higher temperature is because of the low amount of oleoyl group grafted on the cellulose. Double bonds of OA are prone to crosslinking reaction and the lack of fatty acid during heating, which inhibit the decomposition process.

For the thermal profile, the first degradation of cellulose oleate was attributed to the esterified cellulose, while the second degradation indicated the formation of new ordered region, which corresponds to the crystallization of oleoyl chains. Oleic acid contained one double bond between C9 and C10 with the cis configuration, which does not favor the formation of an ordered structure [23]. Hence, only high DS cellulose oleate exhibited two main thermal degradation steps in this study.

3.6. Surface Morphology of Cellulose Oleate (CO) by SEM Analysis

The surface morphologies of pristine cellulose and COs prepared under varying conditions can be clearly observed from their SEM micrographs (Figure 6).

The pristine microcrystalline cellulose composed of smooth rod-like cellulose bundles without small fragment on the surface [34–36]. The images of the modified cellulose Figure 6b–f showed that esterification increased the surface roughness of the cellulose bundles as the reaction temperature and reaction time increased. Pristine MCC was remarkably turned into smaller and irregular particles after mechanical grinding.

COs with milling temperature of 80 and 100 °C exhibited loose and porous structures [37]. Besides, their surface displayed homogenous appearance if compared to pristine MCC and COs with milling temperature of 50 °C. Interruption of the cellulose backbone can explain this observation as a result of the reaction of hydroxy group (–OH) of MCC with OA. Larger surface area could provide a facile access for oleoyl side chain to contact with the hydroxyl groups of cellulose, which significantly enhanced the mechanochemical esterification reaction [38].

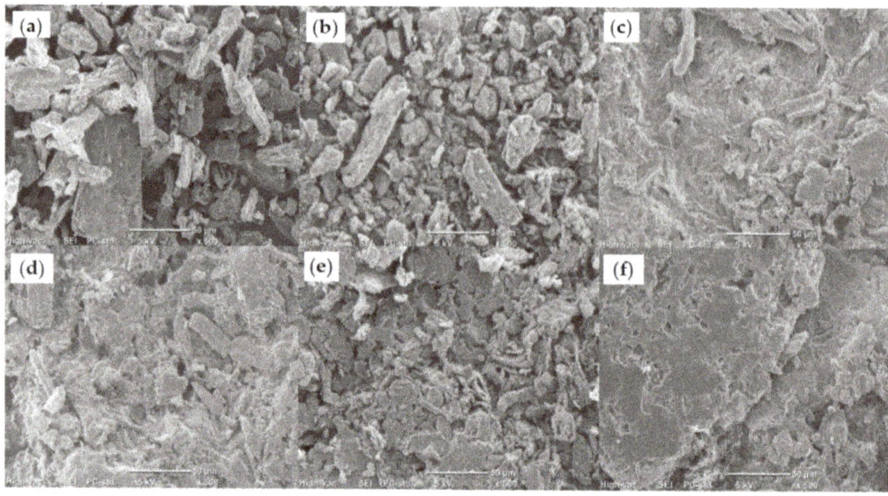

Figure 6. SEM images of (**a**) pristine MCC; (**b**) cellulose oleate 50 °C 12 h; (**c**) cellulose oleate 100 °C 12 h; (**d**) cellulose oleate 80 °C 4 h; (**e**) cellulose oleate 80 °C 12 h; (**f**) cellulose oleate 80 °C 24 h.

4. Conclusions

In this research, cellulose oleates were successfully prepared via a green approach of mechanochemical esterification with a small amount of oleic acid and ionic liquid. By controlling the reaction time and temperature, various DS values (0.001–0.210) of cellulose oleates were obtained and the crystalline structure of modified cellulose was not altered via functionalization. Increasing DS of cellulose oleate promoted decrystallization of cellulose, but it retained the cellulose I structure as possessed by pristine MCC. The fine tuning of the reaction conditions will be beneficial in various industries requiring cellulose with certain properties. This reported method enables the preparation of cellulose ester with different properties for their intended applications. More reaction conditions will be considered in the future, so that high DS of cellulose can be obtained easily using magnetic mortar method. This fundamental study provides a simple yet effective method for functionalizing cellulose, opening up numerous opportunities for future research to address the drawbacks of pristine cellulose.

Supplementary Materials: The following are available online at https://www.mdpi.com/article/10.3390/polym13244397/s1, Figure S1: Solubility Test. Cellulose Oleates (COs) were dispersed by 90 min of sonication and kept static for 1 day. 20 mg of CO was dispersed in 10 mL of dimethyl sulfoxide (DMSO), *N,N*-dimethylformamide (DMF), tetrahydrofuran (THF), chloroform ($CHCl_3$), diethyl ether (Et_2O), toluene (Tol) and hexane (Hx), respectively, Figure S2: Images of comparison of microcrystalline cellulose (MCC) and COs in 10 mL of (a) DMF and (b) THF after 90 min of sonication and kept static for 1 day, Figure S3: DSC thermograms of cellulose oleates with the reaction time and temperature at 50 °C, 12 h (CO_50_12), 80 °C, 12 h (CO_80_12) and 100 °C, 12 h (CO_100_12); Table S1: Polarity index and dielectric constant of various organic solvent.

Author Contributions: Conceptualization, J.L. and Y.A.; Methodology, Y.A. Validation, J.L.; Formal analysis, J.L.; Investigation, J.L. and T.K.; Resources, Y.A.; Data curation, Y.A.; Writing-original draft preparation, J.L.; Writing-review and editing, J.L. and Y.A.; Visualization, J.L.; Supervision, Y.A.; Project administration, Y.A.; Funding acquisition, Y.A. All authors have read and agreed to the published version of the manuscript.

Funding: This research received no external funding.

Institutional Review Board Statement: Not applicable.

Informed Consent Statement: Not applicable.

Data Availability Statement: The data presented are contained within the article.

Acknowledgments: The authors would like to thank Teruhisa Ohno from Kyushu Institute of Technology for the technical support in sample analysis in this research.

Conflicts of Interest: The authors declare no conflict of interest.

References

1. Chen, T.; Liu, W. Highly unsaturated microcrystalline cellulose and its cross-linked soybean-oil-based thermoset composites. *ACS Sustain. Chem. Eng.* **2019**, *7*, 1796–1805. [CrossRef]
2. Zhang, J.; Chen, W.; Feng, Y.; Wu, J.; Yu, J.; He, J.; Zhang, J. Homogeneous esterification of cellulose in room temperature ionic liquids. *Polym. Int.* **2015**, *64*, 963–970. [CrossRef]
3. Wang, Y.; Wang, X.; Xie, Y.; Zhang, K. Functional nanomaterials through esterification of cellulose: A review of chemistry and application. *Cellulose* **2018**, *25*, 3703–3731. [CrossRef]
4. David, G.; Gontard, N.; Angellier-Coussy, H. Mitigating the impact of cellulose particles on the performance of biopolyester-based composites by gas-phase esterification. *Polymers* **2019**, *11*, 200. [CrossRef]
5. Huang, L.; Wu, Q.; Wang, Q.; Wolcott, M. One-step activation and surface fatty acylation of cellulose fibers in a solvent-free condition. *ACS Sustain. Chem. Eng.* **2019**, *7*, 15920–15927. [CrossRef]
6. Jebrane, M.; Terziev, N.; Heinmaa, I. Biobased and sustainable alternative route to long-chain cellulose esters. *Biomacromolecules* **2017**, *18*, 498–504. [CrossRef]
7. Andou, Y.; Lee, H.S.; Kim, D.; Nagasawa, N.; Nishida, H.; Shirai, Y. Enhancement of compatibility based on vapor-phase-assisted surface polymerization (VASP) method for polymer composites with agricultural wastes. *Compos. Interfaces* **2014**, *9*, 773–785. [CrossRef]
8. Vaca-Garcia, C.; Thiebaud, S.; Borredon, M.E.; Gozzelino, G. Cellulose esterification with fatty acids and acetic anhydride in lithium chloride/N,N-dimethylacetamide medium. *J. Am. Oil. Chem. Soc.* **1998**, *75*, 315–319. [CrossRef]
9. Kulpinski, P. Cellulose nanofibers prepared by the N-methylmorpholine-N-oxide method. *J. Appl. Polym. Sci.* **2005**, *98*, 1855–1859. [CrossRef]
10. Onwukamike, K.N.; Grelier, S.; Grau, E.; Cramail, H.; Meier, M.A.R. Sustainable transesterification of cellulose with high oleic sunflower oil in a DBU-CO2 switchable solvent. *ACS Sustain. Chem. Eng.* **2018**, *6*, 8826–8835. [CrossRef]
11. Gericke, M.; Fardim, P.; Heinze, T. Ionic liquids—Promising but challenging solvents for homogeneous derivatization of cellulose. *Molecules* **2012**, *17*, 7458–7502. [CrossRef]
12. Liu, H.R.; Yu, H.W.; Zhou, E.P.; Zhang, X.H.; Zhang, X.C. Influence of anions of imidazole ionic liquids on dissolution of cellulose. *Asian J. Chem.* **2013**, *25*, 8266–8270. [CrossRef]
13. Verma, C.; Mishra, A.; Chauhan, S.; Verma, P.; Srivastava, V.; Quraishi, M.A.; Ebenso, E.E. Dissolution of cellulose in ionic liquids and their mixed cosolvents: A review. *Sustain. Chem. Pharm.* **2019**, *13*, 100162. [CrossRef]
14. Wu, J.; Zhang, H.; He, J.; Ren, Q.; Guo, M. Homogeneous acetylation and regioselectivity of cellulose in a new ionic liquid. *Biomacromolecules* **2004**, *5*, 266–268. [CrossRef]
15. Liu, Y.; Jing, S.; Carvalho, D.; Fu, J.; Martins, M.; Cavaco-Paulo, A. Cellulose dissolved in ionic liquids for modification of the shape of keratin fibers. *ACS Sustain. Chem. Eng.* **2021**, *9*, 4102–4110. [CrossRef]
16. Kakuchi, R.; Ito, R.; Nomura, S.; Abroshan, H.; Ninomiya, K.; Ikai, T.; Maeda, K.; Kim, H.J.; Takahashi, K. A mechanistic insight into the organocatalytic properties of imidazolium-based ionic liquids and a positive co-solvent effect on cellulose modification reactions in an ionic liquid. *RCS Adv.* **2017**, *7*, 9423–9430.
17. Tang, L.; Huang, B.; Yang, N.; Li, T.; Lin, W.; Chen, X. Organic solvent-free and efficient manufacture of functionalized cellulose nanocrystals via one-pot tande reactions. *Green Chem.* **2013**, *15*, 2369–2373. [CrossRef]
18. Eksiler, K.; Andou, Y.; Yilmaz, F.; Shirai, Y.; Ariffin, H.; Hassan, M.A. Dynamically controlled fibrillation under combination of ionic liquid with mechanical grinding. *J. Appl. Polym. Sci.* **2016**, *134*, 44469. [CrossRef]
19. Kuga, S.; Wu, M. Mechanochemistry of cellulose. *Cellulose* **2019**, *26*, 215–225. [CrossRef]
20. Hou, D.F.; Li, M.L.; Yan, C.; Zhou, L.; Liu, Z.Y.; Yang, W.; Yang, M.B. Mechanochemical preparation of thermoplastic cellulose oleate by ball milling. *Green Chem.* **2021**, *23*, 2069–2078. [CrossRef]
21. Wu, Z.; Liu, P.; Wu, Z.; Cravotto, G. In situ modification of activated carbons by oleic acid under microwave heating to improve adsorptive removal of naphthalene in aqueous solutions. *Processes* **2021**, *9*, 391. [CrossRef]
22. Jandura, P.; Kokta, B.V.; Riedl, B. Fibrous long-chain organic acid cellulose esters and their characterization by diffuse reflectance FTIR spectroscopy, solid-state CP/MAS 13C-NMR, and x-ray diffraction. *J. Appl. Polym. Sci.* **2000**, *78*, 1354–1365. [CrossRef]
23. Almasi, H.; Ghanbarzadeh, B.; Dehghannia, J.; Pirsa, S.; Zandi, M. Heterogeneous modification of softwoods cellulose nanofibers with oleic acid: Effect of reaction time and oleic acid concentration. *Fibers Polym.* **2015**, *16*, 1715–1722. [CrossRef]
24. Huang, F.Y.; Yu, Y.; Wu, X.J. Characterization and properties of cellulose oleate. *Adv. Mater. Res.* **2011**, *197–198*, 1306–1309. [CrossRef]

25. Duchatel-Crépy, L.; Joly, N.; Martin, P.; Marin, A.; Tahon, J.F.; Lefebvre, J.M.; Gaucher, V. Substitution degree and fatty chain length influence on structure and properties of fatty acid cellulose esters. *Carbohydr. Polym.* **2019**, *234*, 115912. [CrossRef] [PubMed]
26. Shin, Y.; Winder, E.M.; Han, K.S.; Lee, H.; Bonheyo, G.T. Enhanced capacities of mixed fatty acid-modified sawdust aggregators for remediation of crude oil spill. *ACS Omega* **2019**, *4*, 412–420. [CrossRef]
27. Freire, C.S.R.; Silvestre, A.J.D.; Neto, C.P.; Belgacem, M.N.; Gandini, A. Controlled heterogeneous modification of cellulose fibers with fatty acids: Effect of reaction conditions on the extent of esterification and fiber properties. *J. Appl. Polym. Sci.* **2006**, *100*, 1093–1102. [CrossRef]
28. Liu, C.F.; Sun, R.C.; Zhang, A.P.; Ren, J.L.; Wang, X.A.; Qin, M.H.; Chao, Z.N.; Luo, W. Homogeneous modification of sugarcane bagasse cellulose with succinic anhydride using a ionic liquid as reaction medium. *Carbohydr. Res.* **2007**, *342*, 919–926. [CrossRef] [PubMed]
29. Zhao, G.; Wang, F.; Lang, X.; He, B.; Li, J.; Li, X. Facile one-pot fabrication of cellulose nanocrystals and enzymatic synthesis of its esterified derivative in mixed ionic liquids. *RSC Adv.* **2017**, *7*, 27017–27023. [CrossRef]
30. Low, F.W.; Samsudin, N.A.; Yusoff, Y.; Tiong, S.K. Hydrolytic cleavage of glycosidic bonds for cellulose nanoparticles (CNPs) production by BmimHSO4 ionic liquid catalyst. *Thermochim. Acta* **2020**, *684*, 178484. [CrossRef]
31. Almasi, H.; Ghanbarzadeh, B.; Dehghannya, J.; Entezami, A.A.; Asl, A.K. Novel nanocomposites based on fatty acid modified cellulose nanofibers/poly(lactic acid): Morphological and physical properties. *Food Packag. Shelf Life* **2015**, *5*, 21–31. [CrossRef]
32. Chen, Z.; Zhang, J.; Xiao, P.; Tian, W.; Zhang, J. Novel thermoplastic cellulose esters containing bulky moieties and soft segments. *ACS Sustain. Chem. Eng.* **2018**, *6*, 4931–4939. [CrossRef]
33. Uschanov, P.; Johansson, L.S.; Maunu, S.L.; Laine, J. Heterogeneous modification of various celluloses with fatty acids. *Cellulose* **2011**, *18*, 393–404. [CrossRef]
34. Sang, X.; Qin, C.; Tong, Z.; Kong, S.; Jia, Z.; Wan, G.; Liu, X. Mechanism and kinetics studies of carboxyl group formation on the surface of cellulose fiber in a TEMPO-mediated system. *Cellulose* **2017**, *24*, 2415–2425. [CrossRef]
35. Lubis, M.; Gana, A.; Maysarah, S.; Ginting, M.H.S.; Harahap, M.B. Production of bioplastic from jackfruit seed starch (Artocarpus heterophyllus) reinforced with microcrystalline cellulose from cocoa pod husk (Theobroma cacao L) using glycerol as plasticizer. *IOP Conf. Ser. Mater. Sci. Eng.* **2018**, *309*, 012100. [CrossRef]
36. Chuayjuljit, S.; Su-Uthai, S.; Tunwattaseree, C.; Charuchinda, S. Preparation of microcrystalline cellulose from waste-cotton fabric for biodegradability enhancement of natural rubber sheets. *J. Reinf. Plast. Compos.* **2019**, *15*, 146–154. [CrossRef]
37. Hamdaoui, L.E.; Es-said, A.; Elmarouani, M.; Bouchti, M.E.; Bchitou, R.; Kifani-Sahban, F.; Moussaouiti, M.E. Tosylation optimization, characterization and pyrolysis kinetics of cellulose tosylate. *ChemistrySelect* **2020**, *5*, 7695–7703. [CrossRef]
38. Hu, H.; Li, H.; Zhang, Y.; Chen, Y.; Huang, Z.; Huang, A.; Zhu, Y.; Qin, X.; Lin, B. Green mechanical activation-assisted solid phase synthesis of cellulose esters using a co-reactant: Effect of chain length of fatty acids on reaction efficiency and structure properties of products. *RSC Adv.* **2015**, *5*, 20656–20662. [CrossRef]

polymers

MDPI

Article

Physicomechanical Properties of Rice Husk/Coco Peat Reinforced Acrylonitrile Butadiene Styrene Blend Composites

Nurul Haziatul Ain Norhasnan [1], Mohamad Zaki Hassan [1,*], Ariff Farhan Mohd Nor [2], S. A. Zaki [2], Rozzeta Dolah [1], Khairur Rijal Jamaludin [1] and Sa'ardin Abdul Aziz [1]

1 Razak Faculty of Technology and Informatics, Universiti Teknologi Malaysia, Jalan Sultan Yahya Petra, Kuala Lumpur 54100, Malaysia; ain.kl@utm.my (N.H.A.N.); rozzeta.kl@utm.my (R.D.); khairur.kl@utm.my (K.R.J.); saa.kl@utm.my (S.A.A.)

2 Malaysia-Japan International Institute of Technology, Universiti Teknologi Malaysia, Jalan Sultan Yahya Petra, Kuala Lumpur 54100, Malaysia; ariffmdnor@yahoo.com (A.F.M.N.); sheikh.kl@utm.my (S.A.Z.)

* Correspondence: mzaki.kl@utm.my

Abstract: Utilizing agro-waste material such as rice husk (RH) and coco peat (CP) reinforced with thermoplastic resin to produce low-cost green composites is a fascinating discovery. In this study, the effectiveness of these blended biocomposites was evaluated for their physical, mechanical, and thermal properties. Initially, the samples were fabricated by using a combination of melt blend internal mixer and injection molding techniques. Increasing in RH content increased the coupons density. However, it reduced the water vapor kinetics sorption of the biocomposite. Moisture absorption studies disclosed that water uptake was significantly increased with the increase of coco peat (CP) filler. It showed that the mechanical properties, including tensile modulus, flexural modulus, and impact strength of the 15% RH—5% CP reinforced acrylonitrile-butadiene-styrene (ABS), gave the highest value. Results also revealed that all RH/CP filled composites exhibited a brittle fracture manner. Observation on the tensile morphology surfaces by using a scanning electron microscope (SEM) affirmed the above finding to be satisfactory. Therefore, it can be concluded that blend-agriculture waste reinforced ABS biocomposite can be exploited as a biodegradable material for short life engineering application where good mechanical and thermal properties are paramount.

Keywords: biocomposites; blend; recycle composites; biodegradable

Citation: Norhasnan, N.H.A.; Hassan, M.Z.; Nor, A.F.M.; Zaki, S.A.; Dolah, R.; Jamaludin, K.R.; Aziz, S.A. Physicomechanical Properties of Rice Husk/Coco Peat Reinforced Acrylonitrile Butadiene Styrene Blend Composites. *Polymers* **2021**, *13*, 1171. https://doi.org/10.3390/polym13071171

Academic Editor: Maya Jacob John

Received: 9 March 2021
Accepted: 3 April 2021
Published: 6 April 2021

1. Introduction

Researchers have recently been looking for high-performance materials based on their lightweight capability and assembly features toward the end product's lowest cost. Not all available products in the market fulfill the designers' need. Therefore, the researchers move toward the advanced material called composite materials. The composite material is a combination of two or more constituent materials with different physical and chemical properties. When combined, they produce unique properties compared to individual material. In general, composite materials offer excellent weight to strength ratio, thermal, moisture uptake, and wear properties. Engineering composite is a combination composite materials harder and stronger phase, which is called reinforcement material, and the stiff continuous segment is termed matrices. The matrix can be either thermosetting or thermoplastics, while the reinforcement material could be metals, ceramics, or fibers. Previously, synthetic fibers, including glass [1,2], boron, carbon, and Kevlar, were famously used for reinforcement. However, due to their high cost and non-biodegradable property, scientists and technologies shifted to full usage of natural fibers and agro-waste materials. Recently, vegetable fibers and crop residues like banana [3], kenaf [4], and bamboo [5], coco peat (CP), and rice husk (RH), were evaluated to reinforce polymer composite materials and were the best candidates to replace synthetic fibers. Besides being recyclable, the plant fibers haves offer many advantages, such as low cost, low density, and abundant availability. Due to ecological concern and new rules and regulations, the development of a new biocomposite that consists of thermoplastic polymer and agro-waste fiber as reinforcement is mandatory. Consequently, this study will reduce the carbon cycle, decrease environmental impact, and thus produce a greener product.

Generally, rice (*Oryza sativa*) is one of the essential agricultures product in this world. In 2017, the global annual production of 670 million tons of paddy was harvested from Asia, America, Africa, and Europe [6]. Approximately 20% of the paddy were estimated to be rice husks. In Malaysia, a total of 840 thousand tons of RH are produced annually [7]. The utilization of RH in composite structures is recommended in many engineering applications due to its abrasive nature, low cost, lightweight, renewability, biodegradability, universally available, and weathering resistant. It has been reported that the incorporation of RH into thermoplastic matrices enhances the mechanical [8], flammability [9], and thermal stability [10] properties of biocomposites. Chen et al. [8] reported that the RH filler sample improved up to 58% tensile modulus than neat polymer. Moreover, with the addition of RH in the eco-composite structure, the flammability heat release rates were significantly dropped by approximately 39%, which can be attributed to the presence of silica in RH [9]. Additionally, the degradation rate values for RH reinforced polymer were shifted to a higher temperature, indicating improved thermal stability of the biocomposite as compared to recycled polymer [10]. Many studies also mentioned that RH-filled composite material has a low moisture absorption kinetic and good dimension stability than other natural fiber composites [11]. High moisture absorption capability of the biocomposite results in weak interfacial bonding which reduces microbial resistance, is easy to buckle under compressive loading, and contributes to deterioration of mechanical properties. Therefore, these moisture absorption characteristics are essential and critical factors to be evaluated for potential use in outdoor application. Several research works have assessed the water absorption behavior of RH composites upon immersion in distilled water.

In addition, coco peat is a spongy particle and by-product waste of coconut shell. It is a rich source of lignocellulose consisting of lignin, cellulose, and hemicellulose. Traditionally, this coconut waste is abundantly available and also widely used as soilless potting mix media in agriculture. A coconut mostly contains approximately 100 g of coco peat with a mixture of coarse-to-fine cork particles (83–95% in total) and fibers. Currently, about eight million tons of coco peat are being produced from coconut husks in the world each year. In Malaysia, approximately 5280 kg per hectare of coconut waste, mainly coconut husk, is obtained, but most have not yet been processed and fully utilized. Coco peat is highly plausible with adequate water capacity storage, high water retention, and eco-friendly. However, it has a moderate mechanical property. Borawski [12] stated that unconventional materials like coconut waste are an excellent alternative for many automotive applications, such as brake pads. Besides, hybrid coco peat composites also offer higher wear resistance to manufacture clutch plates lining and superabsorbent capability for desiccant evaporating cooling systems [13].

Based on past findings, even though extensive research has been done to explore composites' physical and mechanical properties, very few involved blend biocomposites. The claims of excellent RH composites on mechanical and especially tensile properties and low moisture absorption mentioned before, together with abundantly available but possess moderate mechanical properties of CP fiber, became the biocomposites' filler selection to run the investigation. Indeed, the aim of this study was to explore the synergistic effects of the blended RH and CP filler reinforced ABS composite on the physical and mechanical properties that has not been conducted previously. However, only a handful of publications raised concerns about the moisture characteristics of the blended composite, especially with coco peat and acrylonitrile butadiene styrene (ABS). Therefore, the moisture absorption, tensile behaviors, flexural test, and impact resistant of these biodegradable composites were tested. Besides, the composites were tested according to ASTM standards, and the surface morphological were finally evaluated experimentally. This investigation direction is beneficial for producing green waste composite materials with a synergistic combination of cost and performance. This topic remains unexplored at the time of writing, and the need to investigate this area of matter is crucial to fill the knowledge gap.

2. Materials and Methods

2.1. Materials

Acrylonitrile Butadiene Styrene (ABS) was supplied by Muleh Zaman Enterprise (Gombak, Selangor, Malaysia) in pallet form and was used as the polymer matrix. The average pellet size was 4 mm. Rice husk (RH) particle and cocoa peat (CP) were purchased from Innovative Pultrusion Sdn. Bhd. (Seremban, Negeri Sembilan, Malaysia). RH has an average density of 0.22 g/cm^3 specifically from *Oryza sativa* species and CP has an average density of 0.10 g/cm^3 solely from *Cocos nucifera* species.

2.2. Sample Fabrication

Initially, RH particle sized ±10 µm, CP particle sized ±0.25 mm, and thermoplastic ABS in pellet form were dried in an electric air circulated oven (CMH Ltd., Lancing, UK) at 80 °C for 48 h. Five sets of RH/CP (20/0, 5/15, 10/10, 15/5, and 0/20) wt.% reinforced thermoplastic ABS were fabricated as tabulated in Table 1. The RH/CP blend composites were prepared via the melt blend internal mixer (Brabender, Duisburg, Germany) at an optimum processing temperature of 190 °C and a rotating speed of 40 rpm. The composites were then oven-dried (CMH Ltd., Lancing, UK) at 80 °C for 2 h [14], followed by chopping (Cheso N3, Loyang Way, Singapore) the composite to form pallets. The composite granules were then fed to a screw-type injection molding (Engel Gmbh, Schwertberg, Austria). It was further mixed, heated, and then extruded through three plate mold dies to create uniform distribution tensile, three-point bending, and impact test samples.

Table 1. Sample composition and designation.

Sample Composition (wt.%)	Designation
ABS (80) + Rice Husk (20) + Coco Peat (0)	RH20/CP0
ABS (80) + Rice Husk (15) + Coco Peat (5)	RH15/CP15
ABS (80) + Rice Husk (10) + Coco Peat (10)	RH10/CP10
ABS (80) + Rice Husk (5) + Coco Peat (15)	RH5/CP15
ABS (80) + Rice Husk (0) + Coco Peat (20)	RH0/CP20

2.3. Physical Properties of RH/CP Reinforced ABS Blend Composites

The density of biocomposites was evaluated according to ASTM D4018. The thickness and weight of the developed blend composites of RH/CP were recorded for tests. The samples were weighed to the nearest 0.001 g by using a close chamber EMS 300-3 precision balance (Kern and Sohn, Balingen, Germany).

Water absorption characteristics were conducted following the ASTM D570. The sample was initially dried in a circulation oven (CMH Ltd., Lancing, UK) at 80 °C for 2 h before the composites' weight was obtained. In this study, five replicates of composite specimens with the dimension of 20 mm × 20 mm × 3 mm were immersed in distilled water at room temperature of 25 °C. The coupons were taken out from the moist condition, and all exposed surfaces were dried by using a microfiber cloth (Spontex, Worcester, Worcestershire). The moisture uptake data were recorded by using a weight balance EMS 300-3 (Kern and Sohn, Balingen, Germany) regularly at every 2 h of water immersion. The moisture absorption characteristic was evaluated by using the following equation:

$$\%M = \frac{W_t - W_0}{W_0} \times 100 \quad (1)$$

where, W_t is the sample's weight at a recorded immersion time, and W_0 is weight of the dried sample. The kinetic and diffusion mechanism was evaluated based on Fickian's theory. Its relation is:

$$\log\left[\frac{M_t}{M_\infty}\right] = \log(k) + n\log(t) \tag{2}$$

where M_t and M_∞ are water absorption at time, t and saturation point, respectively. k and n are constants.

The ability of moisture to penetrate the composite's molecule, also known as diffusion coefficient, D is a primary parameter of Fick's model [15]. The D value is obtained from initial linear portion of the moisture absorption percentage versus the square root of the time curve. The one-dimensional diffusion coefficient, D, can be determined from the following equation:

$$D = \pi\left[\frac{h}{4M_\infty}\right]^2\left[\frac{M_2 - M_1}{\sqrt{t_2} - \sqrt{t_1}}\right]^2 \tag{3}$$

where h is the plate thickness, M_2, M_1 are the moisture content at time t_1 and t_2, respectively.

2.4. Tensile Test

Tensile properties of the RH/CP reinforced ABS blend composites were performed according to the ASTM D618 test standards. The samples were fabricated into flat dog-bone shaped to accommodate the calibrated universal testing machine (AG-X plus 50 kN, Shimadzu, Kyoto, Japan). The speed of the tensile testing was fixed at 1 mm/min. The tests were performed on five samples, and the average reading was taken as the final result.

2.5. Flexural Test

Flexural properties of the fabricated RH-CP/ABS blend composite were evaluated according to the ASTM D790-03 (3-point bending) standard. The testing was executed by using a universal testing machine (AG-X plus 50 kN, Shimadzu, Kyoto, Japan) with a span to depth ratio of 16:1. The flexural testing speed was fixed at 1 mm/min by using a 50 kN of the load cell.

2.6. Impact Test

In this study, the unnotch Charpy impact test was used to measure the impact characteristic of the fabricated RH-CP reinforced ABS blend composite. Samples with a dimension of 65 mm × 12.7 mm × 3.2 mm were evaluated by using a pendulum impact tester (Zwick 5113, Ulm, Germany). The pendulum was released at an energy capacity of 4 J and release angle of 160°. This experimental procedure was conducted according to ASTM D256 test standards. An average of five samples relative to the RH-CP reinforced ABS blend composites' impact strength was evaluated.

2.7. Composite Characterization

The morphological investigations of RH-CP reinforced ABS blend composite were examined by using a field emission scanning electron microscope (FESEM) (JEOL, JSM-7800F, Tokyo, Japan). The composite samples were then analyzed under a magnification of 500× at an accelerating voltage of 2 kV. Prior to the evaluation, the samples were initially coated with platinum by using a fine auto coater (JEOL, JEC-3000FC, Tokyo, Japan).

3. Results and Discussion

3.1. Density of Blend Composites

The measured density values of the RH/CP reinforced thermoplastic ABS composites are illustrated in Figure 1. This composite density was subjected to various particle loading of RH/CP (0/20, 5/15, 10/10, 15/5, and 20/0) in a weight fraction. It was shown that by increasing the rice husk to coco peat in blend composites, the density also increases. The highest density result was obtained from RH20/CP0 as compared to other biocomposite configurations. This result was mainly ascribed to the higher density of RH to the CP filler. Similarly, Hemnath et al. [16] reported that by increasing the RH content the void content decreased due to smaller particle size in blend composites. Higher RH loading reduces the pores content resulting in tighter and pack composites. This void percentage of the composite can be controlled by adjusting the reinforcement amount and molding parameters during fabrication. These lead to improved mechanical properties, sound absorption capabilities, and thermal insulation of the structures.

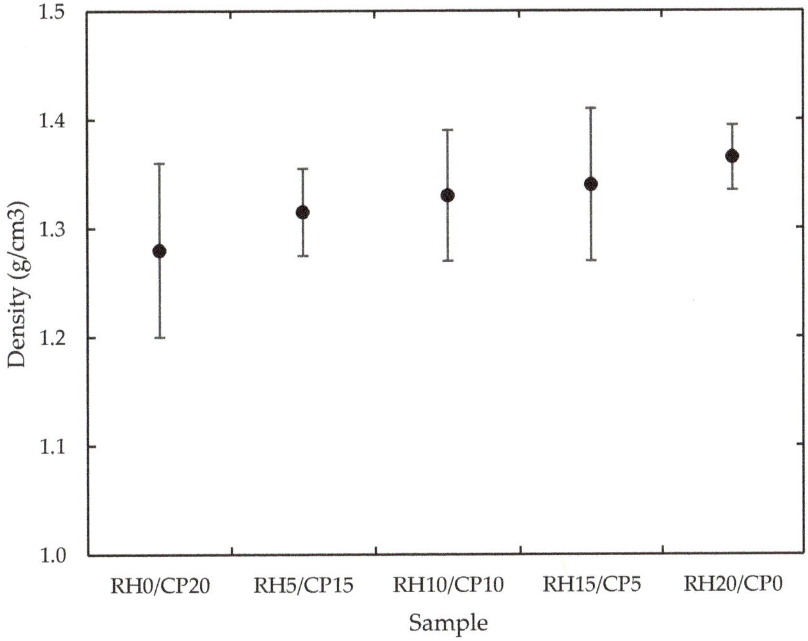

Figure 1. Typical density of rice husk (RH)/coco peat (CP) reinforced acrylonitrile-butadiene-styrene (ABS) blend composites.

3.2. Moisture Absorption Study

The moisture absorption trends of the RH/CP (0/20, 5/15, 10/10, 15/5, and 20/0) reinforced ABS are illustrated in Figure 2a. The plots were deducted from an average value of three specimens. For immersion time lower than 20 h, the moisture absorption (%) rate steadily increased with increasing coco peat content. The maximum moisture absorption is from the RH0/CP20 wt.% of coco peat composite composition. It can be suggested that the hydrophilicity of the coco peat particle was higher than the rice husk filler. Deo et al. [17] reported that weak fiber-matrix adhesion and void content influenced the natural fiber composites' moisture uptake. The high content of coco peat filler increased free hydroxyls (OH) groups in cellulose. These OH groups enhanced the contact with moisture and form several hydrogen bonding, resulting in weight gain in the blend composites. In general, moisture uptake increases with immersion time. However, it remained as a constant plateau after 40 h, as shown in Figure 2a. The time to reach a saturation condition was nearly similar for all tested samples. The extent of blend composites in a humid environment promotes an increase in the swelling dimension and low-stress transfer between particle and matrix, which corresponds to the reduction in mechanical properties [18].

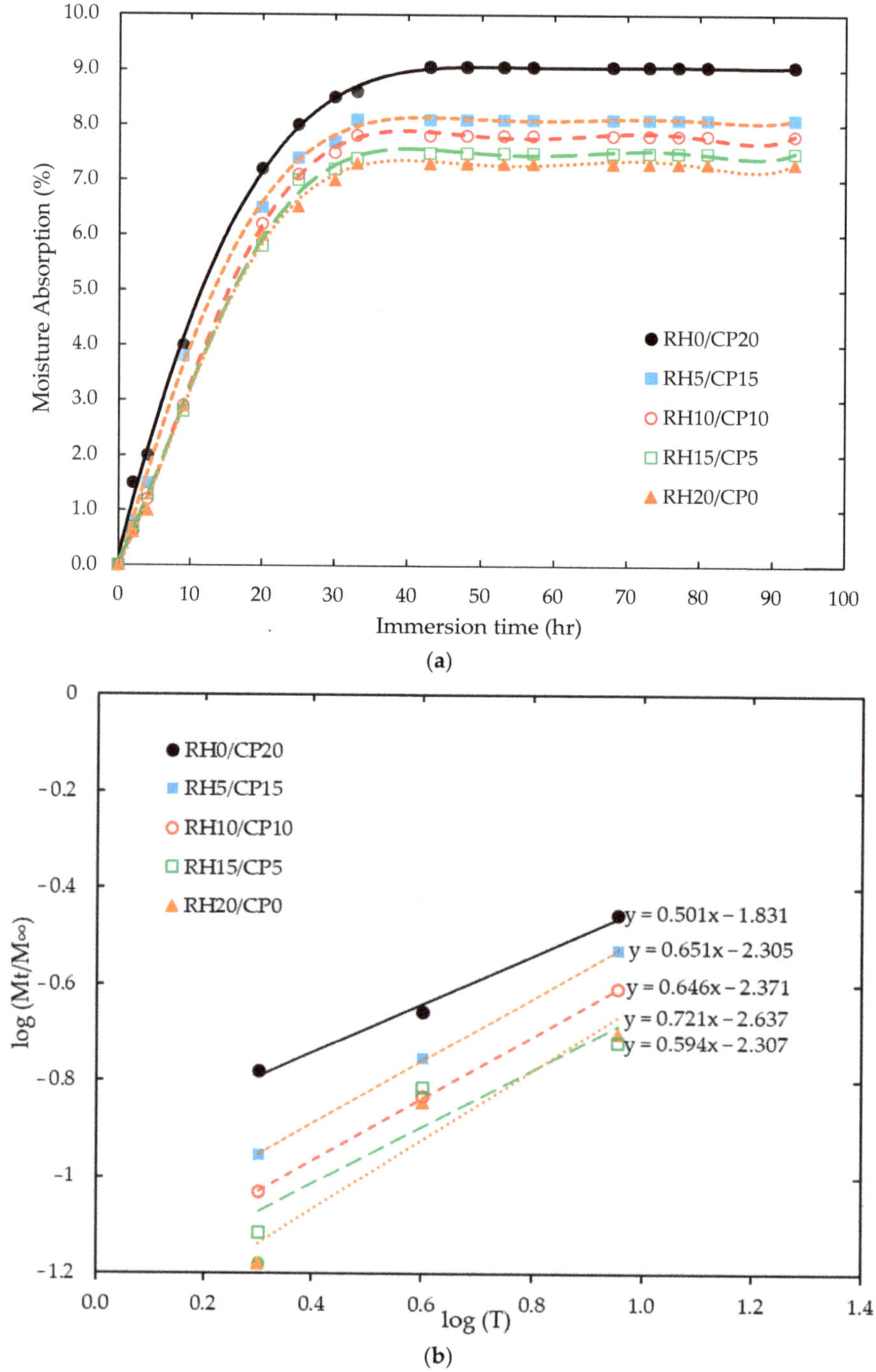

Figure 2. Typical (**a**) moisture absorption characteristics and (**b**) plot of log M_t/M_∞ versus log time of blend RH/CP reinforced ABS composites.

Figure 2b presents a typical log (M_t/M_∞) versus log (t) for the RH/CP reinforced ABS composites. The swelling transport exponent 'n' and characteristic constant (k) of the moisture absorption values were determined from the slopes and intercepts of these plots, respectively. The results indicated that the diffusion exponent for the composite samples lies between 0.56 and 0.59. This can be explained that the RH/CP reinforced ABS was in the range of Fick's model. Similar findings were reported by Razavi et al. [19] and Awasthi et al. [20]. The n values were between 0.5 and 0.7 when the Fickian diffusion property was applied in the studies to determine the rice husk reinforced thermoplastic composites' moisture absorption characteristic. It was claimed that these biocomposites were in an intermediate diffusion process between the penetrant mobility cases I and II. However, Guloglu et al. [21] mentioned that most composites structures were in a range of abnormal water diffusion characteristics. This typical non-Fickian moisture absorption behavior of the thermoplastic-based composites was also reported by Melo et al. [22] and Aziz et al. [23] when the samples were continually exposed to a wet environment at a lower absorption rate and prolonged period.

Moreover, the blend composites k value increased with increasing coco peat content, as tabulated in Table 2. It indicated that the higher the coco peat filler in composites, the higher the biocomposites kinetic water absorption characteristic. A higher value of k explained that the fastest blend composite diffusion time is needed to reach a saturation condition. Findings were also aligned with the result, as illustrated in Figure 2a. Similar results were reported by Guna et al. [24]. It was mentioned a decreased water resistance with the increase up to 40 wt.% of coconut filler content of the hybrid composites, resulting in insufficient resin to impregnate and wet out the fillers. In this study, the D values of the biocomposites also increased with increase in coco peat content. At lower rice husk composition, the water dispersion mobility rate in a composite capillary was low, resulting in lower D values. These D values were in a range reported by Chen et al. [8]. It demonstrated that the pack of hybrid arrangement in rice husk filler would reduce the moisture kinetic absorption due to narrow gaps and voids formation in the biocomposites. A similar finding was reported by Nanthakumar et al. [25], which suggested that an improvement of filler-matrix adhesion by using a surface modification technique could reduce water molecules to diffuse and penetrate the composite structures. Besides, an extra information on the blend composites' glass transition temperature helped to predict the plasticizing effect of solvent on the polymer [26,27].

Table 2. Saturation water absorption, water absorption constant, swelling exponent constant, and diffusion coefficient of RH/CP reinforced ABS composites.

Composition	M_∞ (%)	k	n	$D \times 10^{-6}$ (mm^2/s)
RH0/CP20	9.05	0.124	0.415	0.104
RH5/CP15	8.11	0.095	0.448	0.114
RH10/CP10	7.51	0.085	0.469	0.121
RH15/CP5	7.22	0.055	0.519	0.127
RH20/CP0	6.92	0.039	0.567	0.134

3.3. *Tensile Properties*

Five types of RH/CP fiber weight percentages were used in blend composites. RH and CP fiber used as the fiber content in the composites were fixed at 20 wt.% while the epoxy resin matrix was fixed at 80 wt.%. The tensile study was conducted and showed that the tensile properties were affected by the variation in blend compositions. The tensile properties are shown in Figure 3. It was demonstrated that RH15/CP5 composition showed the highest tensile strength among the blend composites. Tensile strength enhancements of 18.4% and 17.8% were reported for RH15/CP5 blend composites were achieved as compared to its monofiber composites for RH0/CP20 (coco peat monofiber composites) and RH20/CP0 (rice husk monofiber composites), respectively. This phenomenon was due to the blending ability to practically overcome the traditional low strength disadvantages of single type natural fiber-reinforced composites [28,29]. Practically, the RH particle size was much smaller than that of CP. Consequently, the surface area available for wetting by the polymeric matrix was higher with higher RH filler loadings. Under quasi-static tensile loading, stress transfer from matrix to filler may be more efficient in the RH dominant composites, resulting in higher mechanical properties [30]. Moreover, the tensile strength also revealed an increment as the RH content increases. Simultaneously, the CP content was decreased, up to 15 wt.% RH and down to 5 wt.% CP, respectively, on the composites. The composite properties mainly depended on the mechanical behaviors of the individual reinforcing fibers [31]. RH was reported as a kind of natural fiber with higher stiffness in tensile properties as compared to CP fiber [32]. This event was explained by the tensile strength increment as the RH loading extends and not the opposition's extension of CP loading. However, the tensile strength decreased with further addition of 20 wt.% RH content, by as much as 15.1%. Poor interfacial bonding with the evidence from the micrograph of a fractured specimen in Figure 8 was the reason for the tensile deterioration. Aridi et al. [33] studied the mechanical properties of RH polypropylene composites, and the results showed that as the RH composition increased to 55 wt.%, the strength decreased. It was concluded that the weak bonding between the hydrophilic filler and the hydrophobic matrix polymer obstructed the stress propagation, and thus caused the tensile strength to fall when the filler loading increased. Besides, poor dispersion caused agglomeration of the fillers and acted as stress concentration points, which led to composite failure [34]. This caused a decrease in tensile strength.

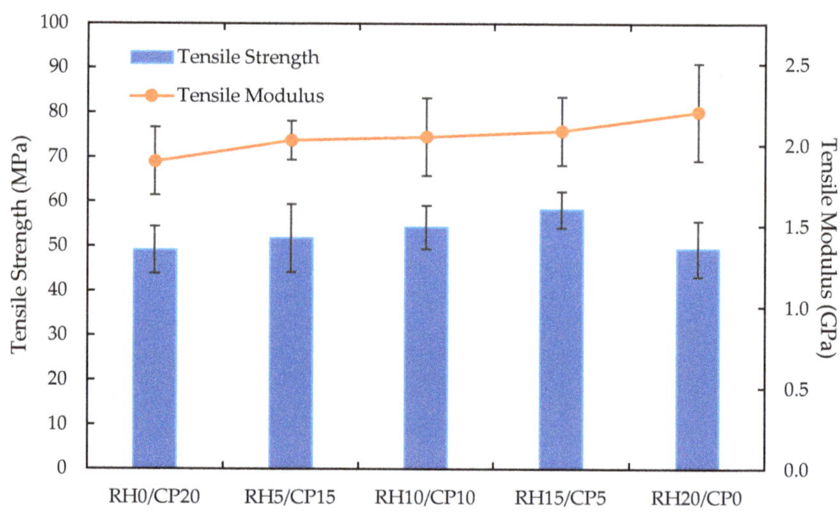

Figure 3. Tensile properties of RH/CP reinforced ABS blend composites.

Nevertheless, the addition of RH fiber loading instead of CP fiber loading proportionally increased the tensile modulus. It was deduced that the tensile modulus was affected by the blending of RH and CP fiber as the reinforcement filler. The highest enhancement of 16% on tensile modulus was detected from RH20/CP0 in contrast with RH0/CP20. This finding remarked that the composites' stiffness came from the stiffness of CP, particularly on top of the ABS itself. Altogether, it was observed that the RH20/CP0 blend composite delivered the maximum Young's modulus, whereas the lowest was delivered by RH0/CP20 filler for the composites. Figure 4 demonstrates the tensile stress-strain curves for RH20/CP0, RH15/CP5, RH10/CP10, RH5/CP15, and RH0/CP20 blend composites from the tensile test. The curve deduced that the 15 wt.% RH and 5 wt.% CP resulted in the highest value in tensile stress. Meanwhile, the tensile strain's highest extent was achieved by the 20 wt.% CP blend composites. The tensile strain recorded that the composites' strain increased as the amount of CP fiber filler increased. The monofiber CP composites showed an improvement of 15.5% strain as compared to the monofiber RH composites. Therefore, the CP fiber upsurged the ductility behavior of the blend composites better than RH fiber.

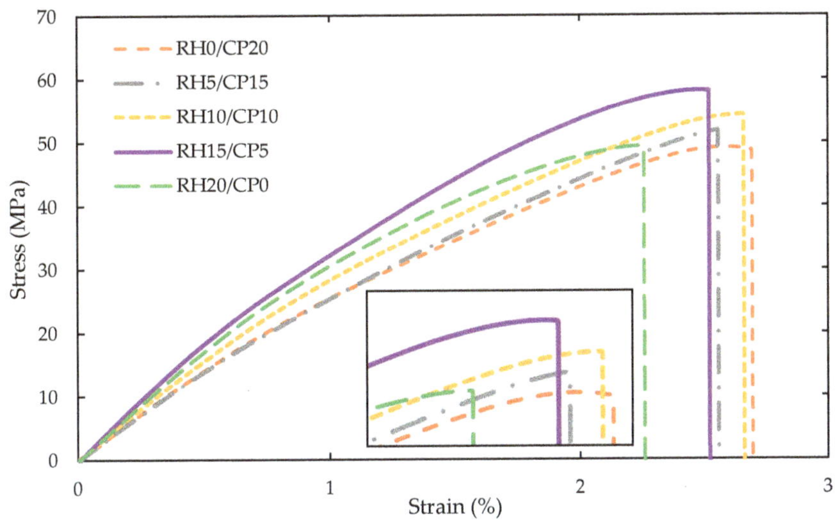

Figure 4. Typical stress-strain curves of RH/CP reinforced ABS blend composites.

3.4. Flexural Properties

In order to further understand the mechanical properties of the blend composites, flexural tests were conducted. Figure 5 presents the flexural strength and flexural modulus results obtained from the tests. It was apparent from this figure that the trend for this test was similar to the tensile test. The finding provided evidence that the flexural strength increased as the RH fiber filler increased, up to 15 wt.%, then decreased after the addition of 20 wt.% RH fiber to the blend composites. Further analysis showed that the RH15/CP5 blend composite revealed the highest value of flexural strength in comparison with its RH and CP monofiber composites, with an improvement of 9.0% and 2.3%, respectively. The flexural strength improvement was not quite as significant as the tensile strength improvement, but still considerably affecting the composites. The lowest flexural strength was RH0/CP20, likewise the tensile strength discussed in the previous section. There were several possible explanations for this result. A possible reason for this might be that a weak link may exist between the interfacial bond between hydrophilic CP fiber and hydrophobic resin [35].

In Figure 5, a clear trend is shown on the increasing flexural modulus as the percentage of introducing RH filler into the composites becomes larger. The optimum flexural modulus was found on the RH20/CP0 at 5.51 GPa, followed by RH15/CP5, RH10/CP10, RH5/CP15, and RH0/CP20. As mentioned in the literature review, no data was found on the blending between RH fiber and CP fiber into a composite for comparison. On the other hand, Figure 6 illustrates the typical flexural stress-strain curves of the blend composites. Contrary to expectations, the result did not show any significant difference between the flexural strain enhancement with the variation of RH and CP fibers filler content. The insignificant difference in the flexural strain could be attributed to the elasticity among the composite, which was almost similar for each of them.

3.5. Impact Response Behavior

The impact response of the rice husk and coco peat filler blend ABS composites is summarized in Figure 7. It corresponds to the combination of RH/CP configuration. The incorporation of RH in the CP-ABS varied the impact behavior of the blend composite. Referring to the plot, the maximum impact energy and impact strength were obtained at RH15/CP5 wt.%. The value of the impact properties significantly increased with the increase of RH from 0 wt.% to 15 wt.%. Further increase in the proportion of biomass in the biocomposite reduced the impact strength due to delamination and insufficient matrix [36] to proper adhesion, and resulting in decreased strength. The overall impact properties of the blend composite can be improved by chemical coupling and surface modification. However, high modification can increase production cost and reduces the agriculture composite residues that are value-added.

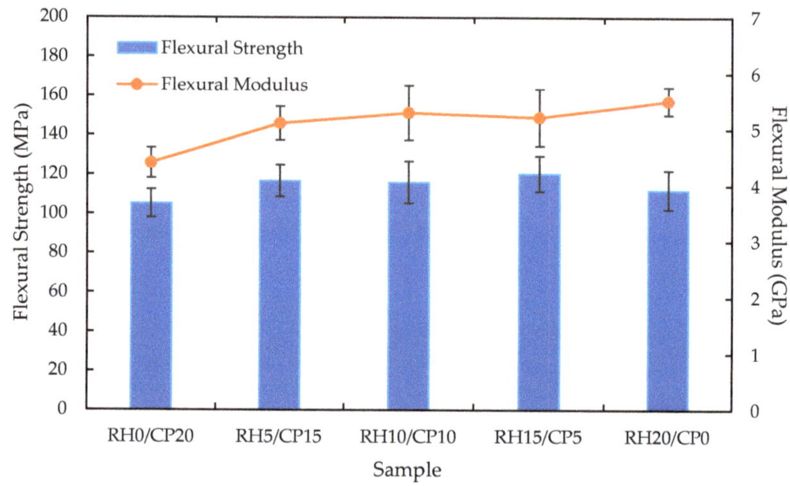

Figure 5. Flexural properties of RH/CP reinforced ABS blend composites.

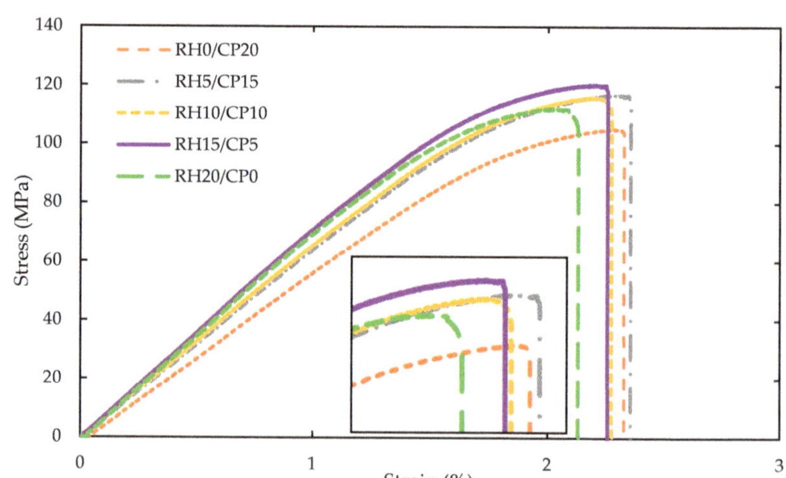

Figure 6. Typical flexural curves of RH/CP reinforced ABS blend composites.

3.6. Fracture Surface Micrograph

Figure 8 demonstrates the differences of fracture surface by using SEM for the composite containing RH15/CP15 and RH0/CP20 filler content. Filler fracture and matrix delamination are predominated in the RH15/CP15 specimen's morphological surface, as illustrated in Figure 8a. Matrix cracking and the presence of void on the interface was also observed. It can be suggested that the stress was well scattered between the filler and thermoplastic matrix due to the clean braking surface. It was also indicated that this composite failed in the brittle mode failure. This means that an excellent interface bonding between filler and matrix was formed, resulting in high tensile strength property. Matrix cracking and fiber breakage were observed as in Figure 8b. As shown from this figure, more cavities were formed in this blend composite resulted from fiber pullout and air trap. It also demonstrated fracture damage occurred at the coco peat filler surface and which may suggest a flaw in these composites. It can lead to stress concentration at this weak point, and it was attributed to the lower mechanical properties. Arslan et al. [37] mentioned that the use of silane coupling agents such as (3-aminopropyl) triethoxysilane (AP) and 3-(trimethoxysilyl) propyl methacrylate (MA) could enhance the tensile properties of the biocomposite due to covalent bond formation between the amino group of coupling agent and the nitrile group of styrene-acrylonitrile (SAN) matrix. However, extra production cost would occur during fabrication.

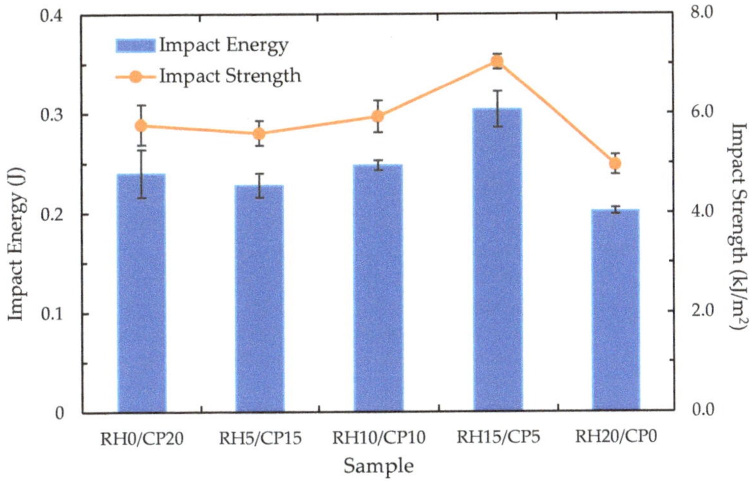

Figure 7. Impact behaviour of RH/CP reinforced ABS blend composites.

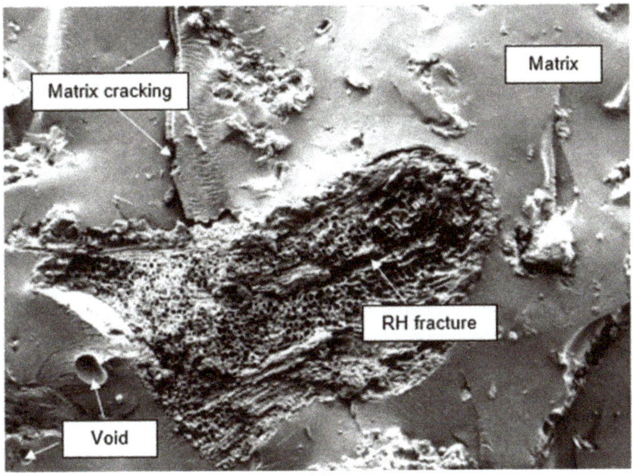

(**a**)

Figure 8. *Cont.*

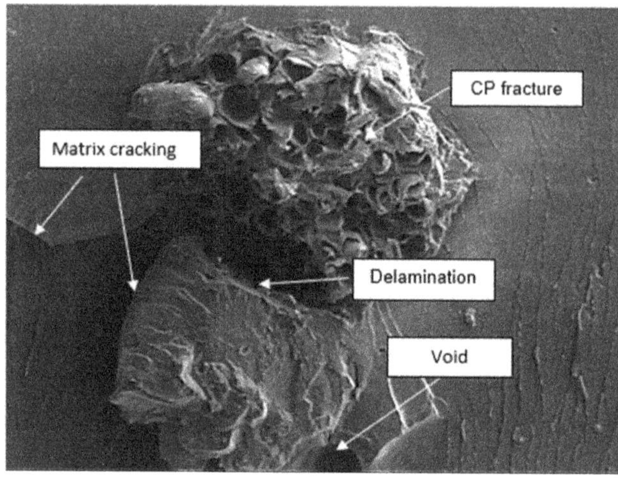

(b)

Figure 8. Scanning electron microscopy (SEM) micrographs of tensile fracture surface from (a) RH15/CP5 and (b) RH0/CP20 wt.%.

4. Conclusions

The blend RH/CP reinforced with ABS was successfully prepared by using two-step processes: melt blend mixer and injection molding. The physicomechanical properties of these biocomposites were then characterized. It was found that the density of the composites increased with increase in RH filler. In contrast, the composite's kinetic water absorption and moisture saturation increased steadily with CP content and obeyed Fick's model. It was also found that the incorporation of RH and CP into the ABS matrix offered better performance in tensile, flexural, and impact strengths. The synergistic effect study clearly shows that the incorporation of RH and CP into the ABS matrix performs better on fiber's interfacial bonding, as compared to the performances of individual composites' components, and thereby enhances the mechanical performances of the overall system.

The highest tensile properties and Young modulus were recorded at the composition of RH15/CP5 wt.%, resulting in lower elongation at break as compared to RH0/CP20 wt.% blend composite. As expected, the incorporation of 15 wt.% RH improved the maximum flexural and modulus properties value more than those blend composites based on the ABS matrix. In addition, this combination was attributed to the highest value of impact strength. The fracture surface morphology for the blend composite dominated by matrix cracking and filler fracture explained the stress was well propagated between filler and matrix thermoplastic. It proved that the excellent adhesion interfaces bonding of the biocomposites resulted in maximum mechanical properties. Therefore, it was interesting to combine these natural waste reinforced ABS composites for a short life engineering application, where the physicomechanical properties of biocomposites are of paramount importance.

Author Contributions: Conceptualization, S.A.Z. and K.R.J.; methodology, A.F.M.N.; validation, S.A.Z. and R.D.; formal analysis, R.D.; investigation, N.H.A.N.; resources, S.A.A.; writing—original draft preparation, N.H.A.N.; writing—review and editing, A.F.M.N.; visualization, K.R.J.; supervision, M.Z.H.; project administration, M.Z.H.; funding acquisition, M.Z.H. All authors have read and agreed to the published version of the manuscript.

Funding: This research was funded by the Ministry of Higher Education (MOHE) with Fundamental Research Grant (FRGS) Scheme FRGS/1/2020/TK0/UTM/02/59 and FRGS/1/2019/TK08/UTM/02/4 for funding this research work. Universiti Teknologi Malaysia also supported this research with 'Geran Universiti Penyelidik' (GUP) UTMFR Scheme Q.K.130000.2656.21H13, and Tier 2 Scheme Q.K.130000.2656.17J66.

Data Availability Statement: The data presented in this study are available on request from the corresponding author.

Conflicts of Interest: The authors declare no conflict of interest.

References

1. Hassan, M.; Cantwell, W. The influence of core properties on the perforation resistance of sandwich structures–an experimental study. *Compos. Part B Eng.* **2012**, *43*, 3231–3238. [CrossRef]
2. Nor, A.F.M.; Sultan, M.T.H.; Jawaid, M.; Talib, A.R.A.; Azmi, A.M.R.; Harmaen, A.S.; Asa'ari, A.Z. The effects of multi-walled CNT in bamboo/glass fibre hybrid composites: Tensile and flexural properties. *BioResources* **2018**, *13*, 4404–4415.
3. Hassan, M.Z.; Sapuan, S.; Roslan, S.A.; Sarip, S. Optimization of tensile behavior of banana pseudo-stem (Musa acuminate) fiber reinforced epoxy composites using response surface methodology. *J. Mater. Res. Technol.* **2019**, *8*, 3517–3528. [CrossRef]
4. Ibrahim, M.I.; Hassan, M.Z.; Dolah, R.; Yusoff, M.Z.M.; Salit, M.S. Tensile behaviour for mercerization of single kenaf fiber. *Malays. J. Fundam. Appl. Sci.* **2018**, *14*, 437–439. [CrossRef]
5. Hassan, M.Z.; Roslan, S.A.; Sapuan, S.; Rasid, Z.A.; Nor, A.F.M.; Md Daud, M.Y.; Dolah, R.; Zuhri, M.Y.M. Mercerization optimization of bamboo (bambusa vulgaris) fiber-reinforced epoxy composite structures using a box–behnken design. *Polymers* **2020**, *12*, 1367. [CrossRef] [PubMed]
6. Quispe, I.; Navia, R.; Kahhat, R. Energy potential from rice husk through direct combustion and fast pyrolysis: A review. *Waste Manag.* **2017**, *59*, 200–210. [CrossRef]
7. Isa, K.M.; Daud, S.; Hamidin, N.; Ismail, K.; Saad, S.A.; Kasim, F.H. Thermogravimetric analysis and the optimisation of bio-oil yield from fixed-bed pyrolysis of rice husk using response surface methodology (RSM). *Ind. Crop. Prod.* **2011**, *33*, 481–487. [CrossRef]
8. Chen, R.S.; Ab Ghani, M.H.; Ahmad, S.; Salleh, M.N.; Tarawneh, M.a.A. Rice husk flour biocomposites based on recycled high-density polyethylene/polyethylene terephthalate blend: Effect of high filler loading on physical, mechanical and thermal properties. *J. Compos. Mater.* **2015**, *49*, 1241–1253. [CrossRef]
9. Das, O.; Hedenqvist, M.S.; Prakash, C.; Lin, R.J. Nanoindentation and flammability characterisation of five rice husk biomasses for biocomposites applications. *Compos. Part A Appl. Sci. Manuf.* **2019**, *125*, 105566. [CrossRef]
10. Chen, R.S.; Ab Ghani, M.H.; Ahmad, S.; Mou'ad, A.T.; Gan, S. Tensile, thermal degradation and water diffusion behaviour of gamma-radiation induced recycled polymer blend/rice husk composites: Experimental and statistical analysis. *Compos. Sci. Technol.* **2021**, 108748. [CrossRef]
11. Boonsuk, P.; Sukolrat, A.; Bourkaew, S.; Kaewtatip, K.; Chantarak, S.; Kelarakis, A.; Chaibundit, C. Structure-properties relationships in alkaline treated rice husk reinforced thermoplastic cassava starch biocomposites. *Int. J. Biol. Macromol.* **2021**, *167*, 130–140. [CrossRef] [PubMed]
12. Borawski, A. Conventional and unconventional materials used in the production of brake pads–review. *Sci. Eng. Compos. Mater.* **2020**, *27*, 374–396. [CrossRef]
13. Singh, A.; Kumar, S.; Dev, R. Studies on cocopeat, sawdust and dried cow dung as desiccant for evaporative cooling system. *Renew. Energy* **2019**, *142*, 295–303. [CrossRef]
14. Nor, A.F.M.; Hassan, M.Z.; Rasid, Z.A.; Sarip, S.; Daud, M.Y.M. Optimization on Tensile Properties of Kenaf/Multi-walled CNT Hybrid Composites with Box-Behnken Design. *Appl. Compos. Mater.* **2021**, 1–26. [CrossRef]
15. Shen, C.-H.; Springer, G.S. Moisture absorption and desorption of composite materials. *J. Compos. Mater.* **1976**, *10*, 2–20. [CrossRef]
16. Hemnath, A.; Anbuchezhiyan, G.; NanthaKumar, P.; Senthilkumar, N. Tensile and flexural behaviour of rice husk and sugarcane bagasse reinforced polyester composites. *Mater. Today Proc.* **2021**. [CrossRef]
17. Deo, C.; Acharya, S. Effect of moisture absorption on mechanical properties of chopped natural fiber reinforced epoxy composite. *J. Reinf. Plast. Compos.* **2010**, *29*, 2513–2521. [CrossRef]
18. Kufel, A.; Kuciel, S. Hybrid Composites Based on Polypropylene with Basalt/Hazelnut Shell Fillers: The Influence of Temperature, Thermal Aging, and Water Absorption on Mechanical Properties. *Polymers* **2020**, *12*, 18. [CrossRef]
19. Razavi, N.M.; Jafarzadeh, D.F.; Ouroumiehei, A.A.; Ershad, L.A. Mechanical properties and water absorption behaviour of chopped rice husk filled polypropylene composites. *Iran. Polym. J.* **2006**, *15*, 757–766.
20. Awasthi, S.; Singhal, R. Mathematical modeling for the prediction of the overall swelling profile from poly (AM-co-AA-co-HEA) hydrogels: Effect of glycidyl methacrylate and ammonium per sulphate. *Int. J. Plast. Technol.* **2015**, *19*, 241–262. [CrossRef]
21. Guloglu, G.E.; Altan, M.C. Moisture absorption of carbon/epoxy nanocomposites. *J. Compos. Sci.* **2020**, *4*, 21. [CrossRef]
22. Melo, R.Q.; Lia Fook, M.V.; de Lima, A.G. Non-Fickian Moisture Transport in Vegetable-Fiber-Reinforced Polymer Composites Using a Langmuir-Type Model. *Polymers* **2020**, *12*, 2503. [CrossRef] [PubMed]
23. Aziz, N.F.A.; Ahmad, Z.; Puaad, M.B.F.M. Study of the behavior of moisture absorption and swelling in high filler loading kenaf core/bast polyethylene composites. *Adv. Civ. Eng. Mater.* **2020**, *9*, 250–261. [CrossRef]
24. Guna, V.; Ilangovan, M.; Rather, M.H.; Giridharan, B.; Prajwal, B.; Krishna, K.V.; Venkatesh, K.; Reddy, N. Groundnut shell/rice husk agro-waste reinforced polypropylene hybrid biocomposites. *J. Build. Eng.* **2020**, *27*, 100991. [CrossRef]
25. Nanthakumar, K.; Yeng, C.M.; Chun, K.S. Tensile and water absorption properties of solvent cast biofilms of sugarcane leaves fibre-filled poly (lactic) acid. *J. Thermoplast. Compos. Mater.* **2020**, *33*, 289–304. [CrossRef]

26. Lu, H.; Du, S. A phenomenological thermodynamic model for the chemo-responsive shape memory effect in polymers based on Flory–Huggins solution theory. *Polym. Chem.* **2014**, *5*, 1155–1162. [CrossRef]
27. Lu, H.; Liu, Y.; Leng, J.; Du, S. Qualitative separation of the physical swelling effect on the recovery behavior of shape memory polymer. *Eur. Polym. J.* **2010**, *46*, 1908–1914. [CrossRef]
28. Yusoff, R.B.; Takagi, H.; Nakagaito, A.N. Tensile and flexural properties of polylactic acid-based hybrid green composites reinforced by kenaf, bamboo and coir fibers. *Ind. Crop. Prod.* **2016**, *94*, 562–573. [CrossRef]
29. Nor, A.F.M.; Sultan, M.T.H.; Jawaid, M.; Azmi, A.M.R.; Shah, A.U.M. Analysing impact properties of CNT filled bamboo/glass hybrid nanocomposites through drop-weight impact testing, UWPI and compression-after-impact behaviour. *Compos. Part B Eng.* **2019**, *168*, 166–174. [CrossRef]
30. Nurihan, O.; Rasid, Z.; Hassan, M.; Nor, A. The elastic properties of unidirectional bamboo fibre reinforced epoxy composites. *Int. J. Recent Technol. Eng.* **2019**, *8*, 7187–7193.
31. Saidane, E.H.; Scida, D.; Assarar, M.; Ayad, R. Damage mechanisms assessment of hybrid flax-glass fibre composites using acoustic emission. *Compos. Struct.* **2017**, *174*, 1–11. [CrossRef]
32. Sundarababu, J.; Anandan, S.S.; Griskevicius, P. Evaluation of mechanical properties of biodegradable coconut shell/rice husk powder polymer composites for light weight applications. *Mater. Today Proc.* **2020**. [CrossRef]
33. Aridi, N.; Sapuan, S.; Zainudin, E.; AL-Oqla, F.M. Mechanical and morphological properties of injection-molded rice husk polypropylene composites. *Int. J. Polym. Anal. Charact.* **2016**, *21*, 305–313. [CrossRef]
34. Olonisakin, K.; Fan, M.; Xin-Xiang, Z.; Ran, L.; Lin, W.; Zhang, W.; Wenbin, Y. Key Improvements in Interfacial Adhesion and Dispersion of Fibers/Fillers in Polymer Matrix Composites; Focus on PLA Matrix Composites. *Compos. Interfaces* **2021**, 1–50. [CrossRef]
35. Singh, Y.; Singh, J.; Sharma, S.; Lam, T.-D.; Nguyen, D.-N. Fabrication and characterization of coir/carbon-fiber reinforced epoxy based hybrid composite for helmet shells and sports-good applications: Influence of fiber surface modifications on the mechanical, thermal and morphological properties. *J. Mater. Res. Technol.* **2020**, *9*, 15593–15603. [CrossRef]
36. Hassan, M.Z.; Sapuan, S.; Rasid, Z.A.; Nor, A.F.M.; Dolah, R.; Md Daud, M.Y. Impact damage resistance and post-impact tolerance of optimum banana-pseudo-stem-fiber-reinforced epoxy sandwich structures. *Appl. Sci.* **2020**, *10*, 684. [CrossRef]
37. Arslan, C.; Dogan, M. The effects of fiber silane modification on the mechanical performance of chopped basalt fiber/ABS composites. *J. Thermoplast. Compos. Mater.* **2020**, *33*, 1449–1465. [CrossRef]

polymers

MDPI

Article

Morphology, Structural, Thermal, and Tensile Properties of Bamboo Microcrystalline Cellulose/Poly(Lactic Acid)/Poly(Butylene Succinate) Composites

Masrat Rasheed [1], Mohammad Jawaid [1,*], Bisma Parveez [2], Aamir Hussain Bhat [3] and Salman Alamery [4]

1 Laboratory of Biocomposite Technology, Institute of Tropical Forestry and Forest Products (INTROP), Universiti Putra Malaysia, Serdang 43400, Selangor, Malaysia; masuqayyuum@gmail.com
2 Kulliyan of Engineering (KOE), Islamic International University Malaysia, Gombak 53100, Kuala Lumpur, Malaysia; mirbisma5555@gmail.com
3 Department of Applied Sciences, Higher College of Technology, University of Technology and Applied Sciences, Al-Khuwair, Muscat 133, Oman; bhataamir@gmail.com
4 Department of Biochemistry, College of Science, King Saud University, P.O. Box 22452, Riyadh 11451, Saudi Arabia; salamery@ksu.edu.sa
* Correspondence: jawaid@upm.edu.my; Tel.: +60-3-97696960

Abstract: The present study aims to develop a biodegradable polymer blend that is environmentally friendly and has comparable tensile and thermal properties with synthetic plastics. In this work, microcrystalline cellulose (MCC) extracted from bamboo-chips-reinforced poly (lactic acid) (PLA) and poly (butylene succinate) (PBS) blend composites were fabricated by melt-mixing at 180 °C and then hot pressing at 180 °C. PBS and MCC (0.5, 1, 1.5 wt%) were added to improve the brittle nature of PLA. Field emission scanning electron microscopy (FESEM), scanning electron microscopy (SEM), X-ray diffraction (XRD), Fourier transform infrared spectroscope (FTIR), thermogravimetric analysis (TGA), differential thermogravimetry (DTG), differential scanning calorimetry (DSC)), and universal testing machine were used to analyze morphology, crystallinity, physiochemical, thermal, and tensile properties, respectively. The thermal stability of the PLA-PBS blends enhanced on addition of MCC up to 1wt % due to their uniform dispersion in the polymer matrix. Tensile properties declined on addition of PBS and increased with MCC above (0.5 wt%) however except elongation at break increased on addition of PBS then decreased insignificantly on addition of MCC. Thus, PBS and MCC addition in PLA matrix decreases the brittleness, making it a potential contender that could be considered to replace plastics that are used for food packaging.

Keywords: poly(lactic acid); microcrystalline cellulose; poly(butylene succinate); scanning electron microscopy; tensile properties

Citation: Rasheed, M.; Jawaid, M.; Parveez, B.; Hussain Bhat, A.; Alamery, S. Morphology, Structural, Thermal, and Tensile Properties of Bamboo Microcrystalline Cellulose/Poly(Lactic Acid)/Poly (Butylene Succinate) Composites. *Polymers* **2021**, *13*, 465. https://doi.org/10.3390/polym13030465

Academic Editor: Maya Jacob John
Received: 26 December 2020
Accepted: 19 January 2021
Published: 1 February 2021

1. Introduction

In current years, due to the higher usage of plastics, their disposals cause environmental pollution that is of great concern. This problem can be solved by the usage of biodegradable materials which are easily disposed due to their microbial action, as a replacement to the synthetic polymers. PLA, PBS, and MCC based composites can fulfil the requirement of replacing commodity synthetic polymers. PLA and PBS are biodegradable polyesters with great degradability and mechanical properties. Since PLA behold amazing thermal, mechanical, and biodegradable properties, it holds a great potential usage in polymer-based applications [1]. In contrast, properties such as flexural properties, impact strength, gas permeability, melt viscosity required for processing, heat distortion temperature (HDT), etc. are not adequate for packaging applications [2]. Additionally, brittleness and higher cost of PLA restricts their potential for commercial usage.

Therefore, combining PLA with other appropriate biodegradable polymer with comparatively better melt processability, flexural properties, and excellent impact strength, can not only improve its properties but also reduces overall material cost. PBA is mostly preferred as it possesses the required properties. Thus, to enhance the properties and minimize the cost of production, numerous works on PLA blended with various biodegradable polymers were executed [3–5]. Bhatia et al. [6] fabricated a blend of PLA and PBS and observed that the blend up to 80/20 wt% (PLA-PBS) composition was partial miscibility. The brittle nature of PLA was reduced by the PBS; therefore, it can be considered as a potential material as a replacement of plastics for packaging applications. However, there was reduction of tensile strength, tensile modulus, and percentage elongation at break as PBS content increased. Liu et al. [7] repaired a blend of PLA and poly (ethylene/butylene succinate) or Binolle and observed that the addition of Bionolle resulted into crystallization of PLA and a small increment in the strain at break, however, there was reduction of ultimate tensile strength and modulus.

Also, Yokohara & Yamaguchi [8] examined the structure and various properties of PLA blended with fibrous PBS and spherical particles of PBS. PBS particle led to nucleation of PLA results into an increase in the crystallinity and it was further enhanced by the heating process. Homklin and Hongsriphan [9] reported the effect of nucleating agents such as sodium benzoate and nano-sized calcium carbonate on mechanical strength and thermal stability of PLA-PBS blends and on filling these blends with the nucleating agents, their tensile strength, energy, and elongation at break reduced. This was because of increased PLA phase crystallinity and the occurrence of stress concentration in these blends.

Microcrystalline cellulose (MCC) is a natural source with added benefits—like low cost, abundance in availability, high strength to weight ratio, renewability, and low density-makes it progressively a potential material as reinforcement for the fabrication of composites [10,11]. Although there are some challenges and further research are carried out to improve their functionalities in order to increase their applications in the materials [12]. PLA and PBS are synthetic polymers(polyesters) but MCC are the micro sized cellulose extracted mostly from plant fibers or other cellulosic waste via acid hydrolysis technique in order to eliminate their amorphous part [13–15]. All of the three biodegradable materials have been commercialized extensively for reduction of the number of plastics in found in the waste disposal. Even though being so beneficial they have got some limitations, low melt strength, brittleness, lower thermal stability, and lower strength that limits their application and their large-scale production [16–18]. However, their blends or composites may compensate their distinct shortcomings because it may lead to enhancement of properties of the developed composites.

Various researchers examined the effect of MCC on the performances of polymers composites [19,20]. Cao et al. [21] examined the effect of the content of a chain-extender on the microstructure and performance of PLA-PBS-MCC composites and observed with the incorporation of chain extender processability and strength of the composites were improved. A few researchers have previously investigated PLA-PBS-MCC composites but there is still a need to explore such composites that can in future replace plastics effectively and thus diminish the level of plastic wastes in waste disposal that otherwise lead to environmental pollution. As from the literature PLA-PBS-MCC composites are not yet studied extensively so far. In addition, these composites may also provide a new possibility of obtaining tailored and more desirable performance of biodegradable plastics, and further enhance their thermal stability and mechanical properties.

In this work, PLA-PBS-MCC composites were fabricated via the hot-pressing technique and the morphology, crystallinity, physiological, and mechanical properties and their thermal behavior were evaluated by carrying out FESEM, XRD, FTIR, tensile testing, TGA, and DSC analysis respectively. This work aims to improve our understanding and endorse the development of multi constituent materials with desirable properties and also explore the potential of MCC in packaging applications.

2. Experimentation

2.1. Materials

Pellet form PLA (7001 IngeoTM of specific gravity 1.24) utilized in this study were obtained from Nature Works LCC, MN, USA. It has melting point of 154 °C and hydrophobic in nature. PBS of density 1.26g/cm^3 in pellet form were obtained from PTT public company limited in Bangkok, Thailand. It has a melting point of 95 °C. MCC were isolated from bamboo fiber through acid hydrolysis technique and further ultrasonicated to prevent agglomeration [22].

2.2. Methods

2.2.1. Extraction of MCC from Bamboo Fiber

Bamboo fiber was used as a source of MCC and were extracted from it by acid hydrolysis technique. The fiber was first bleached to obtain separate lignin and hemicellulose and pre-treated bamboo pulp then was hydrolyzed with sulphuric acid of concentration 85 wt% for 30 min and finally MCC were obtained. These were further vacuum oven dried at 80 for 24 h to get dispersed and high quality MCC. Thus, MCC of higher yield of 80% and crystallinity index of 82.6% were obtained [23].

2.2.2. Fabricated of MCC Reinforced PLA/PBS Composites

MCC, PBS, and PLA were initially oven-dried at a temperature of 60 °C for 24 h. Five compositions, as demonstrated in Table 1, were prepared using hot pressing technique. Initially the constituents of compositions were mixed by Brabender mixer (melt mixer) at the temperature of 180 °C for 15 min at 60 rpm. The mixture was then crushed in a crusher to get it in the form of pellets. Compressed sheets with the thickness of 0.12 mm were obtained by hot compression of these pellets in a hot press at the constant pressing pressure of 150 MPa and the temperature of 180 °C for reheating time of 4 min and then pressing time of 3 min as shown in Table 2. The compressed sheets were then oven-dried for 24 h at a constant temperature of 50 °C and prior to testing stored in desiccator.

Table 1. Composition of PLA/PBS/MCC Composites

S.No	Composite	PLA (wt%)	PBS (wt%)	MCC
1.	M1	100	0	0
2.	M2	80	20	0
3.	M3	80	20	0.5
4.	M4	80	20	1
5.	M5	80	20	1.5

Table 2. Fabrication Process Parameters.

Process	Parameters			
	Temperature (°C)	Pressure (MPa)	Time (min)	Speed (rpm)
Melt mixing	180	-	15	60
Hot pressing	180	150	3	-

2.3. Characterization and Testing

2.3.1. Field Emission Scanning Electron Microscopy (FESEM)

The surface morphologies of composite films were acquired at 10–20 kV (accelerating voltage) using FESEM (JEOL JSM-7000F) from Tokyo, Japan. The films were first coated before examination to avoid electrostatic charging.

2.3.2. Scanning Electron Microscopy (SEM)

The fractured surface of composite films was analyzed by SEM (Jeol, JSM 5410LV) (Hitachi Model S-3400N) from Tokyo, Japan. Specimens were tensile fracture. The samples were fixed to the stub by means of carbon tape and prior to inspection sputter-coating of gold was applied to the fractured surface to avoid electrostatic charging.

2.3.3. X-ray Diffraction (XRD)

XRD analysis of composites was executed to find the crystallinity of composite films at an angular incidence of 5° to 40° using Ni-filtered Cu K-alpha radiation by SHIMADZU XRD-6000 X-ray diffractometer, from Tokyo Japan.

2.3.4. Fourier Transform Infrared Spectroscope (FTIR)

The functional group analysis of each sample was examined at 4 cm^{-1} (resolution) by 32 scans using an imaging microscope (Perkin Elmer 1600 Infrared spectrometer, MA, USA) within a frequency ranging from 500 to 4000 cm^{-1}

2.3.5. Thermal Characterization

Thermal stability of the composite films was estimated by carrying out TGA (Perkinelmer TGA7, MA, USA) at a heat rate of 10 °C/min in range of 10–600 °C in a nitrogen atmosphere. Samples of around 10mg were cut from the films and the change in their weight ratio with respect to temperature was recorded. Thermal analysis was executed using DSC (Perkin-Elmer DSC7, MA, USA) where aluminum pan that was empty was taken as a reference. Samples of weight around 3–5 mg were placed in a pan and a temperature scan was done at a heat rate of 10 °C/min from 0–200 °C under a nitrogen atmosphere. From the thermogram attained, glass transition temperature (T_g), crystallisation temperature (T_c), and melting temperature (T_m) of the composite films were obtained.

2.3.6. Tensile Properties

Tensile properties were measured as per ASTM Standard Method (D638-14 (2014)) via (Lloyds LRX) Universal Testing Machine. Tensile testing was executed using samples of size (10 × 100 mm) cut from the composite films, at 50mm (initial grip separation) and 10 mm/min (crosshead speed). Ten samples of each composition were dried at 50 °C for 24 min a conventional oven prior testing. The tensile strength and modulus properties, and elongation at break were acquired from tensile testing.

3. Results and Discussions

3.1. FESEM

The surface morphology of PLA(M1), PLA-PBS(M2), and PLA-PBS-MCC (M3, M4 and M5) composites was performed as illustrated in Figure 1. PLA composite demonstrated a smoother surface, due to the poor plastic deformation [24] as evident from Figure 1a, while the homogeneously distributed PBS phase exists in all PLA-PBS blends showing the compatibility of the PLA-PBS blends [25]. The addition of PBS up to 20 wt% better compatibility can possibly be attained [1]. It can be clearly seen from Figure 1c–e, MCC was dispersed throughout PLA matrix as well as embedded within PLA matrix because of its smaller size (micro-sized) as shown in Figure 1f–h. By adding MCC and increasing the MCC content in the PLA-PBS composite, the dispersed phase of MCC becomes clearer due to lower dispersion in PLA-PBS matrix [26] and with increasing MCC content the surface topography becomes rougher, as shown in Figure 1c–e. Moreover, MCC in previous works has proved to enhances the interaction at the interfaces of the PLA and PBS blend [27,28].

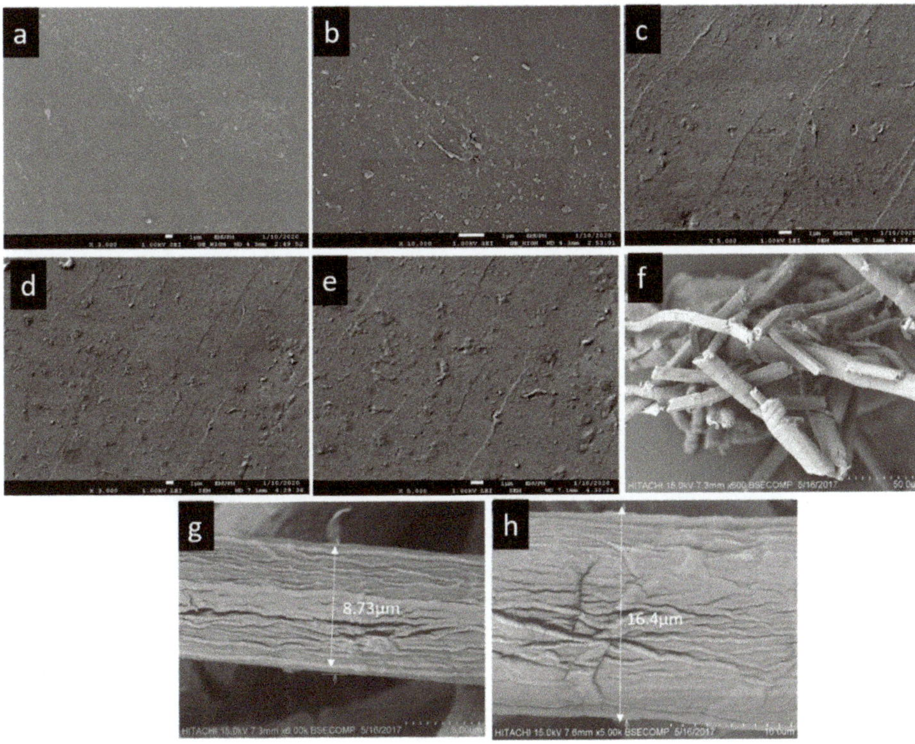

Figure 1. FESEM images of composite films (**a**) M1, (**b**) M2, (**c**) M3, (**d**) M4, (**e**) M5, (**f**,**g**,**h**) SEM images of MCC.

3.2. Thermal Properties

The thermal decomposition temperature of a polymer composite can be determined by its thermal degradation behavior. TGA was executed to examine the influence of PBS and MCC's on the thermal stability of PLA matrix composite films. The thermal stability of the composite films given by various parameters such as maximum degradation (T_{max}), initial degradation (T_i), temperature at 50% weight-loss (T50%), and final degradation (T_f) temperatures are evaluated using TGA. Figure 2 showing TGA curves of composites with varying percentages of MCCs. All the composites exhibited degradation processes in a single step, and the parameters of their thermal stability are given in Table 3. The PLA composites exhibited relatively lower thermal decomposition temperature in comparison to PLA-PBS and PLA-PBS-MCC composites. With addition of PBS the thermal stability of PLA-PBS blend increases. The T_{max}, T_i, T50%, and T_f, all increased on addition of up to 1 wt% MCC also similar behavior was reported by other researchers [21], then on further addition their values decrease as shown in Table 3. Even though MCC has low thermal stability when comparison with both PLA and PBS, still the char residue of MCC is expected to hinder the combustible gases to pass and diffuse into polymer matrix. Certainly, coke formed by decomposed MCC as earlier revealed to be uniformly distributed to the inside and surface of PLA-PBS-MCC composites, that efficiently impeded the discharge of decomposition products of PLA-PBS blends. Similar effect was observed for PLA-PBS-MCC blend [29] and polypropylene (PP)/MCC composites, where MCC effectively hindered the release of decomposition products of the polymers [17,30]. As evident from Table 3, PLA was found to have least residue which increased with the addition of PBS and MCC, due to their lower thermal stabilities as compared to PLA.

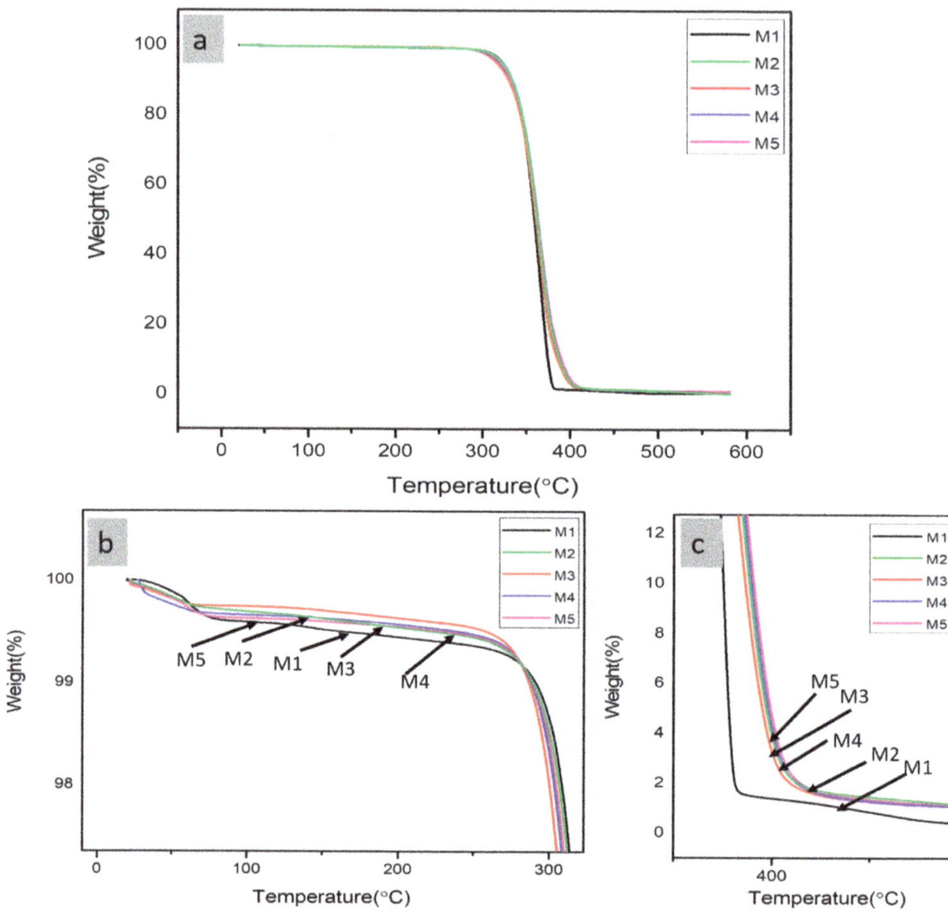

Figure 2. (**a**) TGA curves, (**b**) zoomed initial decomposition curves, (**c**) zoomed final decomposition curves of PLA-PBS-MCC composite films.

Table 3. TGA results of PLA/PBS/MCC composite films

Samples	Ti [a] (°C)	T50 [b]% (°C)	Tmax [c] (°C)	Tf [d] (°C)	Wi [e] (°C)	Wmax [f] (%)	Wfinal [g] (%)	Wresidue [h] (%)
M1	284.81	359.42	365.24	386.09	98.9	27.57	1.698	0.372
M2	286.66	361.197	365.72	416.72	98.95	22.76	2.131	1.044
M3	288.12	362.96	365.10	417.74	97.65	22.04	2.221	1.046
M4	289.22	363.11	366.50	419.65	96.43	21.65	2.163	1.064
M5	289.45	362.48	366.72	420.67	95.88	21.21	2.056	1.099

[a] TGA; initial degradation temperature, [b] TGA;50% degradation temperature, [c] DTG; peak temperature, [d] TGA; final degradation temperature, [e] TGA: initial weight loss, [f] DTG maximum weight loss, [g] TGA; final weight loss, [h] TGA char residue weight.

DTG curves determine the loss of weight and the definite temperature at which material decomposition takes place. The thermal analysis of PLA, PLA-PBS, and PLA-PBS-MCC blend with varying percentages of MCC as evident from the Figure 3a,b. The thermal degradation temperatures increased, and thereby the thermal stability of composites increased as shown in Figure 3. It was found that M1, M2, M3, M4, and M5 displayed single step degradation as evident from Figure 3. From Table 3, W_{max} decreases with the addition of PBS then on addition of MCC it further reduces insignificantly, thus exhibiting improvement of thermal stability with PBS and MCC inclusion in PLA composites.

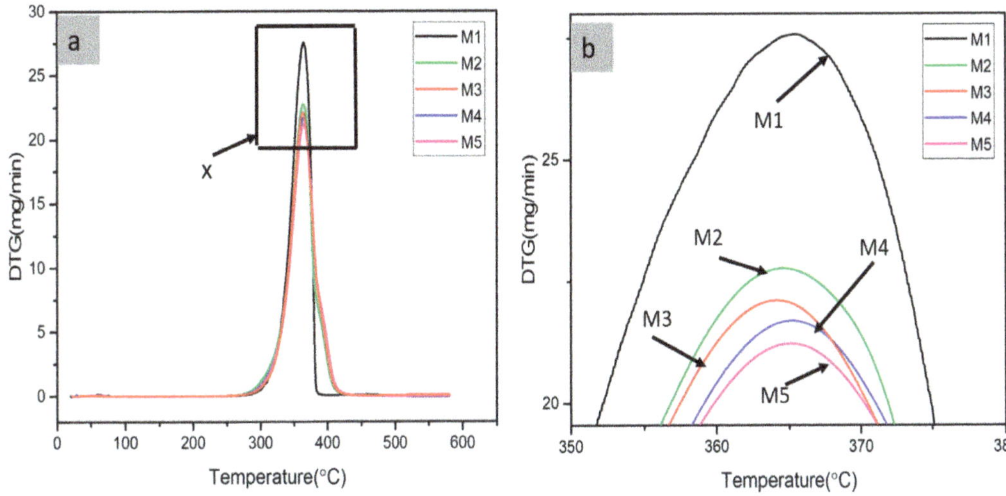

Figure 3. DTG of (**a**) PLA, PLA-PBS, PLA-PBS-MCC composite films, (**b**) zoomed x part.

DSC analysis characterizes the physical nature of the material. The DSC curves displayed two thermal transition temperatures: glass transition temperatures (T_g), and the melting temperatures (T_m) with the absence of crystallization (T_c). The glass transition indicates a change of the degree of freedom of the molecules in the amorphous regions on increasing temperature in the polymer matrix [31]. Figure 4a,b demonstrates the DSC thermograms behavior of the composite films.

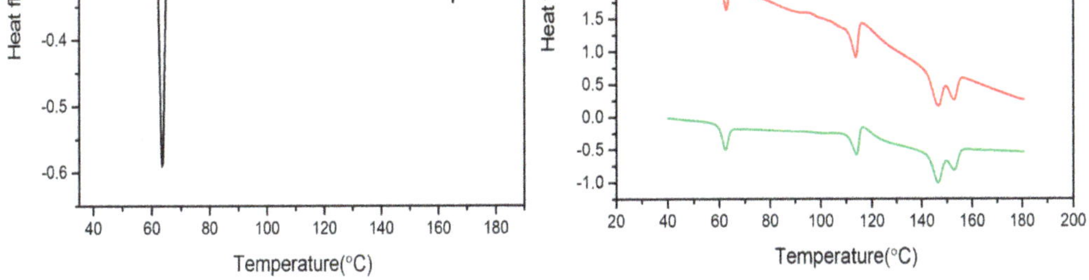

Figure 4. DSC of (**a**) PLA and (**b**) PLA-PBS, PLA-PBS-MCC composite films.

This endothermic peak values for PBS and MCC reinforced composites remain approximately contact however fluctuates to some degrees by the increase in MCC content. This endothermic peak indicates the disorder in the structure. The transition peak of PLA(M1) is obtained at 64 and on further addition of PBS and MCC in case of M2, M3, M4, and

M4 the transition temperature reduces to 62 °C, 63 °C, 63.10 °C and 63.24 °C respectively as mentioned in Table 4. It reveals that the crystals in the pure PLA are more uniformly distributed as in the case of blend and also in the composites due to presence of other components during nucleation leads to crystallites varying in sizes and shapes [32]. This temperature is also relevant to the polymer's transition temperature from glassy state to a rubbery state, i.e., with characteristics such as hard and brittle to a flexible and soft material. The T_g value increases with the addition of the MCC however it reduced with the addition of PBS alone in PLA composites. Also, the melting peak of PBS is obtained around 113 °C in all the composites as shown in Table 4. As per some studies PBS does not induce the crystallization [33] or improve the crystalline nature of PLA [33,34]. The impeding effect of PBS on the crystallinity of PLA was observed during isothermal crystallization around the temperature of 60 °C, indicating that PBS hinders the rate of crystallization of PLA [35]. From Figure 4a neat PLA showed two melting endothermic peaks at 151 °C and 165.14 °C. As evident from the Figure 4, the composites, M2, M3, M4, and M5 also exhibited two separate melting temperature peaks at approximately 146 °C and 153 °C as evident from Table 4 [36].

Table 4. DSC results of PLA, PLA-PBS, PLA-PBS-MCC composites.

Sample	T_g (°C)	T_m (°C)	ΔH (J/g)	T_{m1} (°C)	ΔH (J/g)	T_{m2} (°C)	ΔH (J/g)
M1	64	-	-	151	2.81	165.14	0.32
M2	62	113.94	6.47	146.21	6.58	152.89	4.28
M3	63	113.60	10.68	146.23	6.74	152.73	5.29
M4	63.10	113.51	10.62	146.33	7.77	152.84	2.41
M5	63.24	113.56	9.34	146.01	8.27	152.68	5.61

3.3. *X-ray Diffractometry (XRD)*

The crystalline structures of PLA, PLA-PBS, and PLA-PBS-MCC composite were examined by XRD in order to examine the influence of varying content of MCC on the crystallinity of the PLA matrix. The XRD patterns in 2θ range of 20–30° are evident from Figure 5. The pure PLA and PLA in the compositions displayed wide and indistinct diffraction patterns [37,38], indicating the existence of PLA in amorphous state mostly with small number of crystallites [38]. This may be due to lower rate of crystallization of PLA as a result faster cooling rates during processing [39,40]. Adding the PBS in the PLA resulted into small increment in the peak intensities and also in the crystalline nature of composite blends. Moreover, there was an insignificant co-crystallization among PLA and PBS [41] as also evident from DSC curve in Figure 4. Thus, PLA in the blends was not crystallized as indicated by the absence of PLA diffraction peak in all the four composites (M2, M3, M4, and M5), whereas the diffraction peak of PBS at 22.5° is present in all PLA/PBS composites that increases with increase in MCC content. This can be attributed to the diffraction peaks of both PBS and MCC (2θ = 22.6°) that overlap [42].

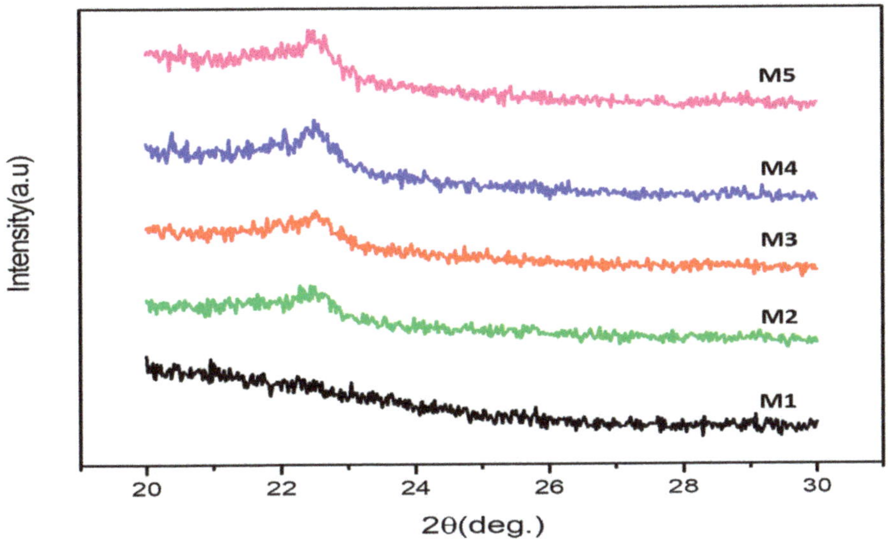

Figure 5. XRD of PLA, PLA-PBS, and PLA-PBS-MCC composite films.

3.4. Fourier Transform Infrared Spectra (FTIR)

FTIR spectra were analyzed to study the chemical structures of composites. The FTIR spectrum of PLA(M1), PLA-PBS(M2), PLA-PBS-MCC (M3, M4, and M5) (Figure 6) displayed characteristic peak at 3750 cm^{-1} (OH stretching vibration) [43,44], and the peak intensity increases with increase in MCC content. The infrared spectra at 2993 represents the stretching vibrations of CH_3 (asymmetric), and 2925 cm^{-1} showed the presence of stretching vibrations of CH_3 (symmetric), respectively in case of PLA and then slowly disappears while as only 2946 cm^{-1}(symmetric stretching vibration of CH_3) peak [44,45], was observed in PLA-PBS and PLA-PBS-MCC (M3, M4, and M5) composites becoming sharper with MCC content. The wide and strong absorption peak at 1753 cm^{-1} can be associated to the presence of C=O stretching of the ester group with the addition of PBS were partially dispersed [29] and it broadens with addition of PBS as well as MCC, while as 1453 cm^{-1} represents bending vibration of CH_3 (asymmetric) in all composites and the asymmetric C–O–C stretching mode at 1184; 1090; and 1047 cm^{-1} [39]. The vibration characteristic of the helical backbone with rocking mode of CH3 was observed at 956.2 cm^{-1} only in case of composites (M2, M3, M4, and M5). The two bands at 869 and 754 cm^{-1} presented the crystalline and amorphous phases of PLA in all compositions. The peak at 869 cm^{-1} could be represents the amorphous phase [46] and the peak at 754 cm^{-1} denotes the crystalline phase [47]. Overall, the peaks (band position) in the spectrum of PLA-PBS blend changed significantly as compared to PLA-PBS-MCC with no significant change, indicating there was no specific chemical reactions between the polymers and MCC. The curves of PLA-PBS-MCC, as shown in Figure 6, were same as that of pure PLA composite. These results show well agreement with the FESEM analysis.

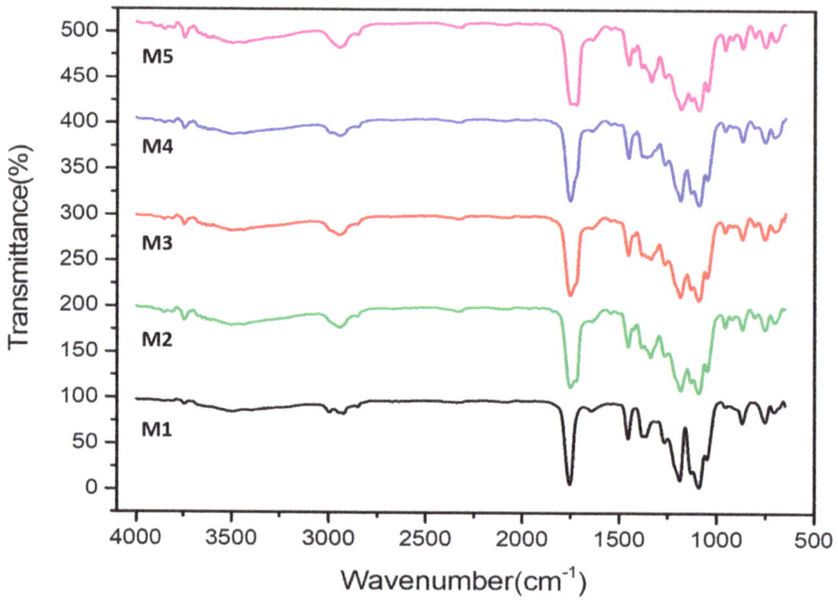

Figure 6. FTIR of PLA, PLA-PBS, and PLA-PBS-MCC composite films.

3.5. Tensile Properties

The tensile strength and tensile modulus of PLA(M1), PLA/PBS(M2), PLA/PBS/MCC (M3, M4, and M5) composites with different MCC content are demonstrated by Figure 7a,b respectively. The tensile strength and the tensile modulus of these composite films decreased on addition of PBS and decreased initially on addition of 0.5 wt% of MCC beyond that it further increases This is due to lower tensile strength and tensile modulus of PBS than PLA. The reduced values of tensile strength may also be attributed to the weak interactions at the interfaces of PLA and PBS, resulting into the inadequate transfer of stress across each polymer phase thereby leading to quicker fracture during the tensile deformation when compared to pure PLA. Also, there was a reduction in brittleness of PLA due to decreased values of modulus from 8500 to 5400 MPa [48]. The decreased value of tensile strength at lower MCC content can be associated with the weak interactions at the interfaces of PLA-PBS matrix and MCC however it increased at higher contents of MCC due to their binding ability. The binder functionality of MCC depends upon its ability to deform plastically on application of compressive force. MCC particles form hydrogen bonds leading to strong compacts and their critical properties corresponding to their functionality as a binder include particle size, moisture content, bulk density, specific surface area, and crystallinity [49]. Similar results of decreased tensile strength were obtained by Ochi [50] in their studies. The increase in tensile strength with higher content of MCC addition in PLA-PBS matrix can be attributed to the interface action of MCC and the penetration of PLA into the network and better interaction between MCC and PLA resin as evident from FESEM images in Figure 1c–e. The increasing of the tensile modulus on addition of MCC beyond 1wt % might be due to the rigid MCC that enhances stiffness of the PLA composites by restricting the molecular movement and the distortion of PLA chains. Luz et al. [51] revealed that insertion of the fibers can increase tensile modulus due to higher modulus of fibers as compared to the modulus of thermoplastics.

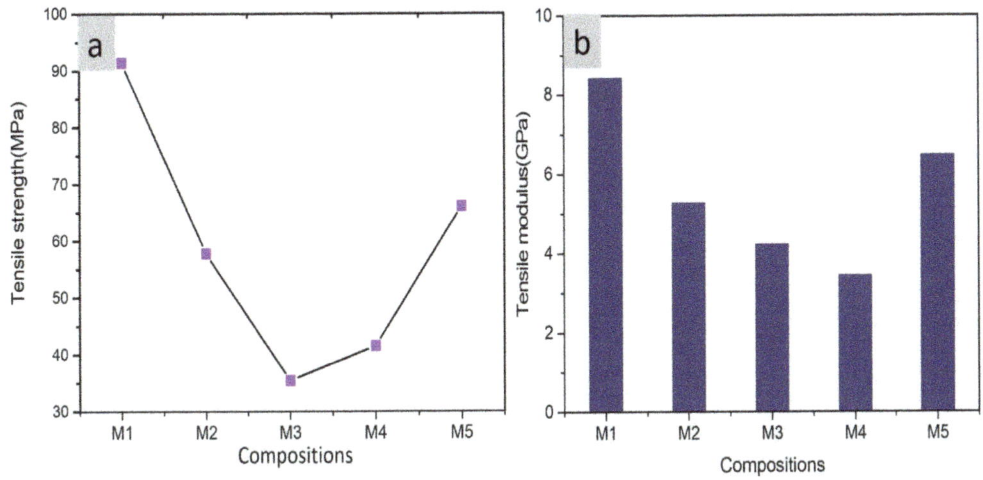

Figure 7. (**a**) Tensile strength (MPa), (**b**) tensile modulus (GPa), of PLA, PLA-PBS, and PLA-PBS-MCC composite films.

The elongation at break of the composites is presented in Figure 8. It reveals that the elongation at break increases on addition of PBS and further decreases with addition of MCC insignificantly. This effect is due to the high flexibility of PBS within the PLA matrix resulting into decrease in the brittleness of the blends [52]. However, MCC inclusion reduces elongation at break but have values greater than pure PLA. Similar effect of MCC extracted from cotton waste on the PLA-PBS blend was reported by Chaiwutthinan et al. [42]. Overall, the elongation at break of the PLA composite films improves.

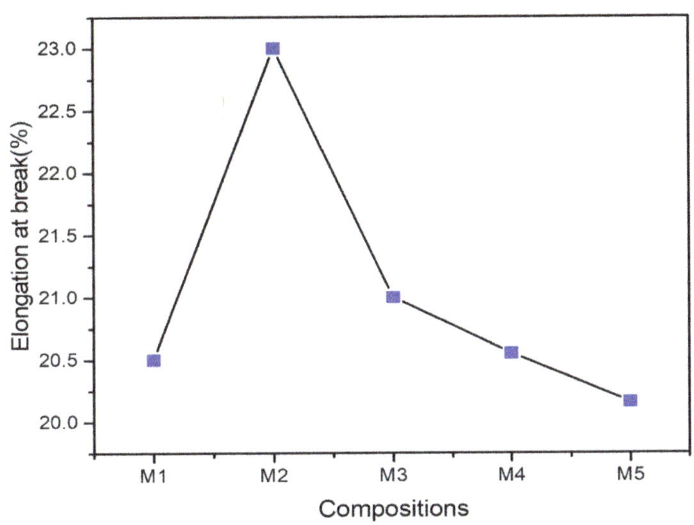

Figure 8. Elongation at break(mm) of PLA, PLA-PBS, and PLA-PBS-MCC composite films

3.6. SEM Analysis of Fractured Surfaces

The tensile properties of the composites also depend upon the compatibility of PLA with PBS and MCC, they also depend on the behavior of composites when loaded. We studied the microstructures of fractured surfaces of the composites as shown in Figure 9. The microstructure indicated the dispersion and compatibility in the composites. SEM morphology of the surfaces(fractured) during tensile testing of the pure PLA as shown in Figure 9a exhibited a clear even-fractured surface and thus showed a nature of brittle material, whereas the fractured surfaces of PLA-PBS blends are rougher as evident from Figure 9b–e, suggesting more flexibility of the samples. This can be due to presence of PBS with higher toughness. The interfaces between PLA and PBS in all the blends were clear as observed in Figure 9. This reveals the immiscibility of the blend [53]. Moreover, the blends with relatively low MCC contents exhibited more interfacial debonding as a result some PBS particles are get removed completely from the composites during the fracture process as evident from Figure 9b–e. In addition, PBS particles in the composites without MCC, exhibited non-uniform distribution within PLA matrix, and thus the gaps present at the interface of PLA and PBS can enhanced water penetration by forming permeation channels that can further facilitate the hydrolytic degradation [54].

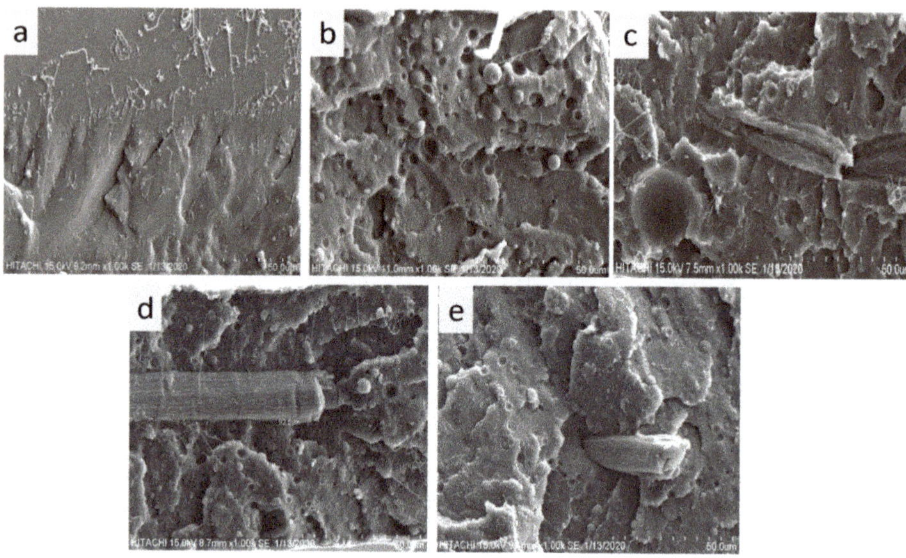

Figure 9. SEM morphology of the fractured surfaces of composite films: (**a**) M1, (**b**) M2, (**c**) M3, (**d**) M4, (**e**) M5.

As evident from Figure 9c, the PLA-PBS-MCC cross section with 0.5 wt% MCC in the PLA-PBS matrix show the rod like structure of MCC was pulled out due to fracturing of the composite leaving a hole on the surface. These exposed MCC rods also indicate poor bonding interface thus reason of deterioration in performance of the composite. The PLA-PBS composite films with MCC exhibited a better compatibility of PLA with PBS in comparison to the composite films without MCC as evident from Figure 9b–e. However, there are MCC pulled outs that are clearly visible. On addition of more MCC (above 0.5 wt%), the MCC rods showed a well-defined and sharp edge as evident from Figure 9d,e. Also, PLA-PBS-MCC composites with MCC content more than 0.5 wt% the cross-sectional surface consists of least number of voids in comparison to composites with lower MCC content. Moreover, observed relatively compact and smooth cross-sectional surface for M4 and M5 as evident from Figure 9d,e. Therefore, there was increase in binding force of the composites on increasing concentration of MCC indicating that MCC modified the surface successfully resulting into enhanced interfacial tension [45]. On addition of 1.5 wt% of MCC, an additional fractured rod-shaped MCC were observed (Figure 9e); indicating the improvement in the cohesive forces at the interface of MCC and the PLA matrix; thus the matrix interface became more miscible as a result increases the compatibility of PLA with

PBS. Thus, the incorporation of MCC improvises the interfacial complications between PLA, PBS, and MCC thereby improving mechanical properties of the composites.

4. Conclusions

In this research work, PLA, PLA-PBS, and PLA-PBS-MCC (0.5, 1, and 1.5 wt%) composites, where MCC extracted from bamboo fiber, were successfully prepared by hot pressing technique. Although PBS revealed insignificant impact on PLA's crystallinity and mechanical properties however it changed PLA from brittle to flexible material that was further enhanced by the addition of MCC. In addition, MCC improved the interfacial bonding between PLA and PBS in the composites. Furthermore, MCC enhanced the crystallinity, thermal stability as well as the tensile strength of the PLA-PBS blends. The tensile strength and tensile modulus of PLA-PBS and PLA-PBS-MCC were lower at low MCC content when compared with the pure PLA composites however the elongation at break of PLA-PBS and PLA-PBS-MCC was found to be higher than that of PLA. The SEM analysis of surfaces (fractured) reveal the better bonding of MCCs in the composites containing MCC and thus exhibit better mechanical properties. Thus, the PBS and MCC-reinforced PLA matrix composites can be regarded as a potential material for packaging applications.

Author Contributions: Conceptualization, M.R., M.J., and A.H.B.; Methodology, M.R. and M.J.; Validation, M.J., A.H.B., and S.A.; Formal analysis and investigation, M.R. and B.P.; Writing—original draft preparation, M.R.; Writing—review and editing, M.J., S.A., B.P., and A.H.B.; Supervision and project administration, M.J.; Funding acquisition, M.J. All authors have read and agreed to the published version of the manuscript.

Funding: The authors are grateful to Malaysian Industry-Government Group for High Technology (MIGHT) for Financial support of this work under Newton-Ungku Omar Fund (grant no. 6300873). The authors would like to extend their gratitude to King Saud University (Riyadh, Saudi Arabia) for the funding of this research through Researchers Supporting Project number (RSP-2020/241).

Informed Consent Statement: Not applicable.

Data Availability Statement: Not applicable.

Conflicts of Interest: The authors declare no conflict of interest.

References

1. Fang, Q.; Hanna, M.A. Rheological properties of amorphous and semicrystalline polylactic acid polymers. *Ind. Crop. Prod.* **1999**, *10*, 47–53. [CrossRef]
2. Ogata, N.; Jimenez, G.; Kawai, H.; Ogihara, T. Structure and thermal/mechanical properties of poly(l-lactide)-clay blend. *J. Polym. Sci. Part B Polym. Phys.* **1997**, *35*, 389–396. [CrossRef]
3. Armentano, I.; Fortunati, E.; Burgos, N.; Dominici, F.; Luzi, F.; Fiori, S.; Jiménez, A.; Yoon, K.; Ahn, J.; Kang, S.; et al. Processing and characterization of plasticized PLA/PHB blends for biodegradable multiphase systems. *Express Polym. Lett.* **2015**, *9*, 583–596. [CrossRef]
4. Bai, H.; Huang, C.; Xiu, H.; Gao, Y.; Zhang, Q.; Fu, Q. Toughening of poly(l-lactide) with poly(ε-caprolactone): Combined effects of matrix crystallization and impact modifier particle size. *Polymer* **2013**, *54*, 5257–5266. [CrossRef]
5. Mittal, V.; Akhtar, T.; Matsko, N. Mechanical, Thermal, Rheological and Morphological Properties of Binary and Ternary Blends of PLA, TPS and PCL. *Macromol. Mater. Eng.* **2015**, *300*, 423–435. [CrossRef]
6. Bhatia, A.; Gupta, R.K.; Bhattacharya, S.N.; Choi, H.J. Compatibility of biodegradable poly (lactic acid) (PLA) and poly (butylene succinate) (PBS) blends for packaging application. *Korea Aust. Rheol. J.* **2007**, *19*, 125–131.
7. Liu, X.; Dever, M.; Fair, N.; Polymer, R.B.-J.U. *Thermal and Nechanical Properties of Poly (Lactic Acid) and Poly (Ethylene/Butylene Succinate) Blends*; Springer: Berlin, Germany, 1997; Volume 5, pp. 225–235.
8. Yokohara, T.; Yamaguchi, M. Structure and properties for biomass-based polyester blends of PLA and PBS. *Eur. Polym. J.* **2008**, *44*, 677–685. [CrossRef]
9. Homklin, R.; Hongsriphan, N. Mechanical and Thermal Properties of PLA/PBS Co-continuous Blends Adding Nucleating Agent. *Energy Procedia* **2013**, *34*, 871–879. [CrossRef]
10. Lin, N.; Huang, J.; Chang, P.R.; Feng, J.; Yu, J. Surface acetylation of cellulose nanocrystal and its reinforcing function in poly(lactic acid). *Carbohydr. Polym.* **2011**, *83*, 1834–1842. [CrossRef]

11. Lu, T.; Liu, S.; Jiang, M.; Xu, X.; Wang, Y.; Gou, J.; Hui, D.; Chen, X. Effects of modifications of bamboo cellulose fibers on the improved mechanical properties of cellulose reinforced poly(lactic acid) composites. *Compos. Part B Eng.* **2014**, *62*, 191–197. [CrossRef]
12. Haldar, D.; Purkait, M. Micro and nanocrystalline cellulose derivatives of lignocellulosic biomass: A review on synthesis, applications and advancements. *Carbohydr. Polym.* **2020**, *250*. [CrossRef] [PubMed]
13. Das, K.; Ray, D.; Bandyopadhyay, N.R.; Sengupta, S. Study of the Properties of Microcrystalline Cellulose Particles from Different Renewable Resources by XRD, FTIR, Nanoindentation, TGA and SEM. *J. Polym. Environ.* **2010**, *18*, 355–363. [CrossRef]
14. Elsakhawy, M.; Hassan, M. Physical and mechanical properties of microcrystalline cellulose prepared from agricultural residues. *Carbohydr. Polym.* **2007**, *67*, 1–10. [CrossRef]
15. Jahan, M.S.; Saeed, A.; He, Z.; Ni, Y. Jute as raw material for the preparation of microcrystalline cellulose. *Cellular* **2010**, *18*, 451–459. [CrossRef]
16. Courgneau, C.; Domenek, S.; Guinault, A.; Avérous, L.; Ducruet, V. Analysis of the Structure-Properties Relationships of Different Multiphase Systems Based on Plasticized Poly(Lactic Acid). *J. Polym. Environ.* **2011**, *19*, 362–371. [CrossRef]
17. Shih, Y.-F.; Wang, T.Y.; Jeng, R.J.; Wu, J.Y.; Teng, C.C. Biodegradable nanocomposites based on poly(butylene succinate)/organoclay. *J. Polym. Environ.* **2007**, *15*, 151–158. [CrossRef]
18. Zhou, J.; Wang, X.; Hua, K.; Duan, C.; Zhang, W.; Ji, J.-H.; Yang, X. Enhanced mechanical properties and degradability of poly(butylene succinate) and poly(lactic acid) blends. *Iran. Polym. J.* **2013**, *22*, 267–275. [CrossRef]
19. Rico, M.; Rodríguez-Llamazares, S.; Barral, L.; Bouza, R.; Montero, B. Processing and characterization of polyols plasticized-starch reinforced with microcrystalline cellulose. *Carbohydr. Polym.* **2016**, *149*, 83–93. [CrossRef]
20. Liu, Y.; Li, Y.; Chen, H.M.; Yang, G.; Zheng, X.; Zhou, S. Water-induced shape-memory poly(d,l-lactide)/microcrystalline cellulose composites. *Carbohydr. Polym.* **2014**, *104*, 101–108. [CrossRef]
21. Cao, Z.; Lu, Y.; Zhang, C.; Zhang, Q.; Zhou, A.; Hu, Y.; Wu, D.; Tao, G.; Gong, F.; Ma, W.; et al. Effects of the chain-extender content on the structure and performance of poly(lactic acid)-poly(butylene succinate)-microcrystalline cellulose composites. *J. Appl. Polym. Sci.* **2017**, *134*, 1–8. [CrossRef]
22. Amado, C.M.; Minahk, C.J.; Cilli, E.; Oliveira, R.G.; Dupuy, F.G. Bacteriocin enterocin CRL35 is a modular peptide that induces non-bilayer states in bacterial model membranes. *Biochim. Biophys. Acta (BBA) Biomembr.* **2020**, *1862*. [CrossRef]
23. Rasheed, M.; Jawaid, M.; Karim, Z.; Abdullah, L.C. Morphological, Physiochemical and Thermal Properties of Microcrystalline Cellulose (MCC) Extracted from Bamboo Fiber. *Molecules* **2020**, *25*, 2824. [CrossRef] [PubMed]
24. Jiang, L.; Zhang, J.; Wolcott, M.P. Comparison of polylactide/nano-sized calcium carbonate and polylactide/montmorillonite composites: Reinforcing effects and toughening mechanisms. *Polymer* **2007**, *48*, 7632–7644. [CrossRef]
25. Hassan, E.A.; Elarabi, S.E.; Wei, Y.; Yu, M. Biodegradable poly (lactic acid)/poly (butylene succinate) fibers with high elongation for health care products. *Text. Res. J.* **2017**, *88*, 1735–1744. [CrossRef]
26. Thiangtham, S.; Runt, J.; Saito, N.; Manuspiya, H. Fabrication of biocomposite membrane with microcrystalline cellulose (MCC) extracted from sugarcane bagasse by phase inversion method. *Cellular* **2019**, *27*, 1367–1384. [CrossRef]
27. Luzi, F.; Fortunati, E.; Jiménez, A.; Puglia, D.; Pezzolla, D.; Gigliotti, G.; Kenny, J.M.; Chiralt, A.; Torre, L. Production and characterization of PLA_PBS biodegradable blends reinforced with cellulose nanocrystals extracted from hemp fibres. *Ind. Crop. Prod.* **2016**, *93*, 276–289. [CrossRef]
28. Zhang, X.; Shi, J.; Ye, H.-M.; Dong, Y.; Zhou, Q. Combined effect of cellulose nanocrystals and poly(butylene succinate) on poly(lactic acid) crystallization: The role of interfacial affinity. *Carbohydr. Polym.* **2018**, *179*, 79–85. [CrossRef]
29. Chaiwutthinan, P.; Pimpan, V.; Chuayjuljit, S.; Leejarkpai, T. Biodegradable Plastics Prepared from Poly(lactic acid), Poly(butylene succinate) and Microcrystalline Cellulose Extracted from Waste-Cotton Fabric with a Chain Extender. *J. Polym. Environ.* **2015**, *23*, 114–125. [CrossRef]
30. Xiuju, Z.; Juncai, S.; Huajun, Y.; Zhidan, L.; Shaozao, T. Mechanical Properties, Morphology, Thermal Performance, Crystallization Behavior, and Kinetics of PP/Microcrystal Cellulose Composites Compatibilized by Two Different Compatibilizers. *J. Thermoplast. Compos. Mater.* **2011**, *24*, 735–754. [CrossRef]
31. Baskaran, R.; Selvasekarapandian, S.; Kuwata, N.; Kawamura, J.; Hattori, T. Conductivity and thermal studies of blend polymer electrolytes based on PVAc–PMMA. *Solid State Ionics* **2006**, *177*, 2679–2682. [CrossRef]
32. Mandal, A.; Chakrabarty, D. Studies on the mechanical, thermal, morphological and barrier properties of nanocomposites based on poly(vinyl alcohol) and nanocellulose from sugarcane bagasse. *J. Ind. Eng. Chem.* **2014**, *20*, 462–473. [CrossRef]
33. Robeson, L.M. Fundamentals of Polymer Blends. *Polym. Blends* **2007**, 11–64. [CrossRef]
34. Diani, J.; Gall, K. Toughening modification of PLLA/PBS blendsviai in situ compatibilization. *Polym. Eng. Sci.* **2009**, 26–33. [CrossRef]
35. Ji, D.; Liu, Z.; Lan, X.; Wu, F.; Xie, B.; Yang, M. Morphology, rheology, crystallization behavior, and mechanical properties of poly(lactic acid)/poly(butylene succinate)/dicumyl peroxide reactive blends. *J. Appl. Polym. Sci.* **2013**, *131*, 1–8. [CrossRef]
36. Silverajah, V.S.G.; Ibrahim, N.A.; Yunus, W.M.Z.W.; Abu Hassan, H.; Woei, C.B. A Comparative Study on the Mechanical, Thermal and Morphological Characterization of Poly(lactic acid)/Epoxidized Palm Oil Blend. *Int. J. Mol. Sci.* **2012**, *13*, 5878–5898. [CrossRef]
37. Li, L.; Song, G.; Tang, G. Novel Biodegradable Polylactide/Poly(butylene succinate) Composites via Cross-Linking with Methylene Diphenyl Diisocyanate. *Polym. Technol. Eng.* **2013**, *52*, 1183–1187. [CrossRef]

38. Yeh, J.; Tsou, C.-H.; Li, Y.-M.; Xiao, H.-W.; Wu, C.-S.; Chai, W.-L.; Lai, Y.-C.; Wang, C.-K. The compatible and mechanical properties of biodegradable poly(Lactic Acid)/ethylene glycidyl methacrylate copolymer blends. *J. Polym. Res.* **2012**, *19*. [CrossRef]
39. Zheng, M.; Luo, X. Phase Structure and Properties of Toughened Poly(L-Lactic Acid)/Glycidyl Methacrylate Grafted Poly(Ethylene Octane) Blends Adjusted by the Stereocomplex. *Polym. Technol. Eng.* **2013**, *52*, 1250–1258. [CrossRef]
40. Zou, H.; Yi, C.; Wang, L.; Xu, W. Crystallization, hydrolytic degradation, and mechanical properties of poly (trimethylene terephthalate)/poly(lactic acid) blends. *Polym. Bull.* **2009**, *64*, 471–481. [CrossRef]
41. Park, J.W.; Im, S.S. Phase behavior and morphology in blends of poly(L-lactic acid) and poly(butylene succinate). *J. Appl. Polym. Sci.* **2002**, *86*, 647–655. [CrossRef]
42. Chaiwutthinan, P.; Song, Z.H.; Leejarkpai, T.; Chuayjuljit, S. Use of Microcrystalline Cellulose Prepared from Cotton Fabric Waste to Prepare Poly(butylene succinate) Composites. *Adv. Mater. Res.* **2011**, *356–360*, 430–434. [CrossRef]
43. Pamuła, E.; Błażewicz, M.; Paluszkiewicz, C.; Dobrzyński, P. FTIR study of degradation products of aliphatic polyesters–carbon fibres composites. *J. Mol. Struct.* **2001**, *596*, 69–75. [CrossRef]
44. Kemala, T.; Budianto, E.; Soegiyono, B. Preparation and characterization of microspheres based on blend of poly(lactic acid) and poly(ϵ-caprolactone) with poly(vinyl alcohol) as emulsifier. *Arab. J. Chem.* **2012**, *5*, 103–108. [CrossRef]
45. Qu, P.; Gao, Y.; Wu, G.F.; Zhang, L.P. Nanocomposites of Poly(lactic acid) reinforced with cellulose nanofibrils. *BioResources* **2010**, *5*, 1811–1823. [CrossRef]
46. Garlotta, D. A Literature Review of Poly(Lactic Acid). *J. Polym. Environ.* **2001**, *9*, 63–84. [CrossRef]
47. Yang, Z.; Peng, H.; Wang, W.; Liu, T. Crystallization behavior of poly(ε-caprolactone)/layered double hydroxide nanocomposites. *J. Appl. Polym. Sci.* **2010**, *116*, 2658–2667. [CrossRef]
48. Chaiwutthinan, P.; Leejarkpai, T.; Kashima, D.P.; Chuayjuljit, S. Poly(Lactic Acid)/Poly(Butylene Succinate) Blends Filled with Epoxy Functionalised Polymeric Chain Extender. *Adv. Mater. Res.* **2013**, *664*, 644–648. [CrossRef]
49. Thoorens, G.; Krier, F.; Leclercq, B.; Carlin, B.; Pestieau, A. Microcrystalline cellulose, a direct compression binder in a quality by design environment—A review. *Int. J. Pharm.* **2014**, *473*, 64–72. [CrossRef]
50. Ochi, S. Mechanical properties of kenaf fibers and kenaf/PLA composites. *Mech. Mater.* **2008**, *40*, 446–452. [CrossRef]
51. Luz, S.; Gonçalves, A.R.; Del'Arco, A. Mechanical behavior and microstructural analysis of sugarcane bagasse fibers reinforced polypropylene composites. *Compos. Part A Appl. Sci. Manuf.* **2007**, *38*, 1455–1461. [CrossRef]
52. Bhatia, A.; Gupta, R.K.; Bhattacharya, S.N.; Choi, H.J. Effect of Clay on Thermal, Mechanical and Gas Barrier Properties of Biodegradable Poly(lactic acid)/Poly(butylene succinate) (PLA/PBS) Nanocomposites. *Int. Polym. Process.* **2010**, *25*, 5–14. [CrossRef]
53. Shibata, M.; Inoue, Y.; Miyoshi, M. Mechanical properties, morphology, and crystallization behavior of blends of poly(l-lactide) with poly (butylene succinate-co-l-lactate) and poly (butylene succinate). *Polymer* **2006**, *47*, 3557–3564. [CrossRef]
54. Wang, Y.-P.; Xiao, Y.-J.; Duan, J.; Yang, J.-H.; Wang, Y.; Zhang, C.-L. Accelerated hydrolytic degradation of poly(lactic acid) achieved by adding poly(butylene succinate). *Polym. Bull.* **2015**, *73*, 1067–1083. [CrossRef]

polymers

MDPI

Article

Characterization of Microcrystalline Cellulose Isolated from Conocarpus Fiber

H. Fouad [1,2], Lau Kia Kian [3], Mohammad Jawaid [3,*], Majed D. Alotaibi [4], Othman Y. Alothman [5,*] and Mohamed Hashem [6]

1 Applied Medical Science Department, Community College, King Saud University, P.O. Box 10219, Riyadh 11433, Saudi Arabia; menhfef@ksu.edu.sa
2 Biomedical Engineering Department, Faculty of Engineering, Helwan University, Cairo 11795, Egypt
3 Institute of Tropical Forestry and Forest Products (INTROP), Universiti Putra Malaysia (UPM), Serdang 43400, Malaysia; laukiakian@gmail.com
4 Life Science and Environment Research Institute, King Abdulaziz City for Science and Technology (KACST), Riyadh 11442, Saudi Arabia; mdalotaibi@kacst.edu.sa
5 Department of Chemical Engineering, College of Engineering, King Saud University, P.O. Box 22452, Riyadh 11451, Saudi Arabia
6 College of Applied Medical Sciences, King Saud University, P.O. Box 10219, Riyadh 11433, Saudi Arabia; mihashem@ksu.edu.sa
* Correspondence: jawaid@upm.edu.my (M.J.); othman@ksu.edu.sa (O.Y.A.)

Received: 6 November 2020; Accepted: 3 December 2020; Published: 7 December 2020

Abstract: Conocarpus fiber is an abundantly available and sustainable cellulosic biomass. With its richness in cellulose content, it is potentially used for manufacturing microcrystalline cellulose (MCC), a cellulose derivative product with versatile industrial applications. In this work, different samples of bleached fiber (CPBLH), alkali-treated fiber (CPAKL), and acid-treated fiber (CPMCC) were produced from Conocarpus through integrated chemical process of bleaching, alkaline cooking, and acid hydrolysis, respectively. Characterizations of samples were carried out with Scanning Electron Microscope (SEM), Energy Dispersive X-ray (EDX), Fourier Transform Infrared-Ray (FTIR), X-ray Diffraction (XRD), Thermogravimetric (TGA), and Differential Scanning Calorimetry (DSC). From morphology study, the bundle fiber feature of CPBLH disintegrated into micro-size fibrils of CPMCC, showing the amorphous compounds were substantially removed through chemical depolymerization. Meanwhile, the elemental analysis also proved that the traces of impurities such as cations and anions were successfully eliminated from CPMCC. The CPMCC also gave a considerably high yield of 27%, which endowed it with great sustainability in acting as alternative biomass for MCC production. Physicochemical analysis revealed the existence of crystalline cellulose domain in CPMCC had contributed it 75.7% crystallinity. In thermal analysis, CPMCC had stable decomposition behavior comparing to CPBLH and CPAKL fibers. Therefore, Conocarpus fiber could be a promising candidate for extracting MCC with excellent properties in the future.

Keywords: microcrystalline cellulose; Conocarpus fiber; morphology; crystallinity; thermal stability

1. Introduction

Development of novel bio-based materials from waste biomass has become the focus of scientists and researchers to mitigate climate change. Agro-wastes are an abundantly available and inexpensive source of cellulose. The utilization of these biomasses is regarded as a dynamic approach that contributes to the establishment of a sustainable and cleaner environment [1,2]. *Conocarpus lancifolius* is a plant found natively in Somalia and Yemen. Today, it is widely cultivated across the Arabian Peninsula, North-East Africa, as well as some tropical coastal regions in North and South America [3,4].

It is commonly cultivated for landscaping and greening purposes in urban areas. Traditionally, Conocarpus had been used for medicinal and textile applications attributing to its anti-hypertension and anti-inflammatory properties. However, large agro-waste can be generated from Conocarpus plant due to improper management [5,6]. Conocarpus trunk fiber generally contains large amounts of lignocellulosic components, serving as the major cellulose source. This endows it with great potential for manufacturing cellulose derivative product such as microfibril cellulose and nanofibril cellulose.

Microcrystalline cellulose (MCC) is a depolymerized cellulose byproduct generated via the hydrolysis of inorganic acid. Generally, it possesses a length of more than 1 μm and diameter of 50–300 μm, as well as fascinating physical properties like stable thermal behavior, excellent reinforcing capability, and biocompatibility [7,8]. These have made it widely applicable in paper production as well as acting filler in composite materials. The thermal stability, particle size, and reinforcing capability of MCC depend significantly of the feedstock characteristics as well as the processing conditions such as acids concentration, acids type, hydrolysis conditions, and pre-treatment reactions [9]. MCC is employed as adsorbent, binder, and excipient in pharmaceutical and cosmetic industry. Moreover, in food and beverage industries, they act as additives, emulsifiers, stabilizers, thickeners, gelling agents, fat substitutes, suspending agents, and anti-caking agents. In addition, they are well known as reinforcing agents in composite production fields [10,11].

From literature studies, there is a lack of reports on the MCC isolation from Conocarpus fibers. Hence, this study aims to extract MCC from Conocarpus fiber through the chemical process of bleaching, alkali, and acid hydrolysis treatments. The obtained chemically treated fibers were examined for their morphology with microscopy and particle size analysis. Meanwhile, the functional chemistry, crystallinity, and thermal stability were also analyzed to comprehensively study the extracted fiber product.

2. Materials and Method

2.1. Materials and Chemicals

Conocarpus fiber was collected from Riyadh, Saudi Arabia. The fiber was grinded into small pieces of about 0.5–1 cm and dried in oven for 24 hr. Chemical reagents of sodium hydroxide (NaOH), acetic acid ($C_2H_4O_2$), sodium chlorite ($NaClO_2$), and hydrochloric acid (HCl) were purchased from R & M Sdn. Bhd, Kajang, Malaysia.

2.2. MCC Extraction

The grounded fiber was firstly treated with 500 mL of 2% $NaClO_2$ (acidified with 5 mL $C_2H_4O_2$) for 2 hrs at 80 °C with constant stirring, and this experimental setting was found optimal to produce clearly white color fiber from Conocarpus with the great removal lignin compound following trial and fail approach. Then, the filtrate containing lignin was removed, and the solid $NaClO_2$-treated fiber (CPBLH) residue was collected by filtration with distilled water through nylon cloth. After that, the fiber was treated with 500 mL of 5% NaOH for 5 hr at 80 °C in der to swell up both cellulose and hemicellulose components which could aid in the following acid hydrolysis for depolymerization process. The collected NaOH-treated fiber (CPAKL) residue was neutralized with distilled water to pH 7 through nylon cloth until showing white color. In further acid hydrolysis treatment, the process was carried out at 80 °C for 30 min using 2.5 M of HCl solution with the aim to depolymerize the fiber into smaller particles. The resulting solution was then quenched for its acidic reaction with 10 times of cold distilled water and subsequently neutralized to around pH 3 through centrifugation. Finally, the acid-treated fiber (CPMCC) with soft pulpy structure was obtained after filtration and drying.

2.3. Characterization

2.3.1. Functional Chemistry Analysis

The functional chemistry of fiber samples was studied with Thermo Nicolet Nexus 670 FTIR analyzer (Thermo Scientific, Woodland, CA, USA) in the range of 400–4000 cm^{-1} wavenumbers, and resolution was maintained at 4 cm^{-1}. The fibers were pelletized after grinding with potassium bromide (KBr) before analysis.

2.3.2. Crystallinity Analysis

A PANalytical Empyrean X-ray diffractometer (XRD) (Malvern Panalytical B.V., Brighton, UK) was employed to examine the fibers with Cu Kα radiation. The X-ray generator was operating at 40 mA and 45 kV with slow scan rate of 2° per min, and the samples were placed on nickel coated steel holder. Crystallinity of fibers was calculated using the Segal method with Equation (1) as below:

$$Crystalinity,\ CrI\ (\%) = \frac{\left(I_{mp} - I_{am}\right)}{I_{mp}} \times 100\% \tag{1}$$

where I_{mp} refers to maximum point intensity at 22.7°, and I_{am} refers to minimum point intensity at 18.50°, following reported work by French and Cintron [12].

2.3.3. Morphology, Particle Size, Elements, and Yield Analyses

Morphology was examined for the samples through JEOL, JSM-7610F Scanning Electron Microscopy (SEM) (JEOL Ltd., Tokyo, Japan) with 1–15 kV voltage ranges as the fibers were non-conductive. Before viewing, the fibers were coated with a metal palladium layer prior to loading onto the stub. Particle size analysis was performed for the fibers via ImageJ software. Moreover, elements were examined through Energy Dispersive X-ray (EDX) analysis under an acceleration voltage of 20 kV with working distance between 14.5–15.5 mm. Meanwhile, the yield (%) of fibers was determined with the given Equation (2) as below:

$$Yield\ (\%) = \frac{(M_2 - M_3)}{M_1} \times 100\% \tag{2}$$

where M_1 is the mass of raw Conocarpus fibers; M_2 is the total mass of treated-fiber in weighing container; M_3 is the mass of weighing container.

2.3.4. Thermal Stability Analysis

Thermal stability of fiber samples was assessed using a Thermogravimetric analyzer model TGA/SDTA 851e (Mettler-Toledo International Inc., Columbus, OH, USA). Thermal analysis was conducted in range of 30–900 °C with heating rate of 10 °C/min under nitrogen purge atmosphere, while Differential scanning calorimetry (DSC) was carried out at 30–600 °C at 10 °C/min heating rate by using DSC 822 analyzer (Mettler-Toledo International Inc., USA).

3. Results and Discussion

3.1. Morphology and Elements Analysis

Figure 1 presents the morphology of extracted fiber samples. CPBLH exhibited fairly smooth structure, indicating impurities of hemicellulose and lignin were substantially removed from the cellulose [13]. Meanwhile, the irregular shape of the bundle structure (red circle) was also observed for CPBLH fiber, further proving that the dissolution of fiber occurred in this bleaching treatment. For CPAKL, it showed a more ruptured and irrigated-like surface feature (blue arrow), evidencing the cellulose had undergone a swelling process during alkaline treatment [9,14].

Furthermore, CPMCC exhibited a smaller size structure compared to CPBLH and CPAKL. During acid hydrolysis treatment, hydronium ions penetrated the internal cellulose structure and broke down the weak amorphous part. Subsequently, depolymerization process separated the fiber into short fibril-like CPMCC product. However, CPMCC showed some aggregated particles (yellow circle), probably driven by the great surface area of MCC particles, which had higher tendency towards agglomeration [10,11]. In this work, the extracted CPMCC had a 60–110 μm diameter and 250–500 μm length. This could favor it with high surface reactivity for polymer fabrication process [1].

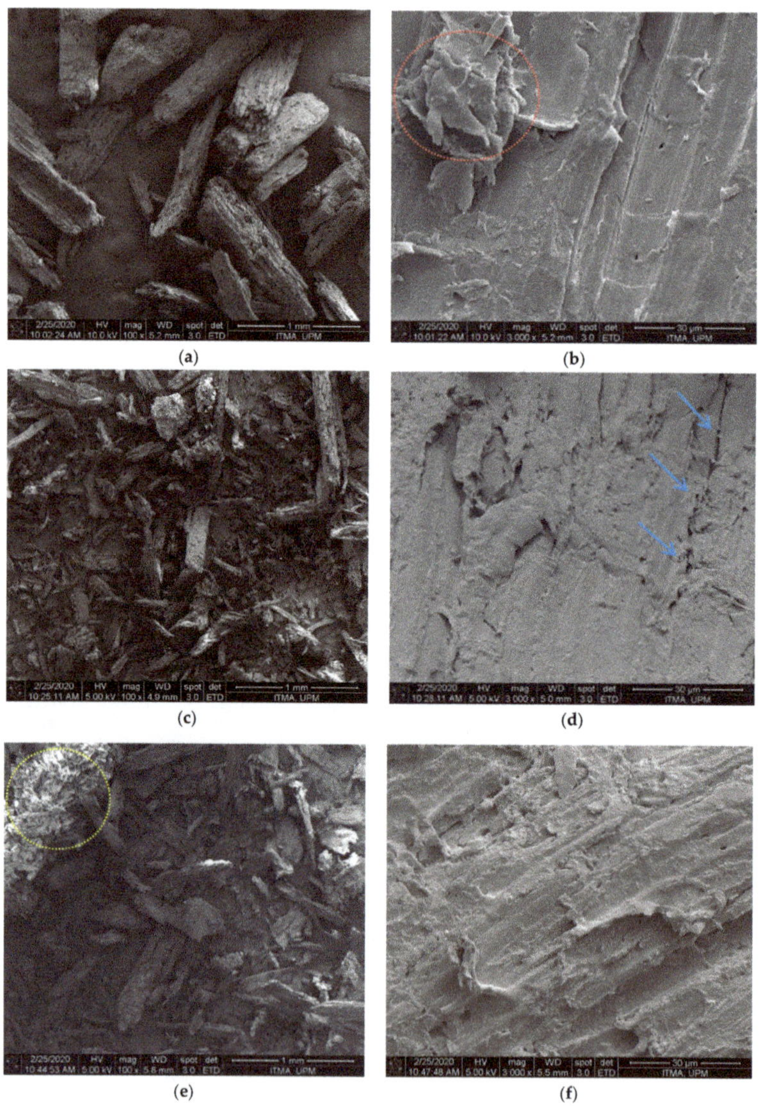

Figure 1. Micrographs of (**a**,**b**) bleached fibrefiber (CPBLH), (**c**,**d**) alkali-treated fibrefiber (CPAKL), and (**e**,**f**) microcrystalline cellulose (CPMCC) with magnifications of ×100 (**a**,**c**,**e**) and ×3000 (**b**,**d**,**f**) under SEM viewing.

From EDX spectra (Figure 2), all samples revealed carbon and oxygen elements, which corresponded to the typical composition of cellulosic compounds. Meanwhile, there was unexpected presence of antimony and fluoride impurities for CPBLH and CPAKL, respectively. It was probably due to the unremoved residual compounds after alkaline and bleaching process that had accumulated the atomic ions through electrostatic attraction. Additionally, the detected zirconium and niobium in both CPAKL and CPMCC samples resulted from the metal coating effect during specimen preparation. As evaluated from Table 1, the built-up of antimony and fluoride impurities was extremely low for CPBLH and CPAKL after excluding zirconium and niobium elements. Moreover, the impurities were not found and could be negligible for CPMCC sample, implying the purity of cellulose content in CPMCC sample was high following acid hydrolysis treatment [2].

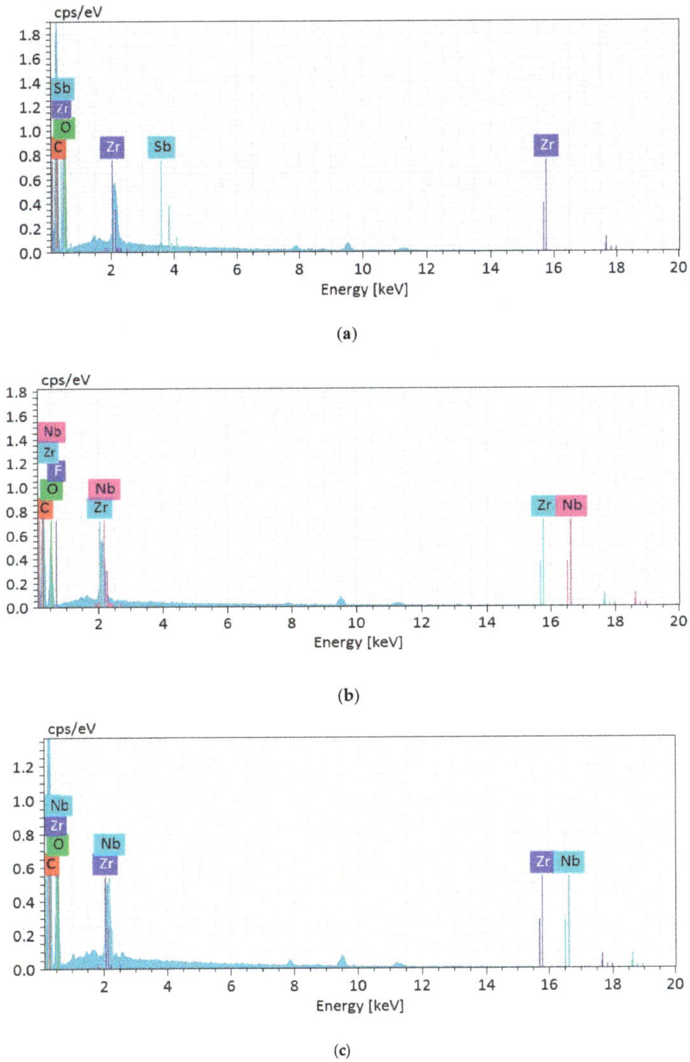

Figure 2. EDX spectra of (**a**) CPBLH, (**b**) CPAKL, and (**c**) CPMCC samples.

Table 1. Elements composition and yield of fiber samples.

Samples	C (%) [a]	O (%) [b]	Sb (%) [c]	F (%) [d]	Zr (%) [e]	Nb (%) [f]	Yield (%)
CPBLH	64.71	34.51	0.02	-	0.75	-	52
CPAKL	64.76	32.48	-	0.91	1.08	0.77	41
CPMCC	66.62	32.37	-	-	0.62	0.40	27

[a] Carbon; [b] Oxygen; [c] Antimony; [d] Fluoride; [e] Zirconium; [f] Niobium.

3.2. FTIR

Figure 3 showed the FTIR spectrum of the samples. Each fiber revealed closely the same pattern of curves, indicating their functional groups were not significantly affected by the chemicals' reaction [13]. The presence of absorption peaks at about 3400–3500, 2920, 1641, 1433, 1372, 1164, and 847 cm^{-1} were correlated to the typical functional groups of cellulosic samples [9,15]. Noticeably, the broad peak at 3408 cm^{-1} for CPBLH had shifted to 3469 cm^{-1} for CPAKL and, subsequently, to 3492 cm^{-1} for CPMCC, implying the increasingly exposed cellulose compartment throughout the treatments [11,14]. Moreover, another peak at 2920 cm^{-1} had reduced intensity for CPMCC. This was likely contributed by the presence of highly pure cellulose within the sample [16]. The peak intensities at 1746 cm^{-1} (C=O stretching of hemicellulose) and 1536 cm^{-1} (C=C vibration of lignin) were also noted decreasing for CPMCC, showcasing the process could remove substantial amorphous residual substances [1,17]. Meanwhile, the observed peaks at around 1641 and 1372 cm^{-1} were related to the cellulose–water interaction. The alteration of these peaks' sharpness from CPBLH to CPMCC was probably influenced by the cellulose crystalline regions that had less affinity towards water absorption. This was supported by the remarkable change of peak intensities at about 1433 cm^{-1}, which corresponded to the rearrangement of cellulose crystals [7]. In addition, the sharpness of peaks at 1164 cm^{-1} (C–O–C asymmetric vibration of ether groups) and 847 cm^{-1} (β-1,4-glycosidic linkages) were reduced possibly owing to the reorientation of cellulose order [2,18].

3.3. XRD

XRD patterns of fiber samples were shown in Figure 4. All fibers presented their main peaks at 2 theta of 15.4°, 16.9°, 22.7°, and 34.6°, which correlate to the (110), (200), and (004) crystal planes [12,19]. These peaks revealed that each fiber possessed Iβ type cellulose structure, evidencing the serial chemical processes did not change the fibers native structure [12,20]. The sharpest crystalline peak at about 22.7° was observed for CPMCC, implying it had the most stable cellulose structure among those samples [1,21]. From calculation, CPMCC possessed the highest crystallinity of 72.7%, followed by the CPBLH with 58.0%, and lastly by CPAKL with 54.9%. The lowest crystallinity for CPAKL was possibly associated to the alkaline cooking treatment that had softened the cellulose structure via swelling process. Besides this, the dramatically increased crystallinity for CPMCC could be explained by the gradually removal of amorphous components during acid hydrolysis, ultimately releasing highly rigid cellulose crystal structure [7,17]. Interestingly, another crystalline peak at 15.2° was noted prominently sharpening out for CPMCC fibers when compared to CPBLH and CPAKL. This might further prove that the acid hydrolysis played a role in eliminating weak residual substances such as surface binding components like lignin and hemicellulose to produce a greatly ordered crystalline cellulose structure [2,16]. Thus, the high crystallinity of CPMCC could endow it as potential load-bearing agent in composite reinforcement application.

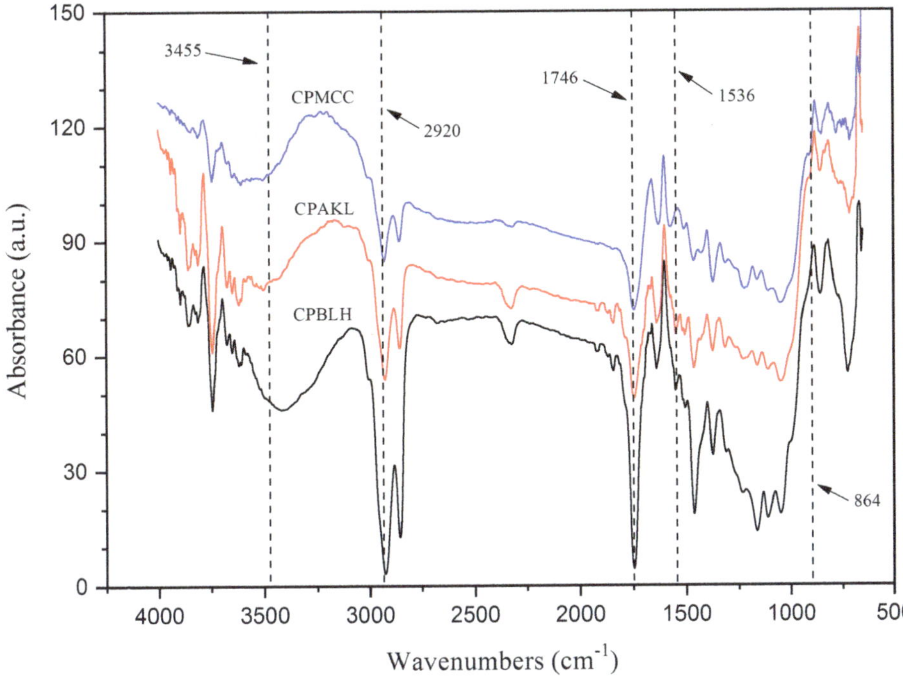

Figure 3. FTIR spectra of CPBLH, CPAKL, and CPMCC samples.

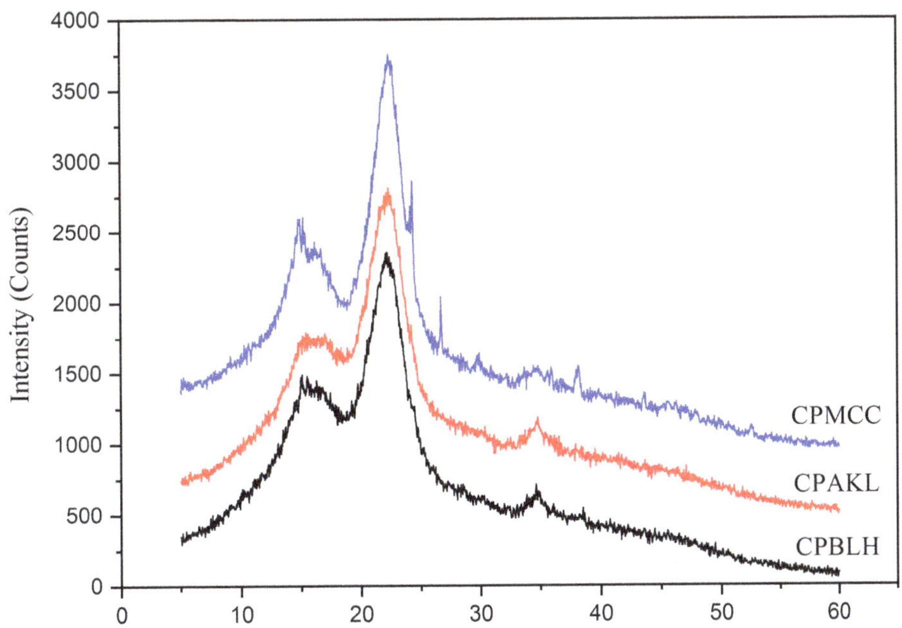

Figure 4. XRD curves of CPBLH, CPAKL, and CPMCC samples.

3.4. Thermostability

Figure 5 demonstrated the TGA curves of fiber samples. Initially, CPBLH showed a well reduced weight curve in the temperature range of 70–130 °C, which related to the evaporation of adsorbed water [22]. However, CPAKL and CPMCC fibers presented more likely a flattened curve, and the weight loss only became prominent in the 150–200 °C temperature range. This was probably due to the high cellulose contents in CPAKL and that CPMCC had absorbed a large amount of water. Consequently, it resulted in increased thermal capacitance degradation and delayed weight loss [15]. In the second degradation step, both CPAKL and CPMCC showed relatively higher decomposition temperature (TD) than that of CPBLH sample (Table 2). It was due to the presence of crystalline cellulose structure having greatly enhanced the thermal endurance towards high temperature [2]. Besides this, CPMCC only exhibited mildly improved decomposition temperature compared to CPAKL. It was possibly due to their same cellulose crystals order as aforementioned having contributed them similar thermal stability trend [16]. Additionally, in this work, the final char residue formation (WCR) was found remarkably low or both CPAKL and CPMCC with less than 2%, whereas the highest was found for CPBLH with more than 7%. This could evidence the exceptionally pure cellulose was obtained at the end of the chemical process [19]. Moreover, from DTG analysis, the maximum decomposition temperature (TMD) was the highest for CPMCC sample at 408.5 °C, suggesting its suitability to be applied in extreme temperature processing application.

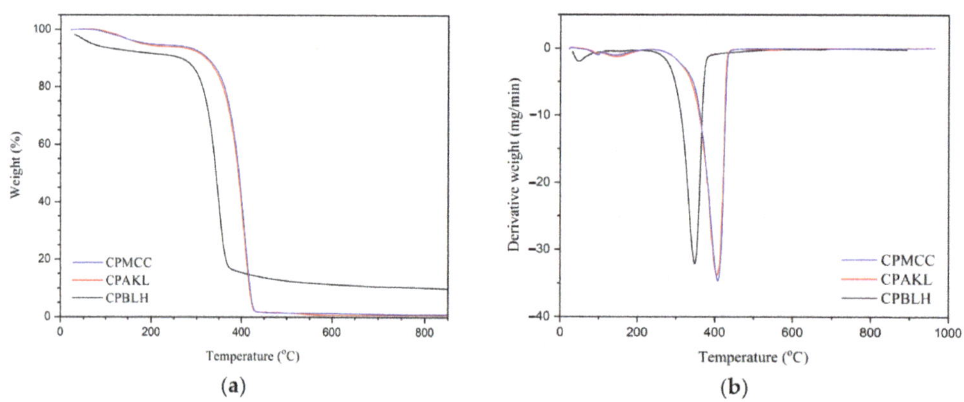

Figure 5. (**a**) TGA and (**b**) DTG curves of CPBLH, CPAKL, and CPMCC samples.

Table 2. Thermal analysis data of CPBLH, CPAKL, and CPMCC.

Samples	TD (°C) [a]	TMD (°C) [b]	WL (%) [c]	WCR (%) [d]
CPBLH	309.5	349.8	89.6	7.4
CPAKL	367.3	402.8	96.0	1.3
CPMCC	372.9	408.5	96.4	1.1

[a] Decomposition temperature; [b] maximum decomposition temperature; [c] weight loss; [d] weight of char residue formation.

As for DSC analysis (Figure 6), an initial broad endothermic band from 60 °C to 140 °C was observed for each sample, where heat energy was absorbed to volatize the water molecules [12,20]. In the region of 150 °C to 200 °C, a small endothermic peak was noticed for CPAKL and CPMCC at 160.5 °C and 164.5 °C, respectively, whereas this peak was less prominent reflected by CPBLH at 154.9 °C and an additional larger peak was presented at 175.3 °C herein. This was probably owing to the containing of amorphous compounds which distorted the heat flow behavior in CPBLH [9]. Beyond 250 °C, CPMCC significantly revealed an exothermic peak near to 288.6 °C temperature. However, CPBLH (at 283.3 °C) and CPAKL (282.9 °C) are more likely to exhibit the exotherms as shoulder bands rather than as sharp peaks. This could indicate that the cellulose degradation process is more uniform in CPMCC as compared to CPBLH and CPAKL fibers [7,14]. Apart from that, another exothermic peak that occurred in the region of 340–360 °C was noted for all fibers, which related to the decarboxylation and depolymerization of cellulose [19].

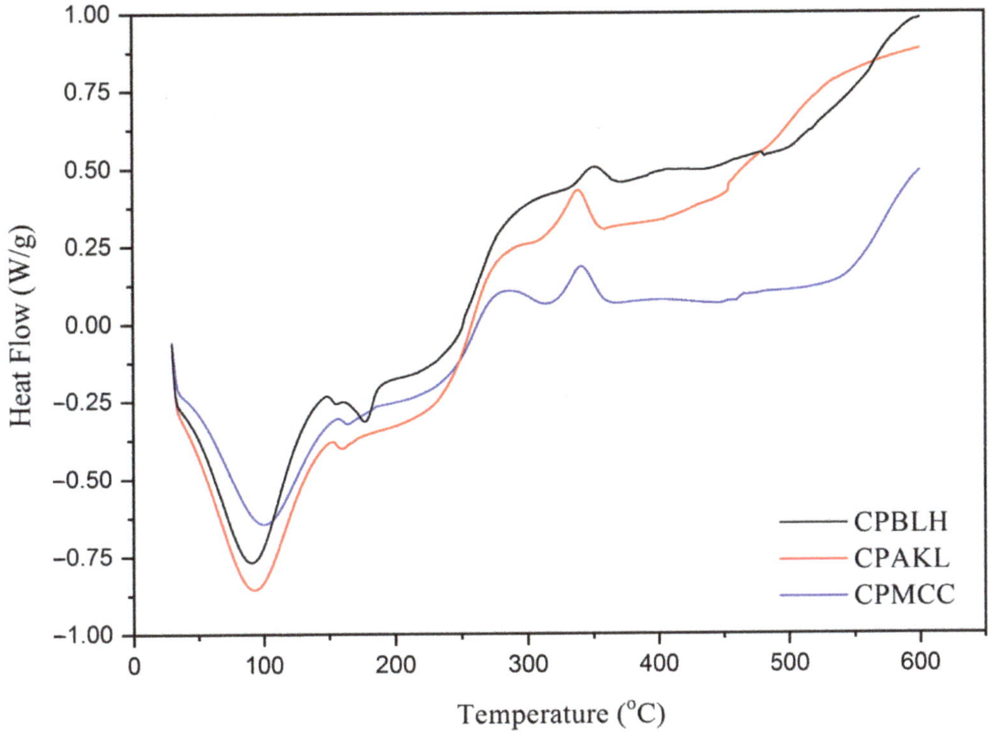

Figure 6. DSC curves of CPBLH, CPAKL, and CPMCC samples.

4. Conclusions

The findings of the present study proved that the sequence of alkaline cooking, bleaching, and acid hydrolysis process is able to isolate Conocarpus fiber for MCC production with excellent properties. A morphology study revealed the MCC had a rough feature and a fibril shape feature which could provide it with high surface reactivity for composite fabrication. Meanwhile, the functional chemistry was not modified throughout the entire chemical treatment, while only lignin and hemicellulose substances were substantially removed. The crystallinity relating to the rigid structure of MCC was greatly improved, suggesting its suitability to be applied as load-bearing agent. Regarding thermal decomposition, the crystal domain of cellulose played a major role in withstanding the MCC towards high temperature. Hence, it was concluded that the MCC product isolated from Conocarpus trunk fiber is a sustainable way to develop a green composite in the future.

Author Contributions: Conceptualization, H.F. and M.H.; methodology, L.K.K. and M.J.; validation, M.J. and H.F.; formal analysis and investigation, L.K.K. and H.F.; writing—original draft preparation, L.K.K.; writing—review and editing, M.J., M.D.A., H.F., and O.Y.A.; supervision and project administration, M.J.; funding acquisition, H.F. and O.Y.A. All authors have read and agreed to the published version of the manuscript.

Funding: The authors extend their appreciation to the Deputyship for Research & Innovation, "Ministry of Education" in Saudi Arabia for funding this research work through the project number IFKSURG-1435-052.

Acknowledgments: The authors extend their appreciation to the Deputyship for Research & Innovation, "Ministry of Education" in Saudi Arabia for funding this research work through the project number IFKSURG-1435-052.

Conflicts of Interest: The authors declare no conflict of interest.

References

1. Katakojwala, R.; Mohan, S.V. Microcrystalline cellulose production from sugarcane bagasse: Sustainable process development and life cycle assessment. *J. Clean. Prod.* **2020**, *249*, 119342. [CrossRef]
2. Liu, Y.; Liu, A.; Ibrahim, S.A.; Yang, H.; Huang, W. Isolation and characterization of microcrystalline cellulose from pomelo peel. *Int. J. Biol. Macromol.* **2018**, *111*, 717–721. [CrossRef] [PubMed]
3. Rehman, S.; Abbas, G.; Shahid, M.; Saqib, M.; Farooq, A.B.U.; Hussain, M.; Farooq, A. Effect of salinity on cadmium tolerance, ionic homeostasis and oxidative stress responses in conocarpus exposed to cadmium stress: Implications for phytoremediation. *Ecotox. Environ. Saf.* **2019**, *171*, 146–153. [CrossRef] [PubMed]
4. Tauqeer, H.M.; Rahman, M.; Hussain, S.; Abbas, F.; Iqbal, M. The potential of an energy crop "Conocarpus erectus" for lead phytoextraction and phytostabilization of chromium, nickel, and cadmium: An excellent option for the management of multi-metal contaminated soils. *Ecotox. Environ. Saf.* **2019**, *173*, 273–284. [CrossRef]
5. Rajabpour, A.; Mashahdi, A.R.A.; Ghorbani, M.R. Chemical compositions of leaf extracts from Conocarpus erectus L. (*Combretaceae*) and their bioactivities against Tribolium castaneum Herbst (Coleoptera: Tenebrionidae). *J. Asia Pac. Entomol.* **2019**, *22*, 333–337. [CrossRef]
6. Santos, N.D.K.D.; Barros, B.R.S.; Aguiar, L.M.S.; Filho, I.J.C.; Lorena, V.M.B.; Melo, C.M.L.; Napoleão, T.H. Immunostimulatory and antioxidant activities of a lignin isolated from Conocarpus erectus leaves. *Int. J. Biol. Macromol.* **2020**, *150*, 169–177. [CrossRef]
7. Tarchoun, A.F.; Trache, D.; Klapötke, T.M.; Derradji, M.; Bessa, W. Ecofriendly isolation and characterization of microcrystalline cellulose from giant reed using various acidic media. *Cellulose* **2019**, *26*, 7635–7651. [CrossRef]
8. Kian, L.K.; Saba, N.; Jawaid, M.; Fouad, H. Characterization of microcrystalline cellulose extracted from olive fiber. *Int. J. Biol. Macromol.* **2020**, *156*, 347–353. [CrossRef]
9. Hou, W.; Ling, C.; Shi, S.; Yan, Z. Preparation and characterization of microcrystalline cellulose from waste cotton fabrics by using phosphotungstic acid. *Int. J. Biol. Macromol.* **2019**, *123*, 363–368. [CrossRef]
10. Yavorov, N.; Valchev, I.; Radeva, G.; Todorova, D. Kinetic investigation of dilute acid hydrolysis of hardwood pulp for microcrystalline cellulose production. *Carbohydr. Res.* **2020**, *488*, 107910. [CrossRef]
11. Zhao, T.; Chen, Z.; Lin, X.; Ren, Z.; Li, B.; Zhang, Y. Preparation and characterization of microcrystalline cellulose (MCC) from tea waste. *Carbohydr. Polym.* **2018**, *184*, 164–170. [CrossRef] [PubMed]
12. French, A.D.; Cintron, M.S. Cellulose polymorphy, crystallite size, and the Segal Crystallinity Index. *Cellulose* **2013**, *20*, 583–588. [CrossRef]
13. Trache, D.; Khimeche, K.; Mezroua, A.; Benziane, M. Physicochemical properties of microcrystalline nitrocellulose from Alfa grass fibres and its thermal stability. *J. Therm. Anal. Calorim.* **2016**, *124*, 1485–1496. [CrossRef]
14. Xiang, L.Y.; Mohammed, M.A.; Baharuddin, A.S. Characterisation of microcrystalline cellulose from oil palm fibres for food applications. *Carbohydr. Polym.* **2016**, *148*, 11–20. [CrossRef]
15. Merci, A.; Urbano, A.; Grossmann, M.V.E.; Tischer, C.A.; Mali, S. Properties of microcrystalline cellulose extracted from soybean hulls by reactive extrusion. *Food Res. Int.* **2015**, *73*, 38–43. [CrossRef]
16. Ferrer, A.; Salas, C.; Rojas, O.J. Physical, thermal, chemical and rheological characterization of cellulosic microfibrils and microparticles produced from soybean hulls. *Ind. Crops Prod.* **2016**, *84*, 337–343. [CrossRef]
17. Owolabi, A.F.; Haafiz, M.K.M.; Hossain, M.S.; Hussin, M.H.; Fazita, M.R.N. Influence of alkaline hydrogen peroxide pre-hydrolysis on the isolation of microcrystalline cellulose from oil palm fronds. *Int. J. Biol. Macromol.* **2017**, *95*, 1228–1234. [CrossRef]
18. Zeni, M.; Favero, D. Preparation of microcellulose (Mcc) and nanocellulose (Ncc) from eucalyptus kraft ssp pulp. *Polym. Sci.* **2016**, *1*, 1–5. [CrossRef]
19. Nomura, S.; Kugo, Y.; Erata, T. 13C NMR and XRD studies on the enhancement of cellulose II crystallinity with low concentration NaOH post-treatments. *Cellulose* **2020**, *27*, 3553–3563. [CrossRef]

20. Hussin, M.H.; Pohan, N.A.; Garba, Z.N.; Kassim, M.J.; Rahim, A.A.; Brosse, N.; Haafiz, M.K.M. Physicochemical of microcrystalline cellulose from oil palm fronds as potential methylene blue adsorbents. *Int. J. Biol. Macromol.* **2016**, *92*, 11–19. [CrossRef]
21. French, A.D. Idealized powder diffraction patterns for cellulose polymorphs. *Cellulose* **2014**, *21*, 885–896. [CrossRef]
22. Sonia, A.; Dasan, K.P. Chemical, morphology and thermal evaluation of cellulose microfibers obtained from Hibiscus sabdariffa. *Carbohydr. Polym.* **2013**, *92*, 668–674. [CrossRef] [PubMed]

Publisher's Note: MDPI stays neutral with regard to jurisdictional claims in published maps and institutional affiliations.

polymers

MDPI

Review

Chitosan: A Sustainable Material for Multifarious Applications

Abdul Zubar Hameed [1], Sakthivel Aravind Raj [2,*], Jayakrishna Kandasamy [2], Majed Abubakr Baghdadi [1] and Muhammad Atif Shahzad [1]

1 Department of Industrial Engineering, Faculty of Engineering, King Abdulaziz University, P.O. Box 80204, Jeddah 21589, Saudi Arabia; aahameed@kau.edu.sa (A.Z.H.); mbaghdadi@kau.edu.sa (M.A.B.); mmoshtaq@kau.edu.sa (M.A.S.)
2 Department of Manufacturing Engineering, School of Mechanical Engineering (SMEC), Vellore Institute of Technology, Vellore 632014, India; mail2jaikrish@gmail.com
* Correspondence: aravindsakthivel@hotmail.com

Abstract: Due to the versatility of its features and capabilities, chitosan generated from marine crustacean waste is gaining importance and appeal in a wide variety of applications. It was initially used in pharmaceutical and medical applications due to its antibacterial, biocompatible, and biodegradable properties. However, as the demand for innovative materials with environmentally benign properties has increased, the application range of chitosan has expanded, and it is now used in a variety of everyday applications. The most exciting aspect of the chitosan is its bactericidal properties against pathogens, which are prevalent in contaminated water and cause a variety of human ailments. Apart from antimicrobial and water filtration applications, chitosan is used in dentistry, in water filtration membranes to remove metal ions and some heavy metals from industrial effluents, in microbial fuel cell membranes, and in agriculture to maintain moisture in fruits and leaves. It is also used in skin care products and cosmetics as a moisturizer, in conjunction with fertilizer to boost plant immunity, and as a bi-adhesive for bonding woods and metals. As it has the capacity to increase the life span of food items and raw meat, it is an unavoidable component in food packing and preservation. The numerous applications of chitosan are reviewed in this brief study, as well as the approaches used to incorporate chitosan alongside traditional materials and its effect on the outputs.

Keywords: chitosan; antibiotic; Gram-positive and -negative bacteria; drug delivery; biodegradable; heavy metal removal; COVID-19; wound healing

Citation: Hameed, A.Z.; Raj, S.A.; Kandasamy, J.; Baghdadi, M.A.; Shahzad, M.A. Chitosan: A Sustainable Material for Multifarious Applications. *Polymers* **2022**, *14*, 2335. https://doi.org/10.3390/polym14122335

Academic Editors: Sabu Thomas and Maya Jacob John

Received: 27 April 2022
Accepted: 30 May 2022
Published: 9 June 2022

1. Introduction

Chitosan is a sugar found mostly in the shells of crustaceans such as crabs, shrimps and lobsters. It is a drug that is used in the pharmaceutical sector. Chitosan, a fibrous substance, may help the body absorb less fat and cholesterol from the foods we eat. It assists in the production of blood clots when applied to wounds. Chitosan is utilized in a variety of applications, including contaminated drinking water, which has a significant impact on human health, and experts are working to develop the best disinfectant to address this issue. Nanocomposites containing chitosan and carbon nanotubes have been identified as viable alternatives to conventional disinfection methods [1]. Nanomaterial-based membranes were discovered to be a sustainable method of treating waste fluids. Membranes and reactors treated with chitosan demonstrated increased resistance to key contaminants and decreased membrane fouling [2]. Chitin and chitosan are formed from fish debris such as heads, tails, skins, scales, and shells, which are unfavorable to humans and the environment [3]. The presence of heavy metals in water is greater, and if consumed in this state, will have a variety of adverse consequences on human health. Chitosan is an excellent absorber of heavy metal ions and is utilized in industry for waste water treatment in fixed-bed column designs [4]. Chitosan makes a greater contribution to biomedical research by acting as an antibiotic, antioxidant, and drug delivery agent.

Due to the additional benefits, such as the non-toxicity and practicality, chitosan is well suited for medical purposes. Due to the deficiencies of chitosan in terms of cell adhesion and biosignaling, peptide chitosan has been developed with enhanced cell adhesion and biosignaling capabilities and is being employed in cell therapy, drug delivery, and as an antimicrobial [5]. Due to decreasing petroleum resources and the environmental damage caused by petroleum products, biomaterials are garnering increased attention. Chitosan is used in a range of industries, including water treatment, medicine, fisheries, and cosmetics; the chemical and packaging industries; as well as food and agriculture. Additionally, it is utilized in the exploration, extraction, refining, and treatment of waste water for petroleum products. Chitosan must undergo particular chemical alterations in order to be used in petroleum field applications [6]. Vacuum-aided filtering, freeze-casting, and biomimetic mineralization are used to make chitosan–calcium phosphate composites. It can be applicable in drug administration, bone implants, wound healing, dental implants, and waste water filtration from heavy and organic metals [7]. Chitosan also possesses antiviral properties and can be utilized as an adjuvant in vaccines against SARS and COVID viruses [8]. Chitosan is a derivative of chitin that exhibits antifungal, biodegradability, biocompatibility, mucoadhesion, and antibacterial properties. It is used in dental and bone implants [9]. Membrane development for water purification is a significant area of research in which a number of researchers are involved. Natural and manmade polymers derived from chitosan have demonstrated their ability to purify water. Chitosan-based nanocomposites are favored for water purification applications because of their low cost, non-toxic nature, biodegradability, and biocompatibility. Chitosan is capable of eliminating heavy metals, dyes, and other hazardous contaminants from water during the purification process [10]. Chitosan is a polysaccharide found naturally in marine crustaceans. Chitosan can also be applied following molecular and chemical changes, including in biomedicine, the pharmaceutical sector, agriculture, gene research, drug delivery, imaging, wound healing, and tissue engineering [11]. The various applications of chitosan are illustrated in Figure 1. The various sources of chitosan, the methods of extraction, the chemical modifications required, the incorporation of chitosan into composites for various applications, and the use of chitosan in various fields are discussed succinctly in this review for the benefit of researchers who work on the application of chitosan for their specific purposes. Table 1 indicates the biological and physicochemical properties of chitosan based on the degree of N-acetylation and the molecular weight.

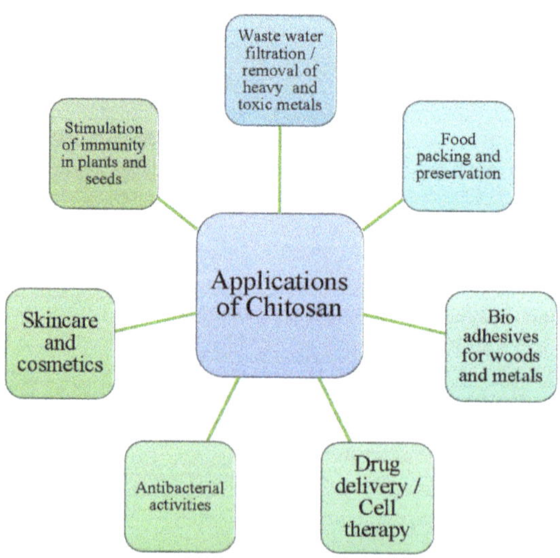

Figure 1. Various applications of chitosan.

Table 1. Various applications of chitosan [12].

Applications	Examples
	Chitosan
Biomedical and pharmaceutical materials	Treating burns, drug delivery systems, dental repair and treatment, surgical structures, artificial skin, lenses for eyes, dialysis of blood, artificial blood vessels, antitumor and antibiotic uses, accelerated wound healing.
Cosmetics	Hair and skin care products
Tissue engineering	Regeneration of bones and tissues, repair of scaffolds, regeneration of sulphate sponges in bone, development of artificial pancreas, diabetes treatment.
Agriculture	Food and seed coating, removal of pesticides and herbicides from soil and water, excellent film coating with antimicrobial activities, preservation of post harvested foods, enhancing plant growth, enhancing soil quality.
Food and feed additives	Food and beverage de-acidification, color stabilization in foods, lipid absorption reduction, extension of natural flavor, antioxidant and food preservation, controlling agent, stabilizing agent, thickening agent, additives in livestock and fish food, manufacture of dietary fibers.
Water engineering	Treatment of waste water, removal of heavy metals from water, removal of pesticides and ions from water, dye removal from water, removal of petroleum products from water, removal of dyes from effluents, color removal from textile waste waters.

2. Chemical Properties and Processing Technologies of Chitosan Based Materials

2.1. Physicochemical and Biological Activities

The deacetylation process is the hydrolysis process of acetamide groups in chitin when strong NaOH solution reacts at temperatures of 100 °C and above, producing the amino groups of the new compound known as chitosan. The formed amino groups in the chitosan decide its biological properties. The deacetylation degree range of 70–85% in chitosan means it can be partly dissolved in water, and above 95 to 100% is the ultrahigh DD range of chitosan, which is a challenging task to achieve. The DD and Mw distributions and average Mw can be determined using the methods mentioned in Figure 2 [13]. Table 2 shows the biological and chemical properties of chitosan.

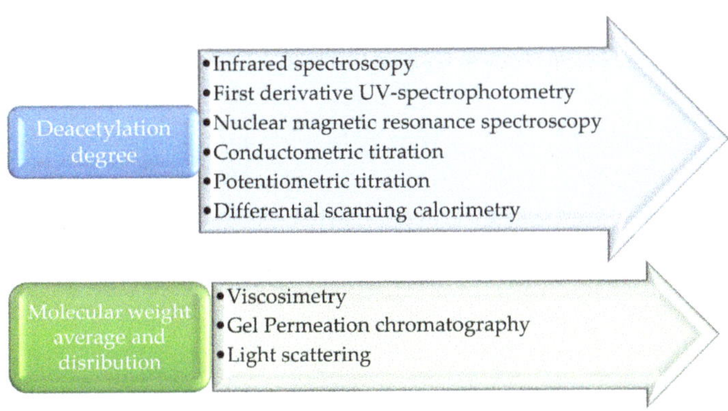

Figure 2. Determination methods followed for Deacetylation degree and Molecular weight distribution and average.

Table 2. Chemical and biological properties of chitosan [12].

Chemical Properties	Biological Properties
Nitrogen content is enhanced	Biocompatible and biodegradable
High hydrophilicity and crystallinity due to structure	Non-toxic to humans
Powerful nucleophile and weak base	Combines with microbial cells quickly
Increases viscosity by forming hydrogen bonds	Regenerates the gum tissues
Has reactive groups for crosslinking and chemical activation	Stops bleeding
Insoluble in water and organic solvents	Enhances bone formation and repair

Table 2. *Cont.*

Chemical Properties	Biological Properties
Soluble in acids	Inhibits the growth of fungi
Leads to salt formation with organic and inorganic acids	Inhibits the growth of tumor cells
Has chelating properties	Enhances birth control
Ionic conductivity	Acts as a cholesterol-reducing agent
Act as a polyelectrolyte in acidic conditions	Anticancer agent
Combines with negatively charged molecules	Act as a nervous depressant
Better adsorption and entrapment properties	Improves the immune response
Better separation and filtration abilities	Combines with mammalians
Ability to form films	Safe for water treatment

Table 3 indicates the physicochemical and biological properties of chitosan with respect to its DA and Mw. The direct and indirect proportionality levels of Mw and DA with the properties of chitosan can be easily understood and can be applied based on the required applications [14]. Crystallinity and hydrophobicity characteristics are influenced by the acetylation degree. Reacetylated chitosan is applied as a coating in cardboard to improve the mechanical properties. Chitosan with a 2% degree of acetylation showed better water and mechanical resistance. This indicates that the molecular groups are well distributed, which increases the hydrophobicity of the polymers [15].

Table 3. Influence of Mw and DA on physicochemical and biological properties [13].

		Physicochemical Properties	
S.No.	**Properties**	**Degree of N-acetylation (DA)**	**Molecular Weight (MW)**
1	Solubility	Indirectly proportional	N/A
2	Crystallinity	Directly proportional	N/A
3	Viscosity	Indirectly proportional	N/A
4	Biodegradability	Directly proportional	Indirectly proportional
5	Biocompatibility	Indirectly proportional	N/A
		Biological properties	
6	Antimicrobial	Indirectly proportional	Directly proportional
7	Analgesic	Indirectly proportional	N/A
8	Anticholestemic	N/A	Indirectly proportional
9	Antioxidant	Indirectly proportional	Indirectly proportional
10	Hemostatic	Indirectly proportional	N/A
11	Mucoadhesion	Indirectly proportional	Directly proportional
12	Permeation enhancing effect	Indirectly proportional	Directly proportional
13	Antitumor	N/A	Indirectly proportional

If the pH value increases, the drug loading, entrapment efficiency, and immune enhancement effects are enhanced, while if the pH value decreases, the antimicrobial properties, antitumor properties, and permeation effect are reduced. The values at which the property changes occur are also mentioned in Figure 3. Chitosan can be embedded, encapsulated, mixed, precipitated, spray-dried, emulsified, crosslinked, and cast into various products such as tablets, capsules, and nano- and microparticles, and can be formed into beads, films, and gels. The processing technologies through which the chitosan can be incorporated into other materials are described in Figure 4 [16].

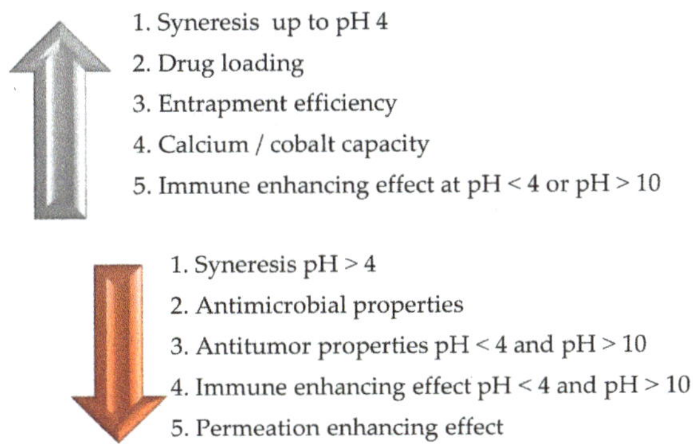

Figure 3. Influence of pH on the biological and physicochemical characteristics of chitosan.

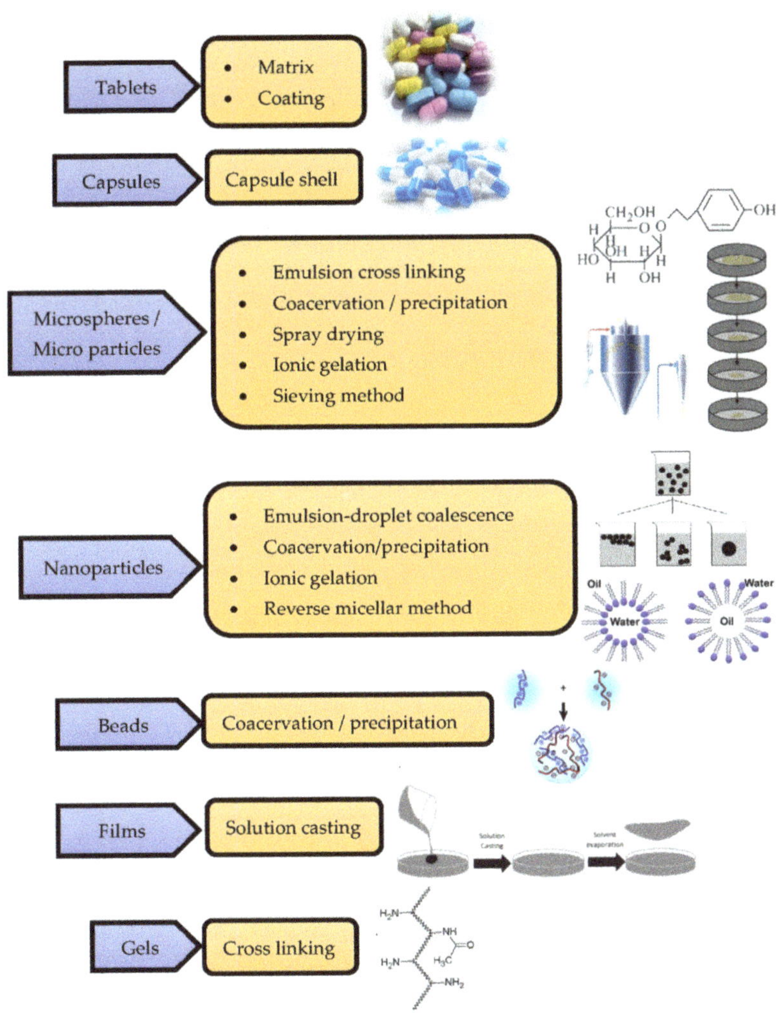

Figure 4. Processing technologies for making products using chitosan.

2.2. Process Technologies for Producing Chitosan Composites

Membranes and films that can be used in air and water filtration processes can be made using the solvent evaporation process. Solvent evaporation is a simple three step process in which the polymer resin is mixed with nano- or micron-sized chitosan fillers; sometimes fibers may also be reinforced to enhance the mechanical properties. The film and cast membrane manufacture process is described in Figure 5. The mixed solution is then poured into a glass container and heated to initiate the evaporation process. After evaporation, the cast membrane or film can be taken out from the container [17].

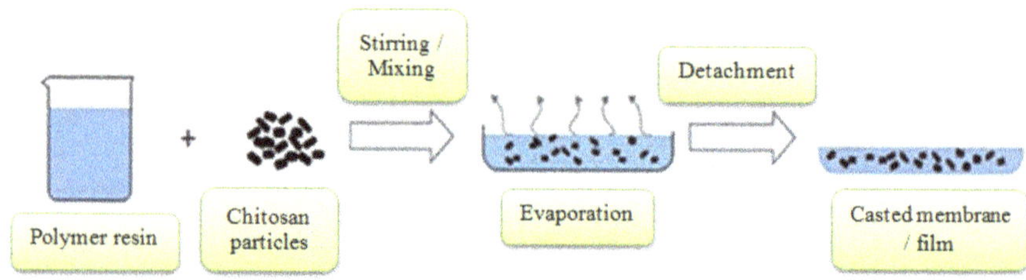

Figure 5. Solvent casting for film and membrane manufacture.

The process of electrospinning and the equipment required are shown in Figure 6. The major differences between conventional fibers and electrospun fibers are their diameter and surface area. The method used for depositing nanochitosan on oppositely charged substrates is shown in Figure 7. The electrospun fibers have a larger surface area and smaller diameter. The polymer solution is subjected to differences in potential generated between the spinneret and collector. Due to the electric filling, the pendant-like droplets are converted into jets, and at a critical value the repulsion of the electricity exceeds the tension offered by the solution on its surface. Due to this phenomenon, the extruded polymer solution is subjected to rapid whipping, which is unstable, while the evaporation leads to nanofibers forming on the collector [18].

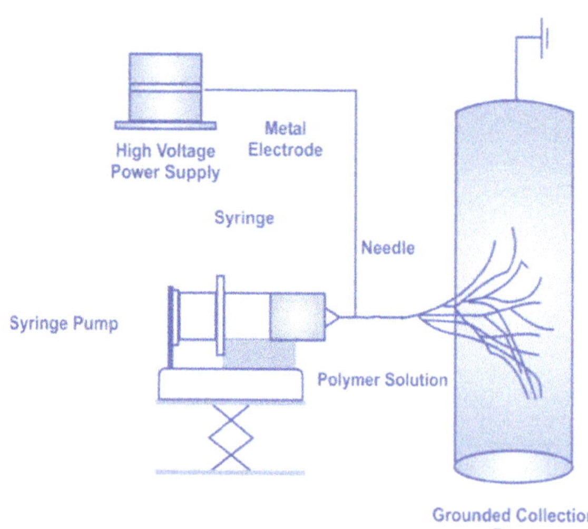

Figure 6. Electrospinning process with rotating mandrel [18].

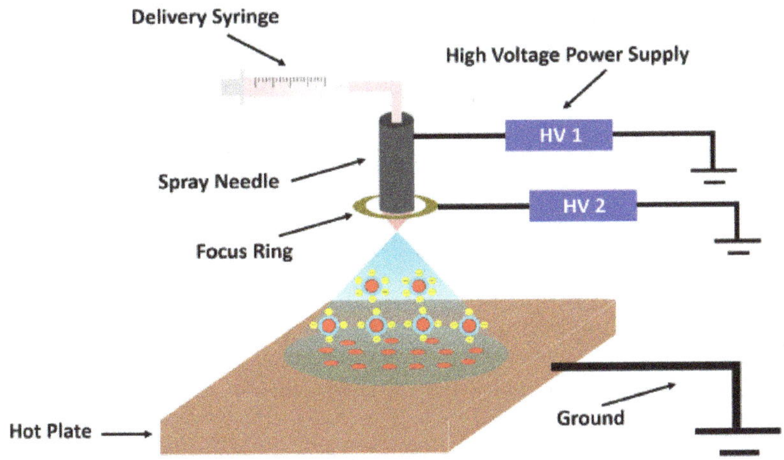

Figure 7. Deposition of chitosan via electrospraying method. Reprinted from [19]. Copyright 2022 with permission from the American Chemical Society.

In this process, the chitosan particles are mixed with a polymer solution and kept in a dispersion needle. A high voltage is supplied to the solution in the needle. Due to the identical charge the droplets repel each other, then due to instability at the needle tip the droplets start dispersing into micron-sized particles and are deposited on oppositely charged surfaces, while the solvent is evaporated rapidly [20]. The method of producing chitosan nanoparticles through emulsion droplet coalescence is described in Figure 8.

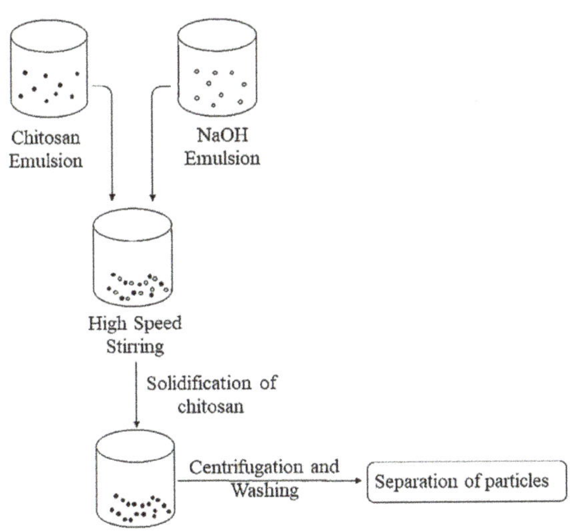

Figure 8. Emulsion droplet coalescence method used to produce chitosan nanoparticles [21].

In Figure 9, the different methods for applying biodegradable coatings on fruits and vegetables are listed. The best alternatives to wax coatings were found, which were chemically formulated. These coatings prevented microbial loading on fruits by resisting against oxidation and reduction. By minimizing the vapor dissipation, decay, and ripening caused by bacteria, the physiological and microbial deterioration is reduced and the shelf lives of the fruits can be extended [22].

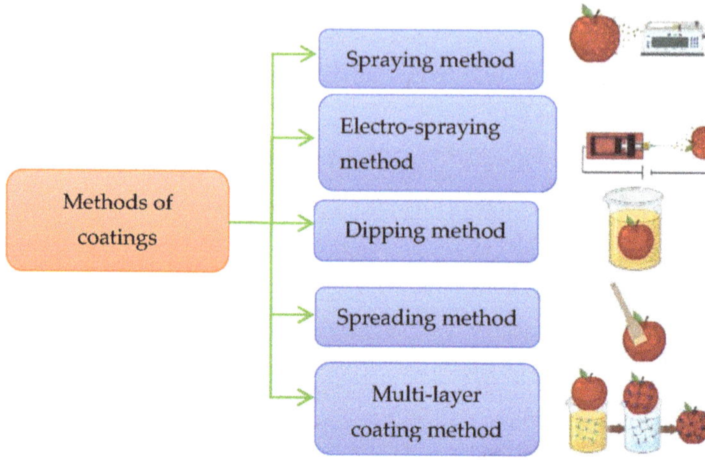

Figure 9. Methods used for coating fruits and vegetables [22].

Post-harvest, the fruits and vegetables are coated with chitosan to maintain their quality. Pure chitosan or a combination of chitosan with citric acid is used for coating. The conditions of post-harvested tomatoes after 15 days with different coating levels and elements are shown in Figure 10. Tomatoes coated with chitosan in combination with citric acid showed less undesirable changes, less weight loss, and less tissue damage, confirming the usage of chitosan as a preservative coating on fruits and vegetables [23]. It is safe to use chitosan similarly to salts as a coating in fruits, as it is non-toxic when consumed by humans.

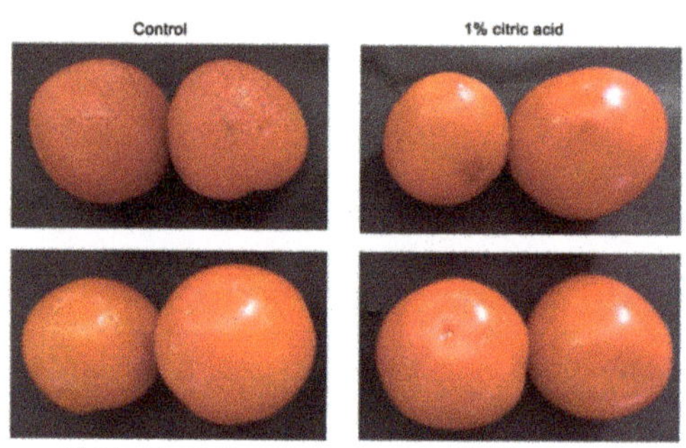

Figure 10. Conditions of tomato samples coated with chitosan and citric acid after 15 days at 28 °C [23].

3. Major Applications of Chitosan

3.1. Water and Air Filtration

By and large, waste water is contaminated with bacteria and microorganisms that cause a variety of ailments. It was discovered that CS with a low Mw inhibits the growth of Gram-positive bacteria, whereas chitosan with a high molecular weight inhibits bacteria. Additionally, it has been demonstrated that doping chitosan with nanoparticles enhances its antibacterial characteristics, and that when coupled with silver nanoparticles, a low concentration of chitosan is sufficient to control bacteria that cause water contamination [24]. Chitosan–cobalt–silica nanocomposites were prepared and utilized for dye absorption and water purification. Additionally, this combination was investigated against bacteria, and the results indicated that it exhibited significant bioactivity toward them [25]. Chitosan composites can be used in special filters in the textile industry, as it has better color absorption properties, along with the ability for metal ion absorption. The PVA-co-PE nanofiber membrane was used to clean waste water and was shown to have a higher water flux and retention rate for nanoparticles and bacterial cells. The membrane's bacteria inactivation rate was also increased from 97.8 to 99.5% against pathogens, and the membranes were capable of being employed for extended periods of time with high stability and efficiency [26]. The filtration of wastewater and industrial effluents using membrane filters has the significant disadvantage of membrane fouling, which prompted the creation of chitosan-based antifouling membranes. Through spinning, chitosan and silver NP's were incorporated into membranes of hollow fibers. The chitosan- and silver-chitosan-based membranes showed superior performance with the highest dye rejection rates, and were determined to be the most suitable for treating industrial effluents without fouling the membrane [27]. To create a multifunctional composite, chitosan, polyethyleneimine, graphene oxide, and glutaraldehyde were combined and coated with membranes capable of removing both positively and negatively charged heavy ions. The glass microfiber filter was chosen and was found to be effective in removing Cr (VI) and Cu (II), demonstrating that this coated membrane may be utilized to remove both positive and negative ions from water [28]. Water flux reduction in membranes used for water treatment is a critical issue owing to the associated bacterial development. To create a PVDF-S/MIL100-CS composite, a recently discovered approach termed solvent-assisted nanoparticle embedding was applied. The new production procedure spread the fillers over an open surface, imparting a hydrophilic quality to the surface. The antibacterial activities of the PVDF-S/MIL100-CS composite are mentioned in Figure 11. The results suggested increases in antibacterial activity and resistance to biofouling, which was validated by a live/dead test for antibacterial activity [29]. The fluoride pollution of groundwater appears to be a significant issue, and numerous experts are striving to develop a cost-effective remedy. A mixed matrix membrane (MMM) based on cellulose acetate was created using the phase inversion approach, together with the use of mixed metal oxide nanoparticles as nanofillers. The results of the tests indicated that the MMM was less susceptible to attack by microbes, and it was discovered that the fluoride ion was rejected due to adsorption, while the membrane surface exhibited electrostatic repulsion, enhancing the defluorination effectiveness [30]. The MWNT was disseminated in an aqueous solution containing varying quantities of chitosan, and the membrane was formed. The bucky paper membranes have excellent mechanical qualities, and their zeta potential improves as the amount of added chitosan increases. Additionally, it was observed that the bucky paper membrane modified with MWNT exhibited improved salt rejection capabilities and smaller interior pores [31]. Bacterial dispersal in the air, combined with particulate matter pollution, is increasing daily, posing a threat to human health. Multilayer membranes with antibacterial properties and excellent air filtration efficiency need to be developed. Sequential electrospinning was used to generate PVA/chitosan membranes with N-halamine that displayed high filtration efficiency and tensile strength in the filtration test. The process of producing multilayer air filters through electrospinning is mentioned in Figure 12 [32]. More efficient water filtration systems are required to address national and global water scarcity challenges. Researchers are focusing their attention on the development of low-cost membranes. PAN membranes were enhanced with nanoparticles of zinc oxide and chitosan to improve the water filtration, mechanical, and antibacterial qualities. It was evident from the results that the created composite membranes possessed excellent antibacterial and self-cleaning qualities [33]. Water filters capable of removing metal ions were created by electrospinning nylon/chitosan fibers. These fibers were tested against lead nitrate and sodium chloride, and the results demonstrated that the membrane was capable of removing metal ions and bacteria from an aqueous solution to a concentration of up to 96% [34]. Numerous contagious diseases are conveyed via air, and numerous concerns have been raised about aerosols and bioaerosols. Electrospinning is used to create polyurethane/chitosan nanofibers, meaning various parameters such as the diameter are changed, the effects of which are examined. The nanofibers demonstrated superior performance when used in filtration processes in industry and in equipment used for respiratory purposes [35]. Superhydrophobic poly (methylmethacrylate)/polydimethylsiloxane fibers with a capture efficiency of 98.23% are used to catch particle matter. On a window screen, continuous particle removal has been proven to filter the particles [36]. Water pathogens and bacteria form biofilms and lead to biofouling, which continue to be significant concerns in many locations. Membranes composed of chitosan, PEG, MWCNT, and iodine were made in three phases. The inclusion of iodine increased the hydrophilicity, porosity, and performance of the membranes. The reduced iodine concentrations killed 99.2 and 100% of *E. coli* and *S. aureus* bacteria, respectively [37].

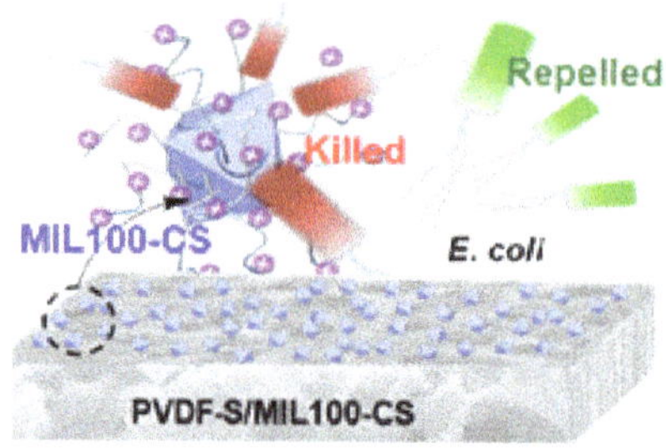

Figure 11. Antibacterial inhibition activity illustration of composite made of PVDF-S/MIL100-CS [29].

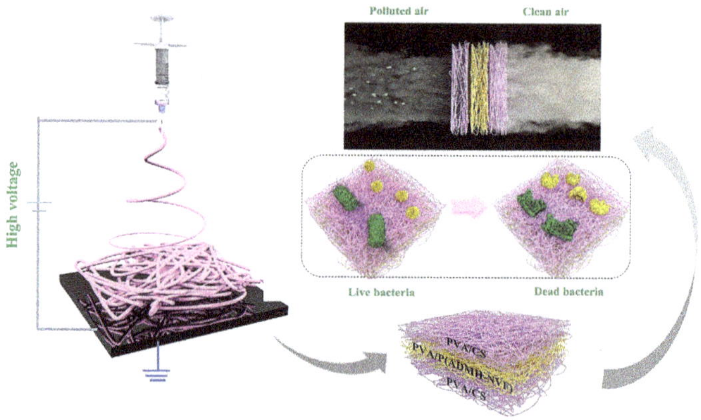

Figure 12. Electrospinning process used to fabricate multilayer air filters. Reprinted with permission from [32]. Copyright 2020 with permission from Elsevier.

Removal of dyes and organic pollutants is a difficult task that can be made possible through incorporating chitosan along with multilayer composite membranes. The production of water filters through the electrospinning process is described in Figure 13. Polyacrylonitrile nanofibers produced via the electrospinning process and supported by polyamide membranes are used to filter waste water and to remove tetracycline from it. These fibers have been laminated and tested and found to be more effective in tetracycline removal [38]. In Figure 14, the filtration process of air pollutants through a fibrous membrane has been illustrated.

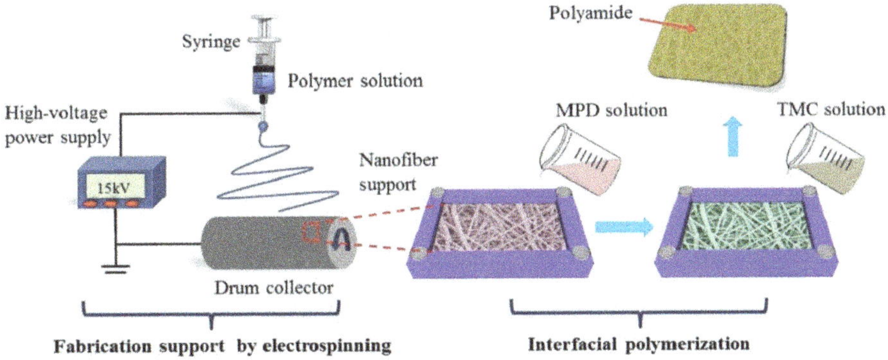

Figure 13. Electrospinning process used to fabricate multilayer water filters [38].

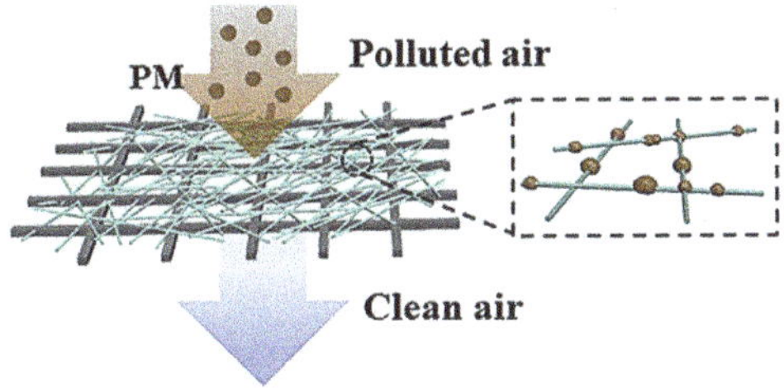

Figure 14. Illustration of fibrous membrane in which the pollutants of air are filtered [36].

Regarding the unique antibacterial mechanism of chitosan-reinforced PVDF-S/MIL100 composites, the negatively charged *E. coli* combine with oppositely charged chitosan. Some *E. coli* bacteria were repelled from the composite surface due to its hydrophillic nature. This property of killing the bacteria or repelling the bacteria from the filter surface is most needed for the air and water filters, meaning the addition of chitosan to the filter membranes is very much needed and effective. When compared to single-layer membranes used for filtering air and water, multilayer membranes have been found to be more effective in filtering the bacteria and other pollutants. The filtration efficiency increases when the number of layers increases. Electrospinning has been found to be the most suitable method for manufacturing filter membranes for both air and water purification purposes. Additionally, the usage of chitosan in the filter membranes has added benefits, such as making the material non-toxic and biodegradable, because when used along with water filters, it may mix with water or be washed away along with the water flow, meaning humans may intake chitosan mixed with water. In such cases, chitosan has no toxicity and is safe. After a certain time period based on the usage of water the filters need to be replaced, and in such cases the used filters may cause pollution issues, which should be avoided as chitosan is biodegradable.

3.2. Metal Removal from Water

Industrial effluents have a high concentration of heavy metals that can cause major diseases and organ damage, meaning they must be separated from drinking water. Metal ions contained in contaminated water are also harmful to health, and their removal from the water is a difficult process that is accomplished by adding chitosan to the filtration membrane. Separating a catalyst from a reaction media is a difficult task. This was accomplished by coating the high surface area of the filter paper with chitosan to increase its affinity for metal ion absorption. The filter paper is composed of cellulose microfibers that act as a support for the catalyst. The catalyst can be recovered and utilized for subsequent chemical reactions using this approach [39]. Chitosan is used to cover cellulose filter paper and is used to adsorb Ni^{2+} ions from aqueous solutions. To transform the adsorbed ions into nanoparticles, they are treated with a NaBH4 solution. FESEM and EDAX are used to characterize the transformed nanoparticles. It has been demonstrated that filter paper containing Ni and CS can be used to detect and catalyze other nanoparticles.

The filter papers containing chitosan and nickel are described in Figure 15. The production process starts with treating the filter paper with a chitosan solution. Then, the paper is dried and kept in a 2 M $NiCl_2.6H_2O$ solution for absorption of Ni^{2+} ions due to the presence of chitosan chains [40]. Here, carboxylated chitosan was deposited onto a membrane and treated with an aqueous copper (II) chloride solution; copper nanoparticles and this thin film membrane were then treated with glutaraldehyde. For 90 days of immersion in water, thin film membranes containing carboxylated chitosan treated with copper (II) chloride demonstrated greater than 99% efficacy against protein fouling. It was demonstrated that chemically modified chitosan acts as an antiprotein fouling agent with increased hydrophilicity [41]. By integrating chitosan and graphene oxide into a polyehtersulfone (PES) membrane, chromium removal and antifouling capabilities can be achieved. The modified membrane demonstrated increased hydrophilicity, a smoother surface, and increased water flux. Additionally, graphene oxide and chitosan-filled membranes exhibit improved antifouling properties. Figure 16 shows a schematic representation of the water filtration process through graphene oxide–chitosan membranes [42].

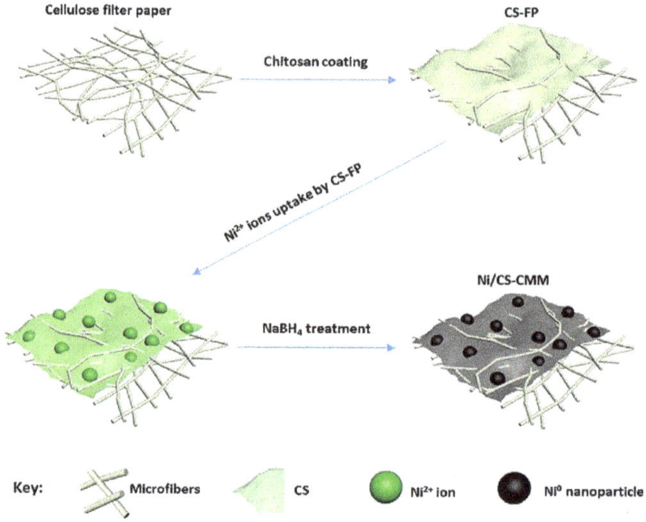

Figure 15. Steps in the preparation of Ni/CS filter paper. Reprinted from [40]. Copyright 2016 with permission from Elsevier.

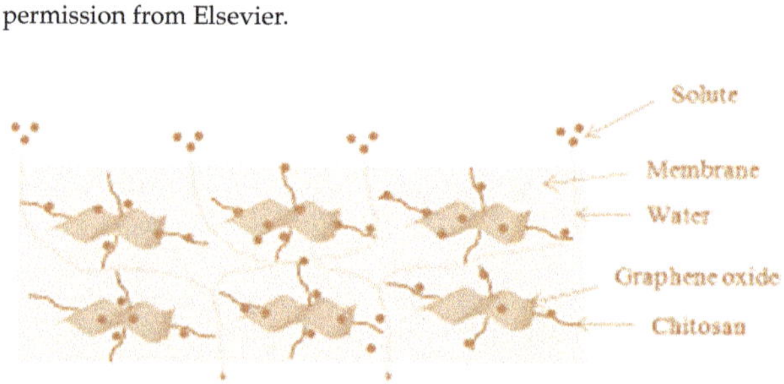

Figure 16. Graphene oxide–chitosan dispersion in water filtration membrane and its feed transport route. Reprinted from [42]. Copyright 2018 with permission from Elsevier.

The SEM images shown in Figure 17 indicate the presence of adsorbed particles, which is proof of the adsorption characteristics of filtration membranes reinforced with chitosan. This image was taken after the salt filtration process. The technique of non-solvent-induced phase inversion was utilized to fabricate a thin membrane made of polyvinyl alcohol, chitosan, and montmorillonite clay. Due to its hydrophilic character, this composite membrane demonstrated a higher rejection rate. The heavy metal chromium removal was verified using EDAX and FT-IR measurements [43].

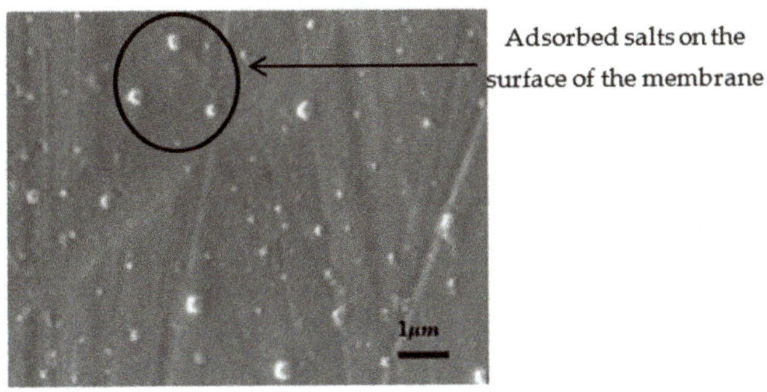

Figure 17. SEM image of membrane with lower graphene–chitosan levels. Reprinted from [42]. Copyright 2018 with permission from Elsevier.

Regeneration is an important process in filter membranes to restore the properties of chitosan after the adsorption of heavy metals. Various agents are used for desorption and regeneration processes, such as alkalis, acids, chelating agents, and salts. These agents must also possess certain other properties, such as being non-toxic, biodegradable, and less expensive. Acidic eluents such as nitric acid, hydrochloric acid, phosphoric acid, and sulfuric acids are used as desorption eluents and for regeneration. The regeneration process through which the base properties of the chitosan are regained by using desorption agents is explained schematically in Figure 18 [44].

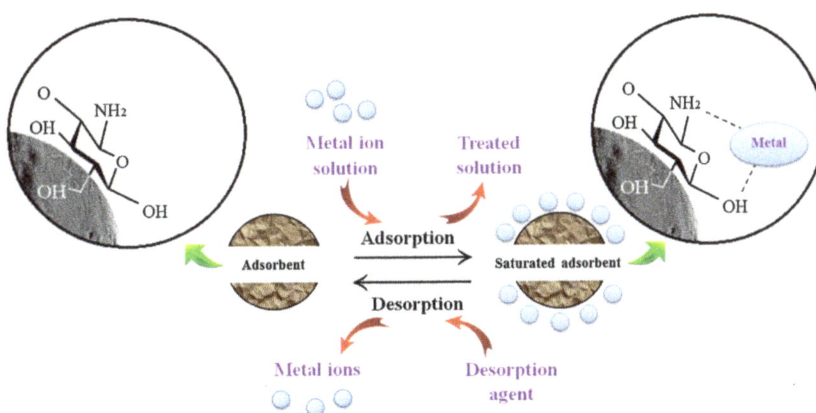

Figure 18. Regeneration process during adsorption. Reprinted from [44]. Copyright 2019 with permission from Elsevier.

3.3. Antibacterial Activities

Chitosan possesses unique antibacterial characteristics, particularly against Gram-positive and -negative microorganisms. It can be coated or electrospun with a filter membrane made of conventional material to increase the membrane's characteristics and performance. The antifouling ability and retention of nanoparticles, as well as the membrane's hydrophilicity and porosity, can be increased. In aqueous solution, chitosan and silver nanoparticles operate as both stabilizers and reductants. Due to the high stability and monodispersity of chitosan-functionalized silver colloids, it has been evaluated for its antibacterial activity against fungi and bacteria and shown to have a higher bactericidal efficacy against them [45]. The bactericidal actions of chitosan, alginate, and silver nanoparticles were evaluated against *E. coli* and *S. aureus*. Chitosan and alginate are employed to create pores, while silver provides antibacterial action. The chitosan–alginate–silver nanoparticle combination demonstrated superior antibacterial activity against bacteria and demonstrated its potential for usage in the treatment of breast cancer. To prepare a chitosan–alginate membrane, alginate solution is slowly added to a chitosan solution drop-by-drop. A polyelectrolyte complex between chitosan and alginate is formed, which is then dispersed by a high-speed stirrer at 500 rpm. The addition of Ag nanoparticles results in a brownish yellow color. A porous scaffold is obtained via the regeneration of chitosan–alginate–Ag NP's using sodium hydroxide and calcium chloride. The production of chitosan–alginate membranes and the step-by-step process is explained schematically in Figure 19 [46].

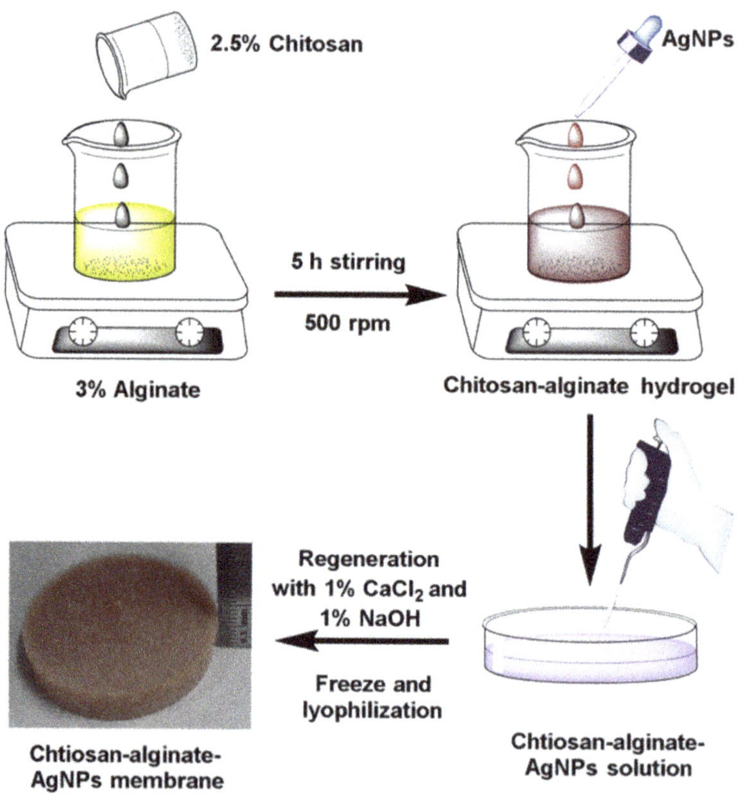

Figure 19. Chitosan–alginate membrane preparation process. Reprinted from [46].Copyright 2017 with permission from Elsevier.

ZnPc-CS composites were made by dispersing ZnPc in chitosan solution and then immersing the composites in salt solution for metal ion adsorption. The composites were treated with sodium borohydride solution for the conversion of metal ions into nanoparticles. Metal-nanoparticle-loaded composite fibers were synthesized in situ and found to be effective against pathogenic bacteria of *E. coli* and to have the ability to reduce 4-nitrophenol, methyl orange, and Cango red [47]. Quaternized chitosan-embedded membranes made in acetic acid medium have a greater effect on *E. coli* and have 99.95% efficiency [48]. Nitric oxide is a free radical that can be engaged in antibacterial actions such as wound healing. S-nitrosoglutathione was added to pluronic F-127 and chitosan as a nitric oxide donor. The GSNO-PL/CS combination was found to release nitric oxide and was proven to be harmless to Vero mammalian cells [49]. Composite membranes composed of chitosan–collogen, chitosan–collogen–montmorillonite, and chitosan–collogen–organomontmorillonite were investigated for their swelling ratio, moisture permeability, and in vitro degradation ratio properties. It was discovered that they have a higher swelling ratio, excellent moisture permeability, and a lower degradation ratio. To increase the antibacterial activity of the composite membrane, Calicarpa nudiflora was added [50]. *T. portulacifolium* leaf extract was utilized as a reducing agent, and the hybrid composite's antibacterial efficacy was determined. The mixture demonstrated improved inhibitory activity against microbes such as *S. marcescens* [51]. Antibiotic treatments are carried out using biodegradable and biocompatible polymer-based Nano capsules, also known as hollow nanoparticles. These are self-assembled polysaccharides that are coated with gold nanoparticles to act as a sacrificial matrix layer. Colloidal gold is removed using cyanide-assisted hydrolysis. The combination of chitosan and alginate is an effective antibacterial substance [52]. Chitosan and chitooligosaccharide were added to the cellulose matrix to increase the antibacterial activity, and tests were performed against Gram-positive and Gram-negative bacteria. The results indicated that they have favorable antibacterial activities, and when compared to pure bacterial cellulose (matrix), the BC-CS and BC-COS exhibited low porosity and a dense structure. This BC-COS composite material has excellent suitability for food and medicinal applications [53]. The antimicrobial activity of chitosan nanoparticles against tomato phytopathogens was assessed via the preparation and testing of chitosan nanoparticles. The testing process involves the use of pathogens such as Phytophthora capsici, Colletotrichum gelosporidies, Sclerotinia sclerotiorum, Gibberella fugikuori, and Fusarium oxysporum. Chitosan nanoparticles had a higher inhibitory impact on phytopathogenic bacteria, and both chitosan and chitosan nanoparticles prohibited the development of Erwinia and Xanthomonas [54]. Silver nanoparticles encapsulated in chitosan–silica scaffolds were synthesized using an electrospinning approach, as well as Ag/CS/silica composites. The tests were conducted with and without the inclusion of silver nanoparticles, and the results indicated that the inhibitory effect against bacteria was strengthened, while the addition of silica improved the composite's mechanical qualities. Biostatic activity was also seen as the diameters varied [55]. To disclose the antibacterial activity of chitosan on fabric made of cotton, a process of layering was used to coat it in its self-assembled form. Through layer-by-layer deposition, silver-loaded chitosan nanoparticles were coated up to 15 bilayers. The fabric displayed effective antibacterial characteristics without compromising its fundamental properties, including its tensile strength, bending stiffness, and air permeability. The base layer in the fabric was formed by dipping it in PSS and PAH aqueous solutions. The body layers were formed after the formation of the base layers by immersing the fabric in 0.6% PSS solution and a nanoparticle suspension consisting of chitosan and Ag. Each immersion was followed by washing with distilled water under sonication. The process can be repeated based on the number of layers required. PSS-CS-Ag layer formation on fabrics is explained in Figure 20 [56].

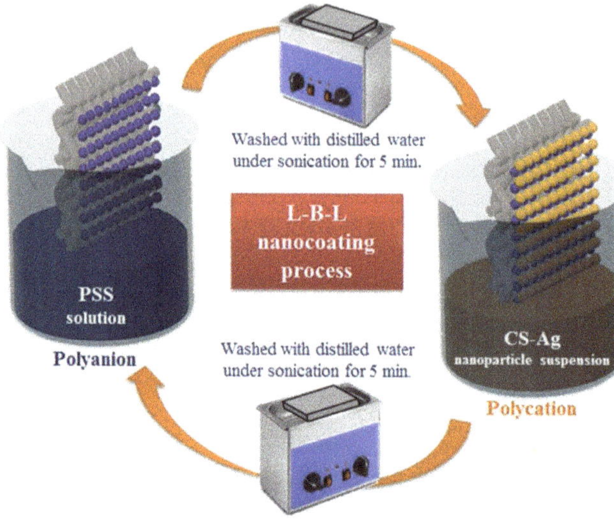

Figure 20. Formation of layers of PSS-CS-Ag on fabric through coating process. Reprinted from [56]. Copyright 2020 with permission from Elsevier.

Chitosan–graphite was produced using silica nanoparticles and the bactericidal action towards bacteria was determined after 18 h at 310 K. The production of radical oxygen species occurred as a result of the UV radiation, which caused damage to the bacteria [57]. The effects of nano cerium oxide particles on chitosan films were examined, and antibacterial activity against *E. coli* and *S. aureus* bacteria was discovered. Additionally, this composite film was recommended for use as a coating and packaging material due to its high mechanical strength, flexibility, and antibacterial efficacy [58]. Antimicrobial scaffolds consisting of cuprous oxide nanoparticles and chitosan nanofibers were produced. The Cu_2O particles became smaller and their shape changed from cubic to irregular as the concentration of $CuSO_4$ increased. The composite exhibited increased hydrophilicity and antimicrobial action against both Gram-negative and -positive bacteria [59]. A starch-based film was created and coated with chitosan nanoparticles for antibacterial activity against *E. coli* and *S. aureus*. Additionally, mechanical, morphological, and biodegradable characteristics were enhanced [60].

3.4. Wound Dressing and Healing

When combined with wound dressing materials and plasters, chitosan acts as an antibacterial agent against germs and viruses and provides the antibodies necessary for rapid wound healing. Incorporating silver nanoparticles, graphene oxide, chitosan, and curcumin into PVA nanofibers resulted in a hybrid composite. The antibacterial activity was shown to be superior to that of other nanoparticles, and the inclusion of graphene oxide boosts the mechanical characteristics. The in vitro test proved its biocompatibility, and it may be used to patch wounds that require both mechanical and antibacterial properties [61]. Chitosan coated with copper oxide and copper nanoparticles is produced for wound dressing applications. Copper oxide and copper nanoparticles with chitosan caps were synthesized via a simple chemical reduction of Cu2+ ions using ascorbic acid and sodium hydroxide. The antibacterial activity of the produced composite was evaluated using the inhibitory zone method against Gram-positive and -negative bacteria and found to be significantly greater than that of other nanomaterials, indicating that it is a superior alternative for wound dressing [62]. Electrospun polyaniline/chitosan nanofiber membranes were created and evaluated for their antibacterial activity in the treatment of chronic wounds and for their ability to minimize wound bioburden. The results suggested that increasing the concentration of polyaniline enhances the bactericidal activity and that it is also efficient against both Gram-positive and -negative bacteria. This property makes the polyaniline/chitosan membrane ideal for wound dressings and other healthcare applications [63]. For antimicrobial and wound healing applications, polyvinyl/chitosan nanofibers were combined with carboxymethyl chitosan nanoparticles encapsulated with an antibacterial peptide. These nanofibers containing varying concentrations of nanoparticles were applied to a mouse's skin wound and demonstrated improved wound healing and antibacterial activity. The improved day wise wound healing progress in the mouse skin with different levels of concentration is shown in Figure 21 [64].

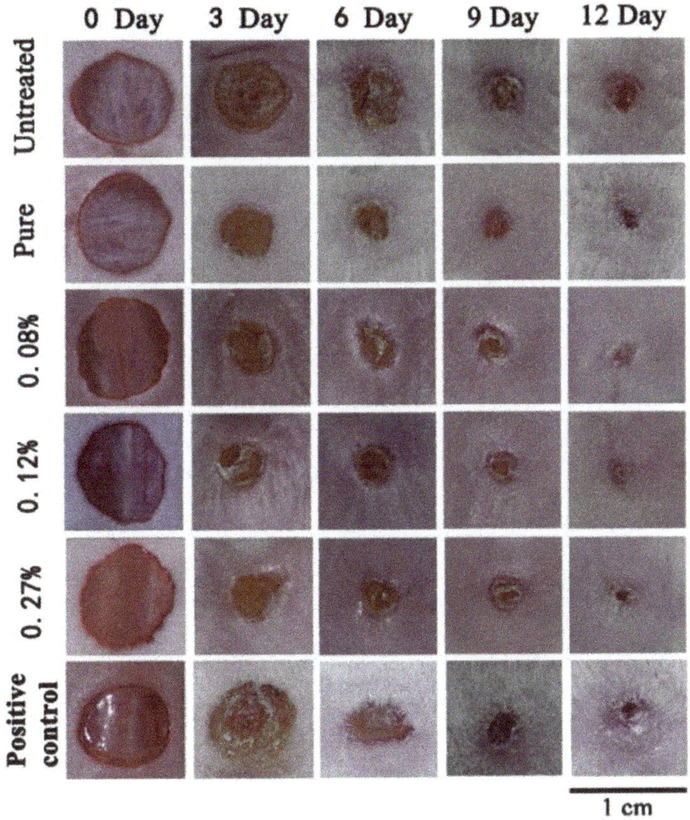

Figure 21. Stages of wound healing effects on mice with different levels of treated groups. Reprinted from [64]. Copyright 2020 with permission from Elsevier.

There are many factors that affect the wound healing process. The process and time taken to heal a wound completely will vary between a normal and a person with diabetes. Factors such as infection, oxygenation, interruption of foreign bodies, wound depth and area, age, gender, obesity, medications, smoking, and alcoholism influence the wound healing process. Figure 22 indicates the various stages of the wound healing process. The wound healing process involves various stages such as coagulation, inflammation, proliferation, and remodeling [65].

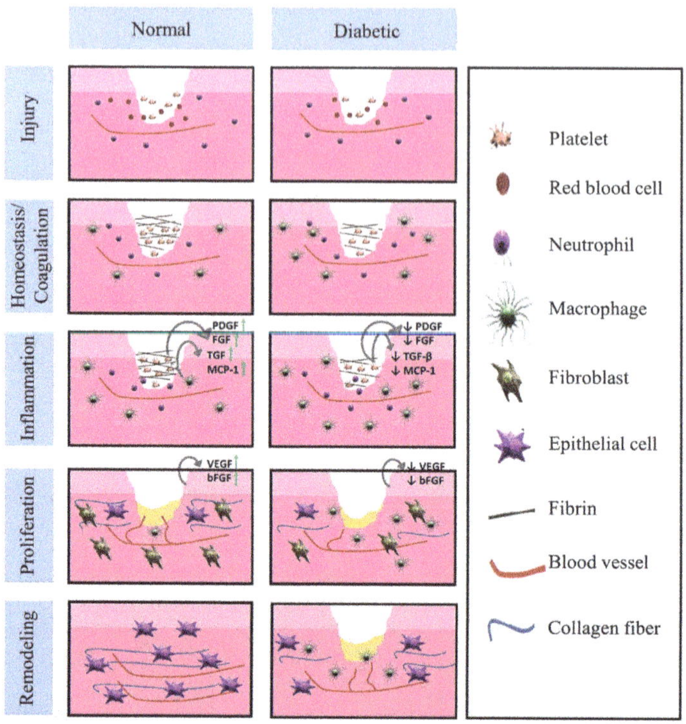

Figure 22. Differences in the wound healing phases between normal and diabetic persons. Reprinted from [65]. Copyright 2013 with permission from Elsevier.

3.5. Food Preservation and Packaging

Chitosan has the capacity to extend the shelf life of raw meat and food items while also minimizing bacterial and viral attack in preserved food. This enables naturally accessible fruits and vegetables to be preserved for longer lengths of time without the addition of preservatives. Chitosan was created in conjunction with zinc oxide coated in gallic acid films to provide an eco-friendly material for food packaging. The addition of gallic acid to chitosan increases its mechanical properties, and SEM images can be used demonstrate that the materials are compatible, suggesting that it could be used as an active material in food packaging [66]. Bacteria such as Escherichia coli and Salmonella enterica serovar typhimurium, which cause food contamination, have been researched using chitosan-based nanofibers. Chitosan nanofibers are created by electrospinning polyethylene oxide. Resistance to germs was determined in vitro, and the shelf life and preservation of red meat were also tested. The results demonstrated that the chitosan membrane was bactericidal, with a 99.9% decrease rate. The fresh meat's shelf life was also increased by seven days, confirming chitosan's contributions to the meat preservation and food packaging industries [67]. The antibacterial activity of chitosan-based nanofiber membranes generated via electrospinning was evaluated against Gram-positive and -negative bacteria. The results suggested that chitosan nanofibers operate as bacterial disruptors and perforators, as when the chitosan membrane comes into contact with negatively charged bacterial cells the membrane ruptures and protein and DNA leakage occur. As a result, it was established that chitosan membranes are good materials for food packing because they help prevent the spread of flora and infections [68].

The fungal growth on the bread pieces packed in various treated packages was observed for 10 days and the results were recorded. It was found that bread pieces packed in LDPE and neat chitosan packages showed fungal growth after 10 days, but packages made of chitosan–AKEO films showed no fungal growth, meaning the antifungal properties were proven. The fungal growth levels on the bread pieces on day 10 with various levels of coating on the packaging material are shown in Figure 23 [69].

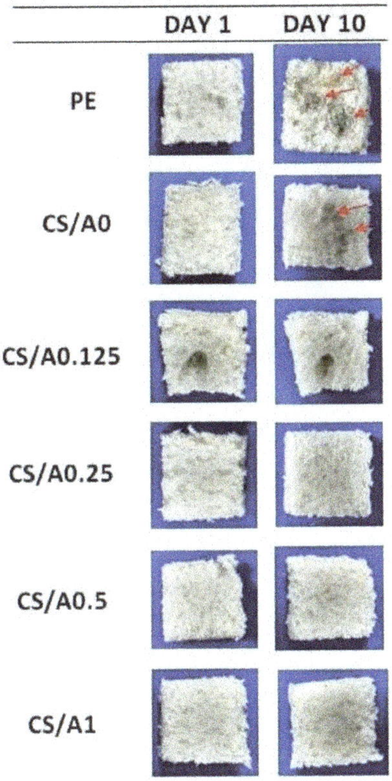

Figure 23. Inhibition of fungal growth on bread pieces packed in modified chitosan films (fungal growth is indicated in red arrows). Reprinted from [69]. Copyright 2018 with permission from Elsevier.

The main reason for the weight loss in the organic substances is the dry matter consumption and stomatal transpiration due to respiration process. A weight loss of 28% was observed after 5 days of storage in chitosan film, which was found to be lower in films made using chitosan LPP groups. In particular, the package with chitosan and 10LPP showed lowest the weight loss of about 6.5%. The hydrogen bond generation between LPP and CS was the reason behind this result, as it reduces the water vapor loss from the package. Apple pieces packed in packages with different levels of chitosan LPP concentrations is shown in Figure 24 [70]. Organic foods and foods subjected to decay are stored in packaging materials embedded or reinforced with chitosan, so that the antimicrobial action of chitosan protects the food product from the attack of bacteria and viruses and extends the shelf life by retaining the moisture content in the food product itself. As the chitosan is non-toxic, if by any chance it mixes with the food items it will not be harmful to humans and will maintain the freshness of the food products. If chitosan is added to any kind of plastic used for packaging purposes, then after usage if the package is removed, the chitosan will take care of the biodegradation process.

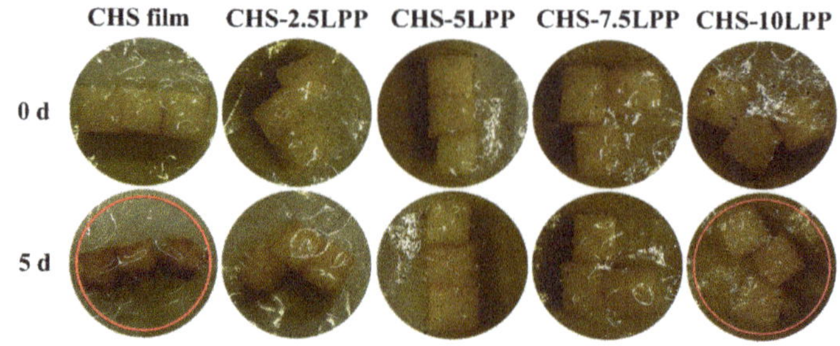

Figure 24. Fresh cut apples stored in chitosan and various concentration of LPP packages after 5 days [70].

3.6. Agricultural Activities

The ability of fruits and leaves to retain moisture can be enhanced by covering them with chitosan. This also increases the shelf life of the fruits and vegetables. Chitosan cerium oxide nanoparticles were synthesized from spherical plant leaves and demonstrated superior antibacterial properties against infections, as well as being effective in biomedical applications. Figure 25 explains the effects of chitin and chitosan on plants [71].

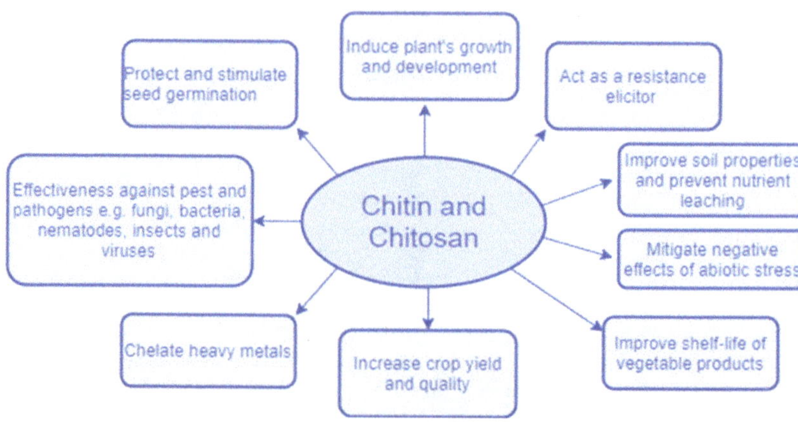

Figure 25. Possible applications of chitin and chitosan in agriculture [72].

To inhibit the growth of *S. aureus* and *E. coli*, chitosan-based coating films with varying titanium dioxide concentrations were produced. Chitosan with 0.05% titanium dioxide nanoparticles displayed the best thermal stability and exhibited superior inhibitory actions against bacteria; additionally, it was suggested that chitosan be utilized as a packaging material for vegetables and fruits to extend their shelf life [73]. To prevent blueberries from bacterial attack, they were preserved using chitosan/silica/nisin sheets. The results indicated that the pH value increased as a result of the addition of nisin, and that the turbidity level was also elevated. When nanoparticles are added to chitosan membranes, their tensile strength and ductility are decreased. The results demonstrated that the fruits lost some moisture and that CH-SN-N films can be utilized to preserve blueberries and extend their shelf life [74].

The maize plants under salt stress were taken for testing and treated with free S-nitroso-mercaptosuccinic acid and compared with another set of plants treated with the same acid encapsulated with chitosan nanoparticles at different concentrations. The 100 μM S-nitroso-MSA-chitosan treatment was found to provide effective relief against salt stress in the plants. The conditions of the plants with different levels of chitosan treatment in combination with NaCl are shown in Figure 26 [75].

Figure 26. Plants under salt stress and under different treatments [75].

Chitosan is a natural growth regulator that improves the defense against diseases. Chitosan improves plant growth by increasing water and nutrient intake. Chitosan helps in generating hydrolytic enzymes, which helps with the mobilization of starch and proteins. Plant hormones such as auxin and cytokinin were activated by chitosan, which promoted the root cells and increased nutrient intake. Seeds that underwent chitosan priming showed stimulated germination and vigor index rates. Genetic activation by chitosan in plants improves the growth of roots and the root biomass, meaning the canopy diameter, leaf area, and number of leaves, and height of the plant are increased. Due to the higher levels of photosynthesis, the fruit size and weight and the overall quality of the fruits are also improved. The influence of chitosan in the growth and quality of output from plants in the form of fruit is explained in Figure 27 [76]. The retention of moisture in fruits and vegetables after harvesting is ensured by coating them with chitosan. This keeps the fruit fresh and enhances the shelf life post-harvesting.

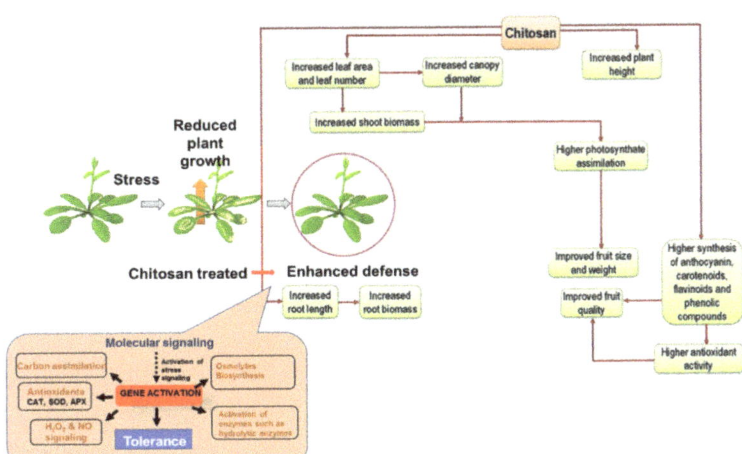

Figure 27. Plant growth conditions under the influence of chitosan [76].

3.7. Drug Delivery

Chitosan has excellent pH sensitivity, and by combining it with hydrophobic groups, more flexible chitosan polymers are formed. Chitosan can be employed in cancer drug delivery because it can release doxorubicin, an anticancer agent, into tumor cells at a lower pH level, resulting in increased antitumor activity [77]. The most critical properties required of medicinal and pharmaceutical materials are biocompatibility and long-term stability. The carbon quantum dots were synthesized using chitosan. Quantum dots with carbon as their core exhibit visible-range luminescence and have been demonstrated to be useful in controlled medication delivery and cell labelling [78]. Curcumin–chitosan–zinc oxide was synthesized using a one-pot technique. It was discovered that it has higher influence against MRSA and *E. coli* than commercially supplied amoxicillin. Following an examination of the CCZ's cytotoxic effects on grown human breast cancer cells, it was determined that it has potential for advanced medicinal applications [79]. For biomedical applications, bioinspired membranes comprised of green nanosilver and chitosan were created using a bottom-up eco-friendly design. This composite material demonstrated improved hemocompatibility, a high antioxidant capacity, and antiproliferative activity against cancer cells, as well as no toxicity against normal cells [80]. Graphene sheets coated in chitosan nanoparticles were tested against multidrug-resistant bacteria and found to be 90% harmful to Artemia franciscana after a 24 h incubation period [81]. Chitosan is frequently employed in the field of medical science due to its biodegradability and compatibility with living organisms. It is used to repair bone, regenerate tissue, and create dental adhesives, as well as to resist oral illnesses. Due to its special features, its use in dentistry is expanding [82]. Chitosan has been shown to be biocompatible, biosafe, and bioactive against the SARS, corona, and AIDS viruses, all of which pose a significant threat to human civilization. The addition of chitosan to ancient medications enhanced their antimicrobial properties [83].

Figure 28 shows the flow process through which a nanocarrier is converted into a functionalized nanocarrier that carries a drug and reacts with the malignant cell.

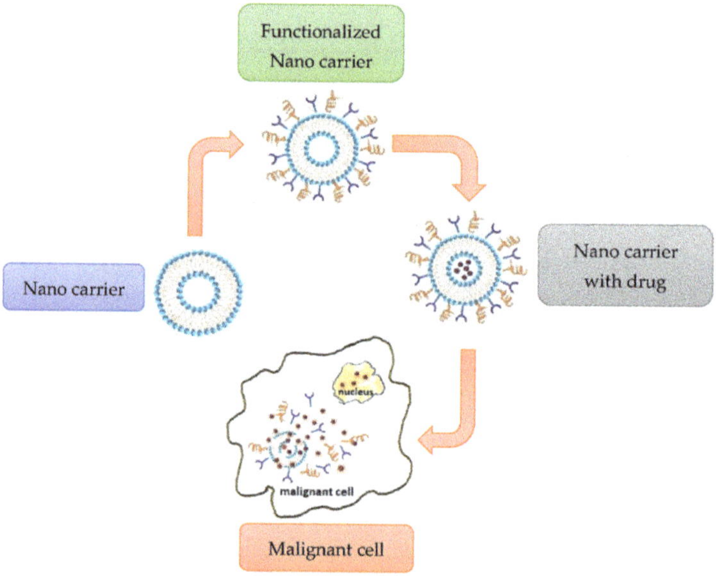

Figure 28. Process of drug delivery. Modified from [84].

The properties and characteristics related to drug delivery for chitosan are described in Figure 29. Polycation enhances absorption and mucoadhesion on dental surfaces, confirming the electrostatic interactions between negatively charged proteins and surfaces. When the drug delivery is increased, this increases the number of positive charges, which improves the mucoadhesion. Chitosan showed 15 to 40% weight reduction rates after 90 days of implantation. No allergy or toxicity was experienced during human trials due to the biocompatibility and biodegradability of chitosan. The adhesion of the drug to the surface must be ensured for a sufficient amount of time, and this was possible due to chitosan's bioadhesion properties. The bacteriostasis properties of chitosan ensures the inhibition of bacteria and other microorganisms. Stimuli responsiveness allows the release of drugs based on changes in the environmental conditions, meaning such systems are termed "intelligent" drug delivery systems. Chitosan is soluble under acidic environments, and to improve the solubility and drug delivery ability, quaternization, carboxylation, and sulfation are performed [85]. Due to the abovementioned properties of chitosan, it is considered to be the best drug carrying and drug delivery agent.

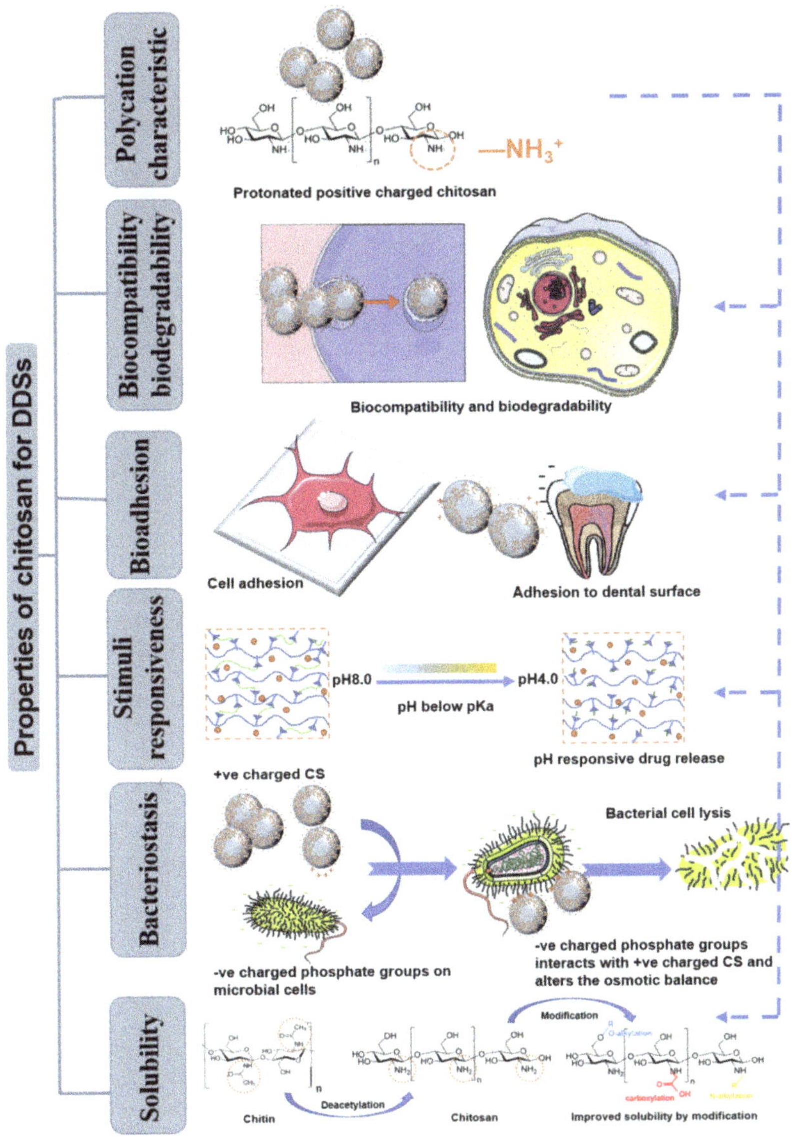

Figure 29. Characteristics and properties of chitosan as the best drug delivery agent. Reprinted from [85]. Copyright 2022 with permission from Elsevier.

3.8. Other Applications

Vacuum filtration was used to add low-viscosity chitosan to carrot cellulose nanofibrils in the range of 9 to 33% on a weight basis. The results indicated that adding chitosan improves the contact angle with water and renders the composite surface hydrophobic. The addition of chitosan increases the thermal stability and decreases the Young's modulus, while the inhibitory effects increase with the increasing chitosan concentration [86]. The silver/chitosan nanoparticles were formed in situ and exhibited superior mechanical characteristics and stability in bodily fluid. The nanoparticles had antibacterial effects against *E. coli* and Staphylococcus aureus bacterial strains [87]. Natural chitosan and gelatin polymers synthesized in ternary solvents have a better water absorption capacity and are referred to as green superabsorbent polymers. Under optimum conditions, the composite reached water saturation in less than 60 min. Additionally, it has a high capacity for water absorption throughout broad temperature, pH, and salt concentration ranges. Without undergoing any chemical reactions, gelatin and chitosan can be mixed [88]. Chitosan/gold nanoparticles efficiently inhibit bacterial activity in human cells. To investigate the interaction of chitosan with bacterial membranes, simulation models were created. The antibacterial activity of the Cs-Au nanoparticles was determined to be satisfactory when compared to the simulated model [89]. The addition of rhamnolipids to chitosan was shown to be successful in developing a nanocomposite that targets Gram-positive bacteria. C/RL nanocomposites exhibit enhanced antibacterial activity while exhibiting less cytotoxicity, making them more suited for pharmaceutical applications [90]. To improve the tensile characteristics and hydrophobicity of chitosan, montmorillonite packed with carboxymethyl cellulose was added, resulting in good dispersion of nanoclay. Increased MMT addition disrupts the biopolymer plasticizer interactions, increasing the surface's wettability [91]. OCMCS-SB, a stabilizer agent synthesized from chitosan and palladium, was tested for stability during Suzuki reactions and found to be satisfactory, with the possibility of further applications in organic transformations [92]. To increase the efficiency of the solar steam generator, semi-conductive in-situ-polymerized MnO2 nanowire–chitosan hydrogels were vertically stacked in macropore water channels. SPM-CH hydrogels enhance the lattice vibrations, while the polymeric network facilitates the creation of intermediate water clusters for steam generation. The solar energy conversion efficiency was determined to be 90.69% while the solar absorption was determined to be 94% using these hydrogels [93]. Separators for microbial fuel cells were built from self-assembled chitosan/montmorillonite. The resistance was lowered by 73.2%, increasing the proton conductivity, while the anode and cathode charge transfer impedances were reduced by 96.44 and 66.14%, respectively [94]. Pure chitosan was also used in the production of wine [95]. Chitosan can also be used as a biodegradable adhesive in woodworking. Synthetic adhesives are hazardous, unsustainable, and volatile. Chitosan-oxidized starch adhesives cure at lower temperatures and exhibit superior bonding and water resistance [96].

Manganese dioxide nanowires in chitosan solution were freeze-cast in the presence of nitrogen and then freeze-dried for 48 h, leading to the formation of MnO_2 hydrogel. The semi-conductive in situ polymerization of polypyrrole films of SPM-CH hydrogel enhances the wettability, tortuosity, and solar absorption. The production process for the SPM–chitosan hydrogel is shown in Figure 30 [93]. The membranes used for the fuel cell are produced via a membrane casting process. The membranes are made of chitin nanowhiskers arranged in a chitosan matrix. Due to the higher proton conductivity and lower methanol permeability, there is great potential for use as electrolyte membranes in fuel cells [97]. Chitosan combined with silica on quartz crystal microbalance sensors provided better film formation abilities in the composite. The modified sensor showed improved sensitivity and reliability, and this type of sensors can be used to detect humidity in the air [98]. Table 4 indicates the various applications of chitin and chitosan and provides recommendations with respect to their capability for drug delivery, as well as their molecular weights.

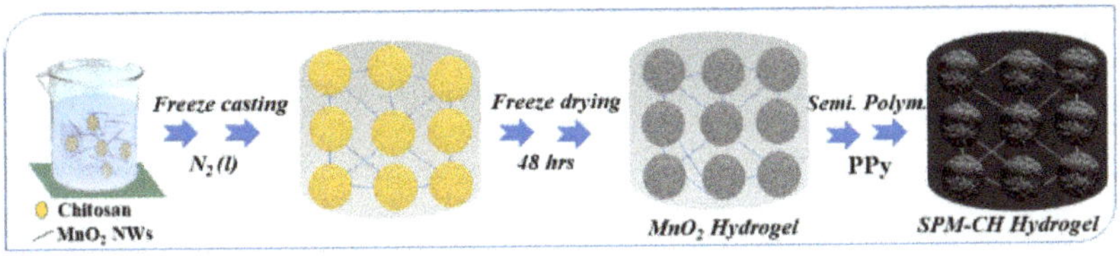

Figure 30. Process involved in making SPM–chitosan hydrogels. Reprinted from [93]. Copyright 2021 with permission from the American Chemical Society.

Table 4. Various applications of chitin and chitosan and general recommendations [99].

Application		General Recommendations
Healing of wounds		Chitosan preferred over chitin due to higher drug delivery capability
DD systems		Higher drug delivery capability and higher Mw
Repairing of Scaffolds		Good proliferation and structure Higher Mw results in prolonged biodegradation
Enzyme immobilization	Adsorption	Chitin used for positively charged and neutral proteins Chitosan used for negatively charged proteins It possess higher drug delivery capability
	Covalent	Chitosan is used for immobilization at multipoints Chitin with higher DD or chitosan with lower DD is used for single-point immobilization
	Encapsulation	Chitosan has higher Mw, higher drug delivery rates, and better retention Chitosan—alginate PECs possess medium Mw and have better stability under different conditions
Food preservative		Higher drug delivery Medium and lower Mw values
Waste water treatment		Depending on pollutants and water conditions such as pH and ionic strength. Chitosan is preferred over chitin due to higher drug delivery ability and lower crystallinity
Metal reduction		Chitosan's characteristics decide the metal reduction rate, higher DD rate, and lower Mw results in the stabilization of nanoparticles

4. Conclusions

This work has concentrated on the numerous applications of chitosan and its dominant traits, including its biocompatibility, antibiotic capabilities, and antibacterial activities. It is readily available as its principal source is the waste from marine species, and because its conversion requires simple chemical processes, the cost of the chitosan is low. It has vast and unique applications in medicine, food preservation and packaging, waste water filtration, dye removal from industrial effluents, wound healing, cancer cell treatment, air filtration, and bone replacements and implants, as well as to enhance the efficiency of solar cells. The water filtration membrane's bioactivity and antifouling qualities are also enhanced. It has also been stated that chitosan has bioactive effects against SARS and COVID-19 viruses, which pose a threat to humans at the current time. Apart from the above-listed applications, chitosan is also used in wine making and is combined with natural adhesives to make it biodegradable. Chitosan also offers moderate mechanical capabilities with good surface hydrophilicity attributes.

5. Future Perspectives

Apart from bactericidal and biocompatibility uses, the material's mechanical and thermal qualities must be enhanced when combined with other materials. Currently, it is employed in conjunction with other materials in water filtration membranes via coating and electrospinning methods. The temporal window within which the membrane bioactivity is most effective has not been well characterized. Chitosan is employed in bone replacements and dental implants due to its non-toxic and biocompatible nature; however, its mechanical qualities such as its strength, corrosion and wear resistance, and toughness are not well defined, which may provide potential opportunities in the development of biomaterials. The addition of chitosan to composite materials used in food packaging and preservation will also have a significant influence, as during natural disasters, food must be preserved for extended periods of time before being consumed, necessitating a longer shelf life.

Moisturizing and retaining moisture on the skin's surface is a difficult task in cosmetics. If chitosan is incorporated into cosmetics and lotions, it will work as an antibiotic and provide antimicrobial resistance against a variety of airborne bacteria and viruses that cause fatal infections. Additional studies must be conducted on waste water filtration to ensure that contaminants are removed efficiently and at a low cost. Incorporating or embedding chitosan into masks, textiles, gloves, and personnel breathing systems will require additional research, and innovative materials resistant to disease-causing substances will need to be developed. In some articles, it has been stated that chitosan can absorb fat from the foods we consume and that it also aids in weight loss. However, no work has shown these aspects empirically, and if established these advantages would represent significant advances in the field of medical science. Nano forms of chitosan can be obtained and filled along with other composites to study the differences in the properties. Catalysts must be developed for use in chitosan conversion processes [100]. Chitosan can be developed and used in dabs, suspensions, wipes, strands, frameworks, drinks, photography, and hydroponics [101].

The step-by-step process of converting raw chitosan into modified chitosan and the sorption, adsorption, and regeneration of chitosan are shown in Figure 31. Chitosan has applications in the recovery of gold from aqueous solutions via modification. Raw chitosan due to its Mw and degree of deacetylation affects the metal absorption ability, meaning the raw chitosan must be modified. Modifications were performed chemically and physically, and the modified chitosan showed better metal absorption and reusability abilities than raw chitosan [102]. Among the available formulations, chitosan-based nanocarriers are promising sources for the treatment of breast cancer and inclusion of chitosan, which when used as a drug carrier will bring about immense changes in the cancer treatment by reducing the cost and increasing the rates of survival and recovery [103]. In dental treatments, the usage of chitosan is also increasing due to its unique characteristics, such as its biocompatibility, biodegradability, hydrophilicity, antifungal activity, and bioactivity. In the future, chitosan will play a major role in dental repairs and in producing teeth with antimicrobial activity [82]. The incorporation of chitosan in medication such as in drug delivery will improve the treatment quality and the patient's recovery rate from illness. Microneedles are under development in drug delivery and the increasing demand for their use will necessitate larger-scale production. The product design process is very important for microneedles if they are to be accepted by patients. Chitosan-based microneedles can be developed so that in future pandemics the diseases can be treated in an effective way [104]. Chitosan can also be used as a catalyst and in the trans-esterification of various oils in the presence of methanol. Furthermore, 90% biodiesel yield can be obtained using chitosan–cryogel beads over 8 to 32 h. In the future, chitosan in combination with other elements could be used for the extraction of biodiesels from various oils by acting as a catalyst. At the same time, it could be used in combination with biodiesel to investigate the pollution created so that newer processes can be implemented to increase efficiency. Chitosan is also used in the polymer electrolyte membrane in the fuel cells, as normal electrolyte fuel cells are expensive. The ionic conductivity can be improved via crosslinking, meaning cost-effective, biodegradable electrolyte membranes can be produced. In the future, the usage of electric and fuel-cell-operated vehicles will increase and the world will be in need of low-cost and ecofriendly electrolyte membranes with high levels of conversion efficiency. Chitosan will be the most suitable material for such purposes [105].

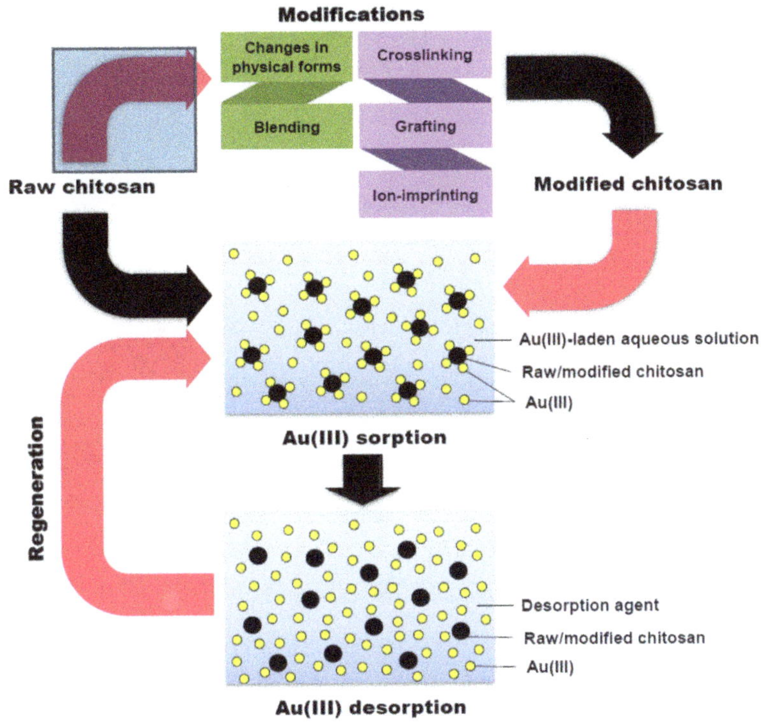

Figure 31. Process flow of gold recovery from aqueous solution using raw and modified chitosan. Reprinted from [102]. Copyright 2021 with permission from Elsevier.

Figure 32 explains the working principles of a surface plasmon resonance sensor and how the quantum dots of chitosan and graphene influence the sensitivity of the sensor. The surface-modified sensor was used for femtomolar detection and it was found that the modification of the sensor chip with chitosan–graphene quantum dots improved its sensitivity. Quantum dots of chitosan and graphene form a thin film that changes the refractive index, thereby shifting the resonance angle [106]. On the whole, chitosan has applications in almost every field, especially in developing fields, being used in fuel cells, electrolytic membranes for battery, sensors with good sensitivity, biodiesel development and extraction, agriculture, post-harvest processing, plant growth enhancement, for the removal of herbicides and pesticides from soil, for the removal of colors from dyes in the textile industry, for the removal of microbes and heavy metal and ions from industrial effluents, for the extraction and separation of gold from aqueous solution, for the removal of contaminants from drinking water and air, and for use in low-cost and effective drug delivery agents, in addition to aiding in wound healing, dentistry repairs, and scaffold development; it even acts as an effective drug delivery agent in the treatment of breast cancer and other chronic diseases. Chitosan has great potential to be the next material to be utilized in multifarious applications.

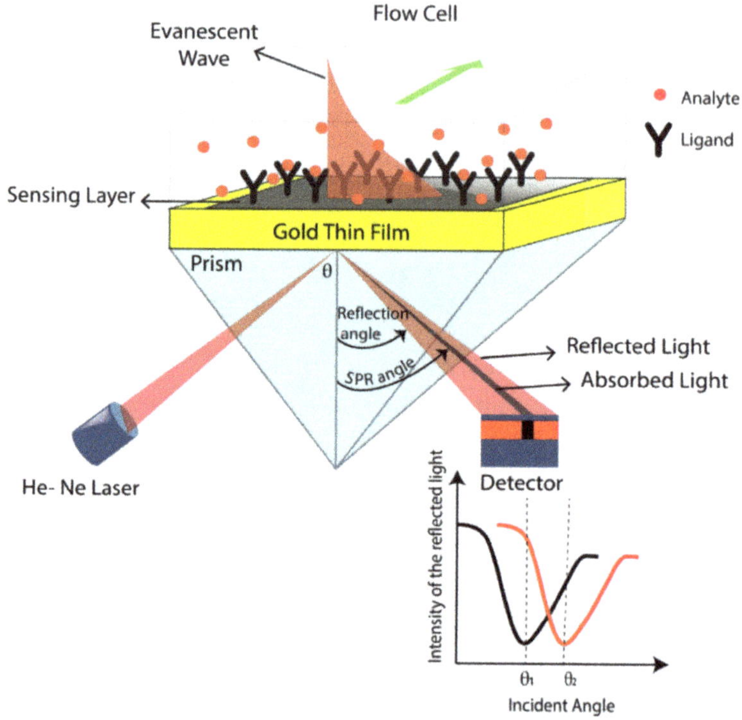

Figure 32. Surface plasmon resonance sensor setup. Reprinted from [106]. Copyright 2021 with permission from Elsevier.

Author Contributions: Conceptualization and original draft preparation, S.A.R. and J.K.; investigation and supervision, S.A.R.; Methodology, A.Z.H., M.A.B. and M.A.S.; final manuscript preparation, S.A.R.; funding acquisition, A.Z.H., M.A.B. and M.A.S. All authors have read and agreed to the published version of the manuscript.

Funding: The authors extend their appreciation to the Deanship for Research and Innovation, Ministry of Education in Saudi Arabia for funding this research work through the project number IFPRC-022-135-2020, and King Abdulaziz University, DSR, Jeddah, Saudi Arabia.

Institutional Review Board Statement: Not applicable.

Informed Consent Statement: Not applicable.

Data Availability Statement: Not applicable.

Acknowledgments: In this section, you can acknowledge any support given which is not covered by the author contribution or funding sections. This may include administrative and technical support, or donations in kind (e.g., materials used for experiments).

Conflicts of Interest: The authors declare no conflict of interest.

Abbreviations

SARS	Severe Acute Respiratory Syndrome
AIDS	Acquired immunodeficiency syndrome
Co/MCM	Chitosan–cobalt–silica
PVA-Co-PE	Poly(vinyl alcohol-co-ethylene)
Cr(IV)	Chromium–IV
Cu(II)	Cupric oxide
PVDF-S/MIL100-CS	Polyvinylidene fluoride metal organic framework chitosan
MWNT	Multiwalled nanotubes
PVA	Polyvinyl alcohol
PAN	Polyacrylonitrile
PEG	Polyethylene glycol
MWCNT	Multiwalled carbon nanotubes
E. coli	Escherichia coli
S. aureus	Staphylococcus aureus
Ni^{2+}	Nickel cation with two positive charges
$NaBH_4$	Sodium borohydride
FESEM	Field emission scanning electron microscope
SEM	Scanning electron microscope
EDAX	Energy-dispersive spectroscopy
Ni	Nickel
CS	Chitosan
FT-IR	Fourier transform infrared spectroscopy
ZnPc-CS	Zinc phthalocyanine chitosan
GSNO-PL/CS	No donor S-nitrosoglutathione-pluronic-chitosan
BC-CS	Bacterial cellulose–chitosan
BC-COS	Bacterial cellulose–chitooligosaccharide
Ag/CS/Silica	Silver–chitosan–silica
UV radiation	Ultraviolet radiation
K	Kelvin
Cu_2O	Cuprous oxide
$CuSO_4$	Copper sulphate
DNA	Deoxyribonucleic acid
CH-SN-N	Chitosan silica nanoparticles
MRSA	Methicillin-resistant staphylococcus aureus
CCZ	Curcumin chitosan zinc oxide
CS-Au	Chitosan gold nanoparticles
C/RL	Chitosan rhamnolipid
MMT	Montmorillonite
OCMCS-SB	O—Carboxymethyl chitosan Schiff base
MnO_2	Manganese dioxide
SPM-CH	Semi-conductive in-situ-polymerized MnO2 nanowire–chitosan hydrogels.
DD	Degree of deacetylation
Mw	Molecular weight

References

1. Kassem, A.; Ayoub, G.M.; Malaeb, L. Antibacterial Activity of Chitosan Nano-Composites and Carbon Nanotubes: A Review. *Sci. Total Environ.* **2019**, *668*, 566–576. [CrossRef] [PubMed]
2. Pervez, M.N.; Balakrishnan, M.; Hasan, S.W.; Choo, K.H.; Zhao, Y.; Cai, Y.; Zarra, T.; Belgiorno, V.; Naddeo, V. A Critical Review on Nanomaterials Membrane Bioreactor (NMS-MBR) for Wastewater Treatment. *npj Clean Water* **2020**, *3*, 43. [CrossRef]
3. Yadav, M.; Goswami, P.; Paritosh, K.; Kumar, M.; Pareek, N.; Vivekanand, V. Seafood Waste: A Source for Preparation of Commercially Employable Chitin/Chitosan Materials. *Bioresour. Bioprocess.* **2019**, *6*, 8. [CrossRef]
4. Upadhyay, U.; Sreedhar, I.; Singh, S.A.; Patel, C.M.; Anitha, K.L. Recent Advances in Heavy Metal Removal by Chitosan Based Adsorbents. *Carbohydr. Polym.* **2021**, *251*, 117000. [CrossRef] [PubMed]
5. Kulkarni, N.; Shinde, S.D.; Jadhav, G.S.; Adsare, D.R.; Rao, K.; Kachhia, M.; Maingle, M.; Patil, S.P.; Arya, N.; Sahu, B. Peptide-Chitosan Engineered Scaffolds for Biomedical Applications. *Bioconjugate Chem.* **2021**, *32*, 448–465. [CrossRef] [PubMed]

6. Negi, H.; Verma, P.; Singh, R.K. A Comprehensive Review on the Applications of Functionalized Chitosan in Petroleum Industry. *Carbohydr. Polym.* **2021**, *266*, 118125. [CrossRef] [PubMed]
7. Salama, A. Recent Progress in Preparation and Applications of Chitosan/Calcium Phosphate Composite Materials. *Int. J. Biol. Macromol.* **2021**, *178*, 240–252. [CrossRef]
8. Jaber, N.; Al-Remawi, M.; Al-Akayleh, F.; Al-Muhtaseb, N.; Al-Adham, I.S.I.; Collier, P.J. A Review of the Antiviral Activity of Chitosan, Including Patented Applications and Its Potential Use against COVID-19. *J. Appl. Microbiol.* **2022**, *132*, 41–58. [CrossRef]
9. Sharifianjazi, F.; Khaksar, S.; Esmaeilkhanian, A.; Bazli, L.; Eskandarinezhad, S.; Salahshour, P.; Sadeghi, F.; Rostamnia, S.; Vahdat, S.M. Advancements in Fabrication and Application of Chitosan Composites in Implants and Dentistry: A Review. *Biomolecules* **2022**, *12*, 155. [CrossRef]
10. Spoială, A.; Ilie, C.I.; Ficai, D.; Ficai, A.; Andronescu, E. Chitosan-Based Nanocomposite Polymeric Membranes for Water Purification—A Review. *Materials* **2021**, *14*, 2091. [CrossRef]
11. Gao, Y.; Wu, Y. Recent Advances of Chitosan-Based Nanoparticles for Biomedical and Biotechnological Applications. *Int. J. Biol. Macromol.* **2022**, *203*, 379–388. [CrossRef] [PubMed]
12. El Kady, E. Chitin, Chitosan and Glucan, Properties and Applications. *World J. Agric. Soil Sci.* **2019**, *3*, 1–19. [CrossRef]
13. Lv, S.H. High-Performance Superplasticizer Based on Chitosan. In *Biopolymers and Biotech Admixtures for Eco-Efficient Construction Materials*; Elsevier: Amsterdam, The Netherlands, 2016; pp. 131–150. [CrossRef]
14. Kumirska, J.; Weinhold, M.X.; Thöming, J.; Stepnowski, P. Biomedical Activity of Chitin/Chitosan Based Materials- Influence of Physicochemical Properties Apart from Molecular Weight and Degree of N-Acetylation. *Polymers* **2011**, *3*, 1875–1901. [CrossRef]
15. Gatto, M.; Ochi, D.; Yoshida, C.M.P.; da Silva, C.F. Study of Chitosan with Different Degrees of Acetylation as Cardboard Paper Coating. *Carbohydr. Polym.* **2019**, *210*, 56–63. [CrossRef]
16. Bansal, V.; Sharma, P.K.; Sharma, N.; Pal, O.P.; Malviya, R. Applications of Chitosan and Chitosan Derivatives in Drug Delivery. *Biol. Res.* **2011**, *5*, 28–37.
17. Ilyas, R.A.; Aisyah, H.A.; Nordin, A.H.; Ngadi, N.; Yusoff, M.; Zuhri, M.; Rizal, M.; Asyraf, M.; Sapuan, S.M.; Zainudin, E.S.; et al. Natural-Fiber-Reinforced Chitosan, Chitosan Blends and Their Nanocomposites for Various Advanced Applications. *Polymers* **2022**, *14*, 874. [CrossRef]
18. Qasim, S.B.; Zafar, M.S.; Najeeb, S.; Khurshid, Z.; Shah, A.H.; Husain, S.; Rehman, I.U. Electrospinning of Chitosan-Based Solutions for Tissue Engineering and Regenerative Medicine. *Int. J. Mol. Sci.* **2018**, *19*, 407. [CrossRef]
19. Green-warren, R.A.; Bontoux, L.; Mcallister, N.M.; Kovacevich, D.A.; Kuznetsova, C.; Tenorio, M.; Lei, L.; Pelegri, A.A.; Jonathan, P. Determining the Self-Limiting Electrospray Deposition Compositional Limits for Mechanically Tunable Polymer Composites. *ACS Appl. Polym. Mater.* **2022**, *4*, 3511–3519. [CrossRef]
20. Islam, S.; Jadhav, A.; Fang, J.; Arnold, L.; Wang, L.; Padhye, R.; Wang, X.; Lin, T. Surface Deposition of Chitosan on Wool Substrate by Electrospraying. *Adv. Mater. Res.* **2011**, *331*, 165–170. [CrossRef]
21. Garg, U.; Chauhan, S.; Upendra, N.; Jain, N. Current Advances in Chitosan Nanoparticles Based Drug Delivery and Targeting. *Adv. Pharm. Bull.* **2019**, *9*, 195–204. [CrossRef]
22. Shiekh, K.A.; Ngiwngam, K.; Tongdeesoontorn, W. Polysaccharide-Based Active Coatings Incorporated with Bioactive Compounds for Reducing Postharvest Losses of Fresh Fruits. *Coatings* **2022**, *12*, 8. [CrossRef]
23. Zhang, J.; Zeng, L.; Sun, H.; Zhang, J.; Chen, S. Using Chitosan Combined Treatment with Citric Acid as Edible Coatings to Delay Postharvest Ripening Process and Maintain Tomato (Solanum Lycopersicon Mill) Quality. *J. Food Nutr. Res.* **2017**, *5*, 144–150. [CrossRef]
24. Fosso-Kankeu, E.; de Klerk, C.M.; van Aarde, C.; Waanders, F.; Phoku, J. Antibacterial Activity of a Synthesized Chitosan-Silver Composite with Different Molecular Weights Chitosan against Gram-Positive and Gram-Negative Bacteria. In Proceedings of the International Conference on Advances in Science, Engineering, Technology and Natural Resources (ICA-SETNR-16), Parys, South Africa, 24–25 November 2016. [CrossRef]
25. Khan, S.A.; Khan, S.B.; Kamal, T.; Yasir, M.; Asiri, A.M. Antibacterial Nanocomposites Based on Chitosan/Co-MCM as a Selective and Efficient Adsorbent for Organic Dyes. *Int. J. Biol. Macromol.* **2016**, *91*, 744–751. [CrossRef] [PubMed]
26. Liu, K.; Cheng, P.; Wang, Y.; Zhong, W.; Lu, Z.; Li, M.; Liu, Q.; Wang, W.; Zhu, Q.; Wang, D. Concurrent Filtration and Inactivation of Bacteria Using Poly(Vinyl Alcohol-Co-Ethylene) Nanofibrous Membrane Facilely Modified Using Chitosan and Graphene Oxide. *Environ. Sci. Nano* **2017**, *4*, 385–395. [CrossRef]
27. Kolangare, I.M.; Isloor, A.M.; Karim, Z.A.; Kulal, A.; Ismail, A.F.; Inamuddin; Asiri, A.M. M. Antibiofouling Hollow-Fiber Membranes for Dye Rejection by Embedding Chitosan and Silver-Loaded Chitosan Nanoparticles. *Environ. Chem. Lett.* **2019**, *17*, 581–587. [CrossRef]
28. Bandara, P.C.; Nadres, E.T.; Rodrigues, D.F. Use of Response Surface Methodology to Develop and Optimize the Composition of a Chitosan-Polyethyleneimine-Graphene Oxide Nanocomposite Membrane Coating to More Effectively Remove Cr(VI) and Cu(II) from Water. *ACS Appl. Mater. Interfaces* **2019**, *11*, 17784–17795. [CrossRef]
29. Cho, K.Y.; Yoo, C.H.; Won, Y.J.; Hong, D.Y.; Chang, J.S.; Choi, J.W.; Lee, J.H.; Lee, J.S. Surface-Concentrated Chitosan-Doped MIL-100(Fe) Nanofiller-Containing PVDF Composites for Enhanced Antibacterial Activity. *Eur. Polym. J.* **2019**, *120*, 109221. [CrossRef]

30. Chaudhary, M.; Maiti, A. Fe–Al–Mn@chitosan Based Metal Oxides Blended Cellulose Acetate Mixed Matrix Membrane for Fluoride Decontamination from Water: Removal Mechanisms and Antibacterial Behavior. *J. Memb. Sci.* **2020**, *611*, 118372. [CrossRef]
31. Alshahrani, A.A.; Algamdi, M.S.; Alsohaimi, I.H.; Nghiem, L.D.; Tu, K.L.; Al-Rawajfeh, A.E.; in het Panhuis, M. The Rejection of Mono- and Di-Valent Ions from Aquatic Environment by MWNT/Chitosan Buckypaper Composite Membranes: Influences of Chitosan Concentrations. *Sep. Purif. Technol.* **2020**, *234*, 116088. [CrossRef]
32. Zhang, L.; Li, L.; Wang, L.; Nie, J.; Ma, G. Multilayer Electrospun Nanofibrous Membranes with Antibacterial Property for Air Filtration. *Appl. Surf. Sci.* **2020**, *515*, 145962. [CrossRef]
33. Makaremi, M.; Lim, C.X.; Pasbakhsh, P.; Lee, S.M.; Goh, K.L.; Chang, H.; Chan, E.S. Electrospun Functionalized Polyacrylonitrile-Chitosan Bi-Layer Membranes for Water Filtration Applications. *RSC Adv.* **2016**, *6*, 53882–53893. [CrossRef]
34. Jabur, A.R.; Abbas, L.K.; Moosa, S.A. Fabrication of Electrospun Chitosan/Nylon 6 Nanofibrous Membrane toward Metal Ions Removal and Antibacterial Effect. *Adv. Mater. Sci. Eng.* **2016**, *2016*, 5810216. [CrossRef]
35. Mohraz, M.H.; Golbabaei, F.; Yu, I.J.; Mansournia, M.A.; Zadeh, A.S.; Dehghan, S.F. Preparation and Optimization of Multifunctional Electrospun Polyurethane/Chitosan Nanofibers for Air Pollution Control Applications. *Int. J. Environ. Sci. Technol.* **2019**, *16*, 681–694. [CrossRef]
36. Liu, H.; Huang, J.; Mao, J.; Chen, Z.; Chen, G.; Lai, Y. Transparent Antibacterial Nanofiber Air Filters with Highly Efficient Moisture Resistance for Sustainable Particulate Matter Capture. *iScience* **2019**, *19*, 214–223. [CrossRef] [PubMed]
37. Khoerunnisa, F.; Rahmah, W.; Seng Ooi, B.; Dwihermiati, E.; Nashrah, N.; Fatimah, S.; Ko, Y.G.; Ng, E.P.; Khoerunnisa, F. Chitosan/PEG/MWCNT/Iodine Composite Membrane with Enhanced Antibacterial Properties for Dye Wastewater Treatment. *J. Environ. Chem. Eng.* **2020**, *8*, 103686. [CrossRef]
38. Nayl, A.A.; Abd-elhamid, A.I.; Awwad, N.S.; Abdelgawad, M.A.; Wu, J.; Mo, X.; Gomha, S.M.; Aly, A.A.; Bräse, S. Review of the Recent Advances in Electrospun Nanofibers Applications in Water Purification. *Polymers* **2022**, *14*, 1594. [CrossRef]
39. Ahmad, I.; Kamal, T.; Khan, S.B.; Asiri, A.M. An Efficient and Easily Retrievable Dip Catalyst Based on Silver Nanoparticles/Chitosan-Coated Cellulose Filter Paper. *Cellulose* **2016**, *23*, 3577–3588. [CrossRef]
40. Kamal, T.; Khan, S.B.; Asiri, A.M. Nickel Nanoparticles-Chitosan Composite Coated Cellulose Filter Paper: An Efficient and Easily Recoverable Dip-Catalyst for Pollutants Degradation. *Environ. Pollut.* **2016**, *218*, 625–633. [CrossRef]
41. Zhang, A.; Zhang, Y.; Pan, G.; Xu, J.; Yan, H.; Liu, Y. In Situ Formation of Copper Nanoparticles in Carboxylated Chitosan Layer: Preparation and Characterization of Surface Modified TFC Membrane with Protein Fouling Resistance and Long-Lasting Antibacterial Properties. *Sep. Purif. Technol.* **2017**, *176*, 164–172. [CrossRef]
42. Bagheripour, E.; Moghadassi, A.R.; Hosseini, S.M.; Van der Bruggen, B.; Parvizian, F. Novel Composite Graphene Oxide/Chitosan Nanoplates Incorporated into PES Based Nanofiltration Membrane: Chromium Removal and Antifouling Enhancement. *J. Ind. Eng. Chem.* **2018**, *62*, 311–320. [CrossRef]
43. Sangeetha, K.; Angelin Vinodhini, P.; Sudha, P.N.; Alsharani Faleh, A.; Sukumaran, A. Novel Chitosan Based Thin Sheet Nanofiltration Membrane for Rejection of Heavy Metal Chromium. *Int. J. Biol. Macromol.* **2019**, *132*, 939–953. [CrossRef]
44. Vakili, M.; Deng, S.; Cagnetta, G.; Wang, W.; Meng, P.; Liu, D.; Yu, G. Regeneration of Chitosan-Based Adsorbents Used in Heavy Metal Adsorption: A Review. *Sep. Purif. Technol.* **2019**, *224*, 373–387. [CrossRef]
45. Biao, L.; Tan, S.; Wang, Y.; Guo, X.; Fu, Y.; Xu, F.; Zu, Y.; Liu, Z. Synthesis, Characterization and Antibacterial Study on the Chitosan-Functionalized Ag Nanoparticles. *Mater. Sci. Eng. C* **2017**, *76*, 73–80. [CrossRef]
46. Venkatesan, J.; Anil, S.; Kim, S.K.; Shim, M.S. Chitosan as a Vehicle for Growth Factor Delivery: Various Preparations and Their Applications in Bone Tissue Regeneration. *Int. J. Biol. Macromol.* **2017**, *104*, 1383–1397. [CrossRef]
47. Ali, F.; Khan, S.B.; Kamal, T.; Anwar, Y.; Alamry, K.A.; Asiri, A.M. Anti-Bacterial Chitosan/Zinc Phthalocyanine Fibers Supported Metallic and Bimetallic Nanoparticles for the Removal of Organic Pollutants. *Carbohydr. Polym.* **2017**, *173*, 676–689. [CrossRef] [PubMed]
48. Cheah, W.Y.; Show, P.L.; Ng, I.S.; Lin, G.Y.; Chiu, C.Y.; Chang, Y.K. *Antibacterial Activity of Quaternized Chitosan Modified Nanofiber Membrane*; Elsevier: Amsterdam, The Netherlands, 2019; Volume 126, ISBN 8862290616.
49. Pelegrino, M.T.; Lima, B.d.A.; do Nascimento, M.H.M.; Lombello, C.B.; Brocchi, M.; Seabra, A.B. Biocompatible and Antibacterial Nitric Oxide-Releasing Pluronic F-127/Chitosan Hydrogel for Topical Applications. *Polymers* **2018**, *10*, 452. [CrossRef] [PubMed]
50. Yu, X.; Guo, L.; Liu, M.; Cao, X.; Shang, S.; Liu, Z.; Huang, D.; Cao, Y.; Cui, F.; Tian, L. Callicarpa Nudiflora Loaded on Chitosan-Collagen/Organomontmorillonite Composite Membrane for Antibacterial Activity of Wound Dressing. *Int. J. Biol. Macromol.* **2018**, *120*, 2279–2284. [CrossRef]
51. Senthilkumar, P.; Yaswant, G.; Kavitha, S.; Chandramohan, E.; Kowsalya, G.; Vijay, R.; Sudhagar, B.; Kumar, D.S.R.S. Preparation and Characterization of Hybrid Chitosan-Silver Nanoparticles (Chi-Ag NPs); A Potential Antibacterial Agent. *Int. J. Biol. Macromol.* **2019**, *141*, 290–297. [CrossRef]
52. Belbekhouche, S.; Bousserrhine, N.; Alphonse, V.; Le Floch, F.; Charif Mechiche, Y.; Menidjel, I.; Carbonnier, B. Chitosan Based Self-Assembled Nanocapsules as Antibacterial Agent. *Colloids Surf. B Biointerfaces* **2019**, *181*, 158–165. [CrossRef]
53. Yin, N.; Du, R.; Zhao, F.; Han, Y.; Zhou, Z. Characterization of Antibacterial Bacterial Cellulose Composite Membranes Modified with Chitosan or Chitooligosaccharide. *Carbohydr. Polym.* **2020**, *229*, 115520. [CrossRef]

54. Oh, J.W.; Chun, S.C.; Chandrasekaran, M. Preparation and in Vitro Characterization of Chitosan Nanoparticles and Their Broad-Spectrum Antifungal Action Compared to Antibacterial Activities against Phytopathogens of Tomato. *Agronomy* **2019**, *9*, 21. [CrossRef]
55. Zienkiewicz-Strzałka, M.; Deryło-Marczewska, A.; Skorik, Y.A.; Petrova, V.A.; Choma, A.; Komaniecka, I. Silver Nanoparticles on Chitosan/Silica Nanofibers: Characterization and Antibacterial Activity. *Int. J. Mol. Sci.* **2020**, *21*, 166. [CrossRef] [PubMed]
56. Gadkari, R.R.; Ali, S.W.; Joshi, M.; Rajendran, S.; Das, A.; Alagirusamy, R. Leveraging Antibacterial Efficacy of Silver Loaded Chitosan Nanoparticles on Layer-by-Layer Self-Assembled Coated Cotton Fabric. *Int. J. Biol. Macromol.* **2020**, *162*, 548–560. [CrossRef] [PubMed]
57. Jamshidi, D.; Sazegar, M.R. Antibacterial Activity of a Novel Biocomposite Chitosan/Graphite Based on Zinc-Grafted Mesoporous Silica Nanoparticles. *Int. J. Nanomed.* **2020**, *15*, 871–883. [CrossRef]
58. Kızılkonca, E.; Torlak, E.; Erim, F.B. Preparation and Characterization of Antibacterial Nano Cerium Oxide/Chitosan/Hydroxyethylcellulose/Polyethylene Glycol Composite Films. *Int. J. Biol. Macromol.* **2021**, *177*, 351–359. [CrossRef]
59. Zhou, X.; Yin, A.; Sheng, J.; Wang, J.; Chen, H.; Fang, Y.; Zhang, K. In Situ Deposition of Nano Cu2O on Electrospun Chitosan Nanofibrous Scaffolds and Their Antimicrobial Properties. *Int. J. Biol. Macromol.* **2021**, *191*, 600–607. [CrossRef]
60. Babaee, M.; Garavand, F.; Rehman, A.; Jafarazadeh, S.; Amini, E.; Cacciotti, I. Biodegradability, Physical, Mechanical and Antimicrobial Attributes of Starch Nanocomposites Containing Chitosan Nanoparticles. *Int. J. Biol. Macromol.* **2022**, *195*, 49–58. [CrossRef]
61. Dubey, P.; Gopinath, P. PEGylated Graphene Oxide-Based Nanocomposite-Grafted Chitosan/Polyvinyl Alcohol Nanofiber as an Advanced Antibacterial Wound Dressing. *RSC Adv.* **2016**, *6*, 69103–69116. [CrossRef]
62. Jayaramudu, T.; Varaprasad, K.; Pyarasani, R.D.; Reddy, K.K.; Kumar, K.D.; Akbari-Fakhrabadi, A.; Mangalaraja, R.V.; Amalraj, J. Chitosan Capped Copper Oxide/Copper Nanoparticles Encapsulated Microbial Resistant Nanocomposite Films. *Int. J. Biol. Macromol.* **2019**, *128*, 499–508. [CrossRef]
63. Moutsatsou, P.; Coopman, K.; Georgiadou, S. Chitosan & Conductive PANI/Chitosan Composite Nanofibers—Evaluation of Antibacterial Properties. *Curr. Nanomater.* **2018**, *4*, 6–20. [CrossRef]
64. Zou, P.; Lee, W.H.; Gao, Z.; Qin, D.; Wang, Y.; Liu, J.; Sun, T.; Gao, Y. Wound Dressing from Polyvinyl Alcohol/Chitosan Electrospun Fiber Membrane Loaded with OH-CATH30 Nanoparticles. *Carbohydr. Polym.* **2020**, *232*, 115786. [CrossRef] [PubMed]
65. Moura, L.I.F.; Dias, A.M.A.; Carvalho, E.; de Sousa, H.C. Recent Advances on the Development of Wound Dressings for Diabetic Foot Ulcer Treatment—A Review. *Acta Biomater.* **2013**, *9*, 7093–7114. [CrossRef] [PubMed]
66. Yadav, S.; Mehrotra, G.K.; Dutta, P.K. Chitosan Based ZnO Nanoparticles Loaded Gallic-Acid Films for Active Food Packaging. *Food Chem.* **2021**, *334*, 127605. [CrossRef] [PubMed]
67. Arkoun, M.; Daigle, F.; Heuzey, M.C.; Ajji, A. Mechanism of Action of Electrospun Chitosan-Based Nanofibers against Meat Spoilage and Pathogenic Bacteria. *Molecules* **2017**, *22*, 585. [CrossRef]
68. Arkoun, M.; Daigle, F.; Heuzey, M.C.; Ajji, A. Antibacterial Electrospun Chitosan-Based Nanofibers: A Bacterial Membrane Perforator. *Food Sci. Nutr.* **2017**, *5*, 865–874. [CrossRef]
69. Priyadarshi, R.; Sauraj; Kumar, B.; Deeba, F.; Kulshreshtha, A.; Negi, Y.S. Chitosan Films Incorporated with Apricot (Prunus Armeniaca) Kernel Essential Oil as Active Food Packaging Material. *Food Hydrocoll.* **2018**, *85*, 158–166. [CrossRef]
70. Jiang, L.; Luo, Z.; Liu, H.; Wang, F.; Li, H.; Gao, H.; Zhang, H. Preparation and Characterization of Chitosan Films Containing Lychee (Litchi Chinensis Sonn.) Pericarp Powder and Their Application as Active Food Packaging. *Foods* **2021**, *10*, 2834. [CrossRef]
71. Senthilkumar, R.P.; Bhuvaneshwari, V.; Ranjithkumar, R.; Sathiyavimal, S.; Malayaman, V.; Chandarshekar, B. Synthesis, Characterization and Antibacterial Activity of Hybrid Chitosan-Cerium Oxide Nanoparticles: As a Bionanomaterials. *Int. J. Biol. Macromol.* **2017**, *104*, 1746–1752. [CrossRef]
72. Shahrajabian, M.H.; Chaski, C.; Polyzos, N.; Tzortzakis, N.; Petropoulos, S.A. Sustainable Agriculture Systems in Vegetable Production Using Chitin and Chitosan as Plant Biostimulants. *Biomolecules* **2021**, *11*, 819. [CrossRef]
73. Xing, Y.; Li, X.; Guo, X.; Li, W.; Chen, J.; Liu, Q.; Xu, Q.; Wang, Q.; Yang, H.; Shui, Y.; et al. Effects of Different Tio2 Nanoparticles Concentrations on the Physical and Antibacterial Activities of Chitosan-Based Coating Film. *Nanomaterials* **2020**, *10*, 1365. [CrossRef]
74. Sami, R.; Soltane, S.; Helal, M. Microscopic Image Segmentation and Morphological Characterization of Novel Chitosan/Silica Nanoparticle/Nisin Films Using Antimicrobial Technique for Blueberry Preservation. *Membranes* **2021**, *11*, 303. [CrossRef] [PubMed]
75. Bandara, S.; Du, H.; Carson, L.; Bradford, D.; Kommalapati, R. Agricultural and Biomedical Applications of Chitosan-Based Nanomaterials. *Nanomaterials* **2020**, *10*, 1903. [CrossRef] [PubMed]
76. Chakraborty, M.; Hasanuzzaman, M.; Rahman, M.; Khan, M.A.R.; Bhowmik, P.; Mahmud, N.U.; Tanveer, M.; Islam, T. Mechanism of Plant Growth Promotion and Disease Suppression by Chitosan Biopolymer. *Agriculture* **2020**, *10*, 624. [CrossRef]
77. Tan, L.; Huang, R.; Li, X.; Liu, S.; Shen, Y.M.; Shao, Z. Chitosan-Based Core-Shell Nanomaterials for PH-Triggered Release of Anticancer Drug and near-Infrared Bioimaging. *Carbohydr. Polym.* **2017**, *157*, 325–334. [CrossRef]
78. Janus, Ł.; Piatkowski, M.; Radwan-Pragłowska, J.; Bogdał, D.; Matysek, D. Chitosan-Based Carbon Quantum Dots for Biomedical Applications: Synthesis and Characterization. *Nanomaterials* **2019**, *9*, 274. [CrossRef]
79. Karthikeyan, C.; Varaprasad, K.; Akbari-Fakhrabadi, A.; Hameed, A.S.H.; Sadiku, R. Biomolecule Chitosan, Curcumin and ZnO-Based Antibacterial Nanomaterial, via a One-Pot Process. *Carbohydr. Polym.* **2020**, *249*, 116825. [CrossRef]

80. Barbinta-Patrascu, M.E.; Badea, N.; Bacalum, M.; Antohe, S. Novel Bio-Friendly Nanomaterials Based on Artificial Cell Membranes, Chitosan and Silver Nanoparticles Phytogenerated From Eugenia Caryophyllata Buds: Eco-Synthesis, Characterization And Evaluation Of Bioactivities. *Rom. Rep. Phys.* **2020**, *72*, 601.
81. Muthuchamy, M.; Govindan, R.; Shine, K.; Thangasamy, V.; Alharbi, N.S.; Thillaichidambaram, M.; Khaled, J.M.; Wen, J.L.; Alanzi, K.F. Anti-Biofilm Investigation of Graphene/Chitosan Nanocomposites against Biofilm Producing P. Aeruginosa and K. Pneumoniae. *Carbohydr. Polym.* **2020**, *230*, 115646. [CrossRef]
82. Zhang, C.; Hui, D.; Du, C.; Sun, H.; Peng, W.; Pu, X.; Li, Z.; Sun, J.; Zhou, C. Preparation and Application of Chitosan Biomaterials in Dentistry. *Int. J. Biol. Macromol.* **2021**, *167*, 1198–1210. [CrossRef]
83. Meng, Q.; Sun, Y.; Cong, H.; Hu, H.; Xu, F.J. An Overview of Chitosan and Its Application in Infectious Diseases. *Drug Deliv. Transl. Res.* **2021**, *11*, 1340–1351. [CrossRef]
84. Sharifi-Rad, J.; Quispe, C.; Butnariu, M.; Rotariu, L.S.; Sytar, O.; Sestito, S.; Rapposelli, S.; Akram, M.; Iqbal, M.; Krishna, A.; et al. Chitosan Nanoparticles as a Promising Tool in Nanomedicine with Particular Emphasis on Oncological Treatment. *Cancer Cell Int.* **2021**, *21*, 318. [CrossRef] [PubMed]
85. Liu, Z.; Wang, K.; Peng, X.; Zhang, L. Chitosan-Based Drug Delivery Systems: Current Strategic Design and Potential Application in Human Hard Tissue Repair. *Eur. Polym. J.* **2022**, *166*, 110979. [CrossRef]
86. Szymańska-Chargot, M.; Chylińska, M.; Pertile, G.; Pieczywek, P.M.; Cieślak, K.J.; Zdunek, A.; Frąc, M. Influence of Chitosan Addition on the Mechanical and Antibacterial Properties of Carrot Cellulose Nanofibre Film. *Cellulose* **2019**, *26*, 9613–9629. [CrossRef]
87. Sun, D.; Turner, J.; Jiang, N.; Zhu, S.; Zhang, L.; Falzon, B.G.; McCoy, C.P.; Maguire, P.; Mariotti, D.; Sun, D. Atmospheric Pressure Microplasma for Antibacterial Silver Nanoparticle/Chitosan Nanocomposites with Tailored Properties. *Compos. Sci. Technol.* **2020**, *186*, 107911. [CrossRef]
88. Peng, J.; Wang, X.; Lou, T. Preparation of Chitosan/Gelatin Composite Foam with Ternary Solvents of Dioxane/Acetic Acid/Water and Its Water Absorption Capacity. *Polym. Bull.* **2020**, *77*, 5227–5244. [CrossRef]
89. Fuster, M.G.; Montalbán, M.G.; Carissimi, G.; Lima, B.; Feresin, G.E.; Cano, M.; Giner-Casares, J.J.; López-Cascales, J.J.; Enriz, R.D.; Víllora, G. Antibacterial Effect of Chitosan–Gold Nanoparticles and Computational Modeling of the Interaction between Chitosan and a Lipid Bilayer Model. *Nanomaterials* **2020**, *10*, 2340. [CrossRef]
90. Marangon, C.A.; Martins, V.C.A.; Ling, M.H.; Melo, C.C.; Plepis, A.M.G.; Meyer, R.L.; Nitschke, M. Combination of Rhamnolipid and Chitosan in Nanoparticles Boosts Their Antimicrobial Efficacy. *ACS Appl. Mater. Interfaces* **2020**, *12*, 5488–5499. [CrossRef]
91. Chen, P.; Xie, F.; Tang, F.; McNally, T. Influence of Plasticiser Type and Nanoclay on the Properties of Chitosan-Based Materials. *Eur. Polym. J.* **2021**, *144*, 110225. [CrossRef]
92. Dong, Y.; Bi, J.; Ming, S.; Zhang, S.; Zhu, D.; Meng, D.; Li, T. Functionalized Chitosan as a Novel Support for Stabilizing Palladium in Suzuki Reactions. *Carbohydr. Polym.* **2021**, *260*, 117815. [CrossRef]
93. Irshad, M.S.; Wang, X.; Abbasi, M.S.; Arshad, N.; Chen, Z.; Guo, Z.; Yu, L.; Qian, J.; You, J.; Mei, T. Semiconductive, Flexible MnO2NWs/Chitosan Hydrogels for Efficient Solar Steam Generation. *ACS Sustain. Chem. Eng.* **2021**, *9*, 3887–3900. [CrossRef]
94. Yousefi, V.; Mohebbi-Kalhori, D.; Samimi, A. Start-up Investigation of the Self-Assembled Chitosan/Montmorillonite Nanocomposite over the Ceramic Support as a Low-Cost Membrane for Microbial Fuel Cell Application. *Int. J. Hydrogen Energy* **2020**, *45*, 4804–4820. [CrossRef]
95. Castro Marín, A.; Colangelo, D.; Lambri, M.; Riponi, C.; Chinnici, F. Relevance and Perspectives of the Use of Chitosan in Winemaking: A Review. *Crit. Rev. Food Sci. Nutr.* **2021**, *61*, 3450–3464. [CrossRef] [PubMed]
96. Xi, X.; Pizzi, A.; Lei, H.; Zhang, B.; Chen, X.; Du, G. Environmentally Friendly Chitosan Adhesives for Plywood Bonding. *Int. J. Adhes. Adhes.* **2022**, *112*, 103027. [CrossRef]
97. Nasirinezhad, M.; Ghaffarian, S.R.; Tohidian, M. Eco-Friendly Polyelectrolyte Nanocomposite Membranes Based on Chitosan and Sulfonated Chitin Nanowhiskers for Fuel Cell Applications. *Iran. Polym. J.* **2021**, *30*, 355–367. [CrossRef]
98. Qi, P.; Xu, Z.; Zhou, T.; Zhang, T.; Zhao, H. Study on a Quartz Crystal Microbalance Sensor Based on Chitosan-Functionalized Mesoporous Silica for Humidity Detection. *J. Colloid Interface Sci.* **2021**, *583*, 340–350. [CrossRef]
99. Aranaz, I.; Mengibar, M.; Harris, R.; Panos, I.; Miralles, B.; Acosta, N.; Galed, G.; Heras, A. Functional Characterization of Chitin and Chitosan. *Curr. Chem. Biol.* **2009**, *3*, 203–230. [CrossRef]
100. Chadha, U.; Selvaraj, S.K.; Ashokan, H.; Hariharan, S.P.; Mathew Paul, V.; Venkatarangan, V.; Paramasivam, V. Complex Nanomaterials in Catalysis for Chemically Significant Applications: From Synthesis and Hydrocarbon Processing to Renewable Energy Applications. *Adv. Mater. Sci. Eng.* **2022**, *2022*, 1552334. [CrossRef]
101. Chadha, U.; Bhardwaj, P.; Selvaraj, S.K.; Kumari, K.; Isaac, T.S.; Panjwani, M.; Kulkarni, K.; Mathew, R.M.; Satheesh, A.M.; Pal, A.; et al. Advances in Chitosan Biopolymer Composite Materials: From Bioengineering, Wastewater Treatment to Agricultural Applications. *Mater. Res. Express* **2022**, *9*, 052002. [CrossRef]
102. Chang, S.H. Gold(III) Recovery from Aqueous Solutions by Raw and Modified Chitosan: A Review. *Carbohydr. Polym.* **2021**, *256*, 117423. [CrossRef]
103. Wang, G.; Li, R.; Parseh, B.; Du, G. Prospects and Challenges of Anticancer Agents' Delivery via Chitosan-Based Drug Carriers to Combat Breast Cancer: A Review. *Carbohydr. Polym.* **2021**, *268*, 118192. [CrossRef]

104. Gorantla, S.; Dabholkar, N.; Sharma, S.; Rapalli, V.K.; Alexander, A.; Singhvi, G. Chitosan-Based Microneedles as a Potential Platform for Drug Delivery through the Skin: Trends and Regulatory Aspects. *Int. J. Biol. Macromol.* **2021**, *184*, 438–453. [CrossRef] [PubMed]
105. Pal, P.; Pal, A.; Nakashima, K.; Yadav, B.K. Applications of Chitosan in Environmental Remediation: A Review. *Chemosphere* **2021**, *266*, 128934. [CrossRef] [PubMed]
106. Kamal Eddin, F.B.; Fen, Y.W.; Omar, N.A.S.; Liew, J.Y.C.; Daniyal, W.M.E.M.M. Femtomolar Detection of Dopamine Using Surface Plasmon Resonance Sensor Based on Chitosan/Graphene Quantum Dots Thin Film. *Spectrochim. Acta—Part A Mol. Biomol. Spectrosc.* **2021**, *263*, 120202. [CrossRef] [PubMed]

polymers

MDPI

Review

Alginate-Induced Disease Resistance in Plants

Roohallah Saberi Riseh [1], Mozhgan Gholizadeh Vazvani [1], Marzieh Ebrahimi-Zarandi [2] and Yury A. Skorik [3,*]

1 Department of Plant Protection, Faculty of Agriculture, Vali-e-Asr University of Rafsanjan, Imam Khomeini Square, Rafsanjan 7718897111, Iran; r.saberi@vru.ac.ir (R.S.R.); mgholizadehvazvani@yahoo.com (M.G.V.)
2 Department of Plant Protection, Faculty of Agriculture, Shahid Bahonar University of Kerman, Kerman 7618411764, Iran; ebrahimimarzieh@gmail.com
3 Institute of Macromolecular Compounds of the Russian Academy of Sciences, Bolshoi VO 31, 199004 St. Petersburg, Russia
* Correspondence: yury_skorik@mail.ru

Abstract: Plants are continuously exposed to a wide range of pathogens, including fungi, bacteria, nematodes, and viruses; therefore, survival under these conditions requires a sophisticated defense system. The activation of defense responses and related signals in plants is regulated mainly by the hormones salicylic acid, jasmonic acid, and ethylene. Resistance to pathogen infection can be induced in plants by various biotic and abiotic agents. For many years, the use of abiotic plant resistance inducers has been considered in integrated disease management programs. Recently, natural inducer compounds, such as alginates, have become a focus of interest due to their environmentally friendly nature and their ability to stimulate plant defense mechanisms and enhance growth. Polysaccharides and the oligosaccharides derived from them are examples of eco-compatible compounds that can enhance plant growth while also inducing plant resistance against pathogens and triggering the expression of the salicylic acid-dependent defense pathway.

Keywords: induced disease resistance; sodium alginate; polysaccharides; plant growth-promoting bacteria

Citation: Saberi Riseh, R.; Gholizadeh Vazvani, M.; Ebrahimi-Zarandi, M.; Skorik, Y.A. Alginate-Induced Disease Resistance in Plants. *Polymers* **2022**, *14*, 661. https://doi.org/10.3390/polym14040661

Academic Editors: Sabu Thomas and Maya Jacob John

Received: 9 January 2022
Accepted: 7 February 2022
Published: 9 February 2022

1. Introduction

Plant pathogens cause diseases with different pathogenicity mechanisms in various parts of plants, resulting in great economic loss [1]. Pathogens successfully infect plants through mechanisms involving the recognition of plant surface receptors, production of virulence and effector proteins, and overcoming plant defense barriers [1,2]. During their evolution, pathogens maintain their attacks on plants by emerging new races, while plants develop mechanisms to cope with and adapt to these new pathogen variants. When a pathogen attacks a plant, it also triggers signal pathways that elicit the expression of the plant's defense genes [3], which activate defense responses against the pathogen [4]. As soon as the plant initiates this immune response, the pathogen contamination process is disrupted, and its gene expression is suppressed. These defense responses can be activated by both endogenous (plant structure) and exogenous (biotic and abiotic) elicitors [5]. Exogenous agents are now broadly used in agriculture worldwide to control the losses caused by different pathogens. The use of genetic resistance to pathogens, the identification of resistance genes, and the transfer of these genes to plants require long-term breeding and genetic engineering programs. By contrast, due to the complex interaction that occurs between the pathogen, plant, biological inducer components, and the environment (combined biotic and abiotic factors), the application of biotic inducers to control plant diseases in field conditions may lead to changes in the mechanism of induction of resistance and the form of the defense signals (Figure 1). The interactions among these three biotic agents (plants, pathogens, and biological inducers) and with the environment must, therefore, be controlled to achieve maximum disease control. Plants, pathogens, and biological factors may respond to different environmental conditions through changes that may affect the mechanism of resistance. Unfortunately, the consequence of the complex quadruple effects (pathogen × plant × inducer agents × environment) that occur between these factors is a less successful biological control in the field, compared with the control achieved in the controlled environments of the greenhouse and laboratory.

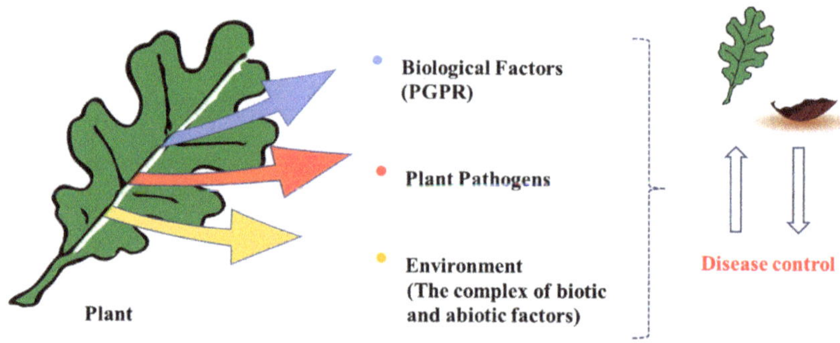

Figure 1. Complex interactions between the plant, the pathogen, plant growth-promoting rhizobacteria as biological factors (inducers), and the environment.

The role of abiotic agents in the induction of resistance against phytopathogens has been reported in many studies [6–8]. One example is β-aminobutyric acid, which has been successfully exploited in practical agriculture for defense priming in different crops [9]. Many natural compounds have now been demonstrated to enhance the defense priming response in plants, ranging from oligosaccharides, glycosides, and amides to vitamins, carboxylic acids, and aromatic compounds [10]. A simple compound, hexanoic acid, shows a potent natural priming capability to protect plants against a wide range of pathogens by inducing callose deposition and activating salicylic acid (SA) and jasmonic acid (JA) pathways [7]. Zhou et al. [11] reported that thiamine (vitamin B1) can modulate cellular redox status to protect Arabidopsis against infection by *Sclerotinia sclerotiorum*. Para-aminobenzoic acid, another member of the vitamin B group, was able to enhance resistance against the cucumber mosaic virus and *Xanthomonas axonopodis* by inducing systemic acquired resistance (SAR). In the same study, benzothiadiazole was also shown to reduce disease severity, but it also caused adverse effects on the plant, as shoot lengths were shortened and cucumber fruit lengths were significantly reduced, compared with plants treated with para-aminobenzoic acid or untreated control plants [12]. Chitosan, a deacetylated derivative of chitin, can enhance plant defenses by various mechanisms, including scavenging reactive oxygen species (ROS), upregulating antioxidant activities, and activating the octadecanoid pathway that leads to the production of phytoprotective fatty acids [13].

The oligosaccharide products arising from the hydrolysis of natural polysaccharides can also serve as elicitors that induce resistance and alter the expression of plant defense genes [4]. For example, exogenous application of oligogalacturonide can induce plant defense responses, such as accumulation of phytoalexin, β-1,3-glucanase, and chitinase, and generation of ROS, by triggering nitric oxide (NO) production [14]. Oligochitosan can protect plants against fungi, bacteria, and viruses by activating the SA and jasmonic acid–ethylene (JA–ET) pathways, while also protecting against abiotic stresses by the induction of an abscisic acid (ABA)-dependent pathway [15]. Microbial products can also induce defense responses in plants; an example is an *Agrobacterium* spp. fermentation product, oligocurdlan, which has been shown to induce defense responses against *Phytophthora infestans* in potatoes [16]. Other examples are the oligosaccharides that naturally occur in green and brown algae and that can activate defense signals in plants [17,18]. These compounds are also used as fertilizers and soil conditioners in agricultural and horticultural industries [19].

Several commercial products containing oligosaccharides are now successfully marketed for plant protection. One example is FytoSave® (LIDA Plant Research, Valencia, Spain), a complex mixture of oligochitosans and oligopectates that is active against downy mildew infection in grape and cucumber [20]. The active component of FytoSave® (LIDA Plant Research, Valencia, Spain), COS–OGA, can induce resistance against *Phytophthora infestans,* the causal agent of potato late blight, by enhancing pathogenesis-related (PR) proteins, such as PR-1 and PR-2. The induction of resistance in plants by COS–OGA is reported as a cumulative process involving SA. COS–OGA combines cationic chitosan oligomers, chitooligosaccharides (COSs), with anionic pectin oligomers, oligogalacturonides (OGAs) [20,21]. In 2018, FytoSave® product (LIDA Plant Research, Valencia, Spain) as the first plant phytovaccine with phytosanitary registration was admitted by the European Commission for use in organic agriculture (https://www.infoagro.com, accessed on 8 January 2022). Another commercial product is Stemicol® (LIDA Plant Research, Valencia, Spain), a mixture of chitooligosaccharides that causes the reduction in fruit rot in tomatoes, strawberries, and grapes (https://www.lidaplantresearch.com/phytovaccines/stemicol, accessed on 8 January 2022). Thus, natural compounds, such as oligosaccharides, are now promising alternatives to chemical fungicides for controlling pathogen diseases in the field [22].

Another plant defense elicitor of considerable interest is sodium alginate (ALG), a polysaccharide derived from seaweeds. ALG oligosaccharides or oligoalginates (AOS) are recognized as new types of functional material and are used to enhance seed germination, shoot elongation, root growth, and resistance against plant pathogens [23–27]. AOS can activate the production of phosphodiesterase in suspension cultures of plant cells by modulating the production of ROS and by activating PR proteins and defense enzymes, such as peroxidase (POD), catalase (CAT), polyphenol oxidase (PPO), and phenylalanine ammonia-lyase (PAL) [28].

Induced resistance is a suitable alternative strategy for chemical pesticides to control plant diseases. Finding new natural sources of elicitors and exploring their effects on plant defense is a significant issue. Recently, natural inducer compounds, such as ALG, have become a focus of interest due to their environmentally friendly nature and their ability to stimulate plant defense mechanisms and enhance growth. In this review, we discuss the main defense pathways invoked by plants to combat pathogen attacks, with a more intense focus on the role of ALG and AOS in the induction of resistance against plant diseases.

2. Plant Immune System against Pathogens

Plant cells are capable of sensing evolutionarily conserved microbial molecular signals, termed pathogen-associated or microbe-associated molecular patterns (PAMPs or MAMPs), through plant pattern recognition receptors [29–31]. The PAMP molecules are essential for pathogen fitness; therefore, they represent an efficient form that plants exploit to sense the presence of pathogens. The perception of PAMPs by plant pattern recognition receptors activates an immune response, referred to as PAMP-triggered immunity, which provides protection against nonhost pathogens and limits diseases caused by virulent pathogens [32].

However, pathogens also adapt to their host plants and evolve mechanisms for the suppression of plant defenses induced by pathogenicity signals and genes [33–36]. In return, plants evolve resistance proteins (R proteins) that can detect, either directly or indirectly, the effector proteins of the pathogen and trigger a different form of disease resistance, known as effector-triggered immunity, which is highly specific and often accompanied by the appearance of the hypersensitive response (HR) and SAR in the plant. Damage-associated molecular patterns, which include plant cell walls and cutin fragments characteristically released by the enzymatic actions of pathogens, can serve as triggers of immune responses in plants [31,37,38]. The effector-triggered immunity and PAMP-triggered immunity pathways activate a set of downstream defense responses, including signaling pathways and transcription factors that limit pathogen proliferation or disease symptom expression [39]. Further, ROS accumulate, cell wall defense mechanisms are activated, and defense hormones such as SA, ET, and JA accumulate. Crosstalk between the SA and JA–ET signaling pathways has also emerged as an important regulatory mechanism in plant immunity [32,40–43].

Plants are equipped with various defense genes, but the expression of these genes is often latent in healthy conditions. Intriguingly, these defense genes can be induced in plants by the application of any type of inducer in a process known as induced resistance [44]. The inducer triggers the plant's defense system against a subsequent pathogen attack, thereby suppressing the occurrence of disease. Induced resistance activates a wide range of defense mechanisms, and the defense signals in this pathway lead to two types of resistance: SAR and induced systemic resistance (ISR) [45].

2.1. Systemic Acquired Resistance (SAR)

SAR describes a type of plant defense response that provides long-term protection against various plant pathogens. The systemic signals involved in SAR include SA, lipid-based signal molecules, and ROS; these molecules transport the systemic signal that is activated by the plant–pathogen interaction [46]. SAR is related to the production of SA as a signaling molecule and the accumulation of PR proteins [46]. SAR can be activated in many plant species by different pathogens that cause necrosis or hypersensitive reactions in plants. This type of resistance is long-lasting and effective against a broad spectrum of pathogens [47,48]. SA is a defense hormone, and pathogen infections induce SA synthesis by upregulating the expression of isochorismate synthase 1 (*ICS1*), a gene that encodes a key enzyme in the SA synthesis pathway [49]. The enhancement of another defense signal for SAR—namely, the increased expression of palmitic acid and its derivatives, has been observed in the primed guard cells of Arabidopsis plants [50]. NO and ROS, which are both early chemical signals in systemic immunity, operate in a feedback loop in SAR. ROS also act additively to mediate the chemical hydrolysis of unsaturated fatty acids to induce SAR in plants [51]. During SAR, SA binds the H_2O_2-scavenging enzymes, CAT, and ascorbate peroxidase, and inhibits their activities, thereby promoting an increase in H_2O_2 levels. This increase is then responsible for the signal transduction that leads to the induction of pathogenesis-related genes and pathogen resistance [46].

2.2. Induced Systemic Resistance (ISR)

Plant growth-promoting rhizobacteria (PGPR) colonize the root surface, thus preventing the penetration of pathogens while inducing systemic resistance in plants. A specific recognition response is needed between the plant and the rhizobacteria for the onset of ISR [52]. Rhizobacterial determinants, such as flagellar proteins, lipopolysaccharides, antibiotics, quorum-sensing molecules, volatile organic compounds, and siderophores, can elicit ISR [53,54]. When this type of resistance occurs, the plant's immune system is strengthened against other invaders [55].

ISR is a nonspecific response, as indicated by its broad action against different pathogens [56]. ISR is generally activated by a pathway in which JA and ET are central players [57]. Although beneficial rhizobacteria often trigger JA–ET-dependent pathways, several PGPR have been reported to trigger SA-dependent pathways [58]. Some of the signal pathways that regulate ISR are similar to those of SAR [45,57,58]. One example is NPR1, a common regulator of both SAR and ISR pathways that functions as a transcriptional coactivator of SA-responsive pathogenesis-related genes. However, the role of NPR1 in ISR has not yet been established [58,59].

Immune responses are induced in plants by many biological and chemical stimuli that trigger defense priming and increase the plant's defense capacity. Priming is defined as enhanced sensitivity and responsiveness to stress that results from prior experience and leads to increased resistance. Primed plants respond faster and have stronger defense responses against subsequent stresses [60]. Table 1 shows examples of biological priming agents and the mechanisms by which they induce resistance against pathogens in different plants.

Table 1. Examples of biological agents that induce resistance against plant pathogens.

Biological Inducer	Pathogens	Host	Mechanism	Reference
Pseudomonas spp.	*Botrytis cinerea*	Grapevine	Oxidative burst and phytoalexin accumulation in grape cells and leaves.	[61]
	Clavibacter michiganensis	Tomato	Increase in levels of PR1a and ACO transcripts and SA signaling pathways.	[62]
	Meloidogyne spp.	Tomato	SA production by bacteria.	[63]
	Pythium aphanidermatum	Cucumber	Reduced pathogen spread.	[64]
Bacillus spp.	*Heterodera glycines*	Soybean	Expression of defense-related genes involved in the SA and JA pathways.	[65]
	Fusarium sp.	Tomato	Production of phthalic acid methyl ester by Bacillus.	[66]
	Botrytis cinerea	Arabidopsis	Activation of the JA–ET signaling pathway.	[67]
Trichoderma spp.	*Botrytis cinerea*	Tomato	Activation of the JA, SA, and ABA signaling pathways.	[68]
			Enhanced activation of jasmonate-responsive genes.	[69]
	Sclerotinia sclerotiorum	*Brassica napus*	Induction of SA- and JA–ET-dependent defenses and decreased disease symptoms.	[70]
Mycorrhizal fungi	*Botrytis cinerea*	Lettuce	Provision of biotic stress protection with no nutritional or growth benefits.	[71]
	Blumeria graminis f.sp. *tritici*	Wheat	Accumulation of phenolic compounds and H_2O_2, upregulation of genes encoding several defense markers (POD, PAL, chitinase 1)	[72]

3. Abiotic Inducers of Disease Resistance in Plants

Abiotic inducers include chemicals that act at various points in the signaling pathways involved in disease resistance and against biotic and abiotic stress. One compound, 2,6-dichloroisonicotinic acid, and its methyl ester were the first synthetic compounds shown to prime defense responses in plants [73]. A wide range of cellular responses, including alterations in ion transport across the plasma membrane, synthesis of antimicrobial secondary metabolites (e.g., phytoalexins, cell wall phenolics, and lignin-like polymers), and activation of defense genes, are potentiated by these chemical inducers [6]. The resistance induced by chemical elicitors is broad spectrum and long-lasting, and many of these elicitors provide disease control ranging between 20% and 85% [74]. For instance, exposure of plants to β-aminobutyric acid, probenazole, benzothiadiazole, and SA can all induce resistance against a broad range of pathogens [75]. Durable induced resistance, based on priming of gene expression, was reported after treatment of tomato seeds with β-aminobutyric acid or JA [76]. Similarly, the treatment of faba beans with acibenzolar-S-methyl induced SAR against rust and ascochyta blight diseases in both greenhouse and field conditions, and this protection was still evident several weeks after acibenzolar-S-methyl application [77]. Table 2 shows examples of abiotic components known to induce pathogen resistance in plants.

Table 2. Abiotic components that induce pathogen resistance in plants.

Abiotic Component	Pathogen/Plant Disease	Type of Plant	Mechanism	Reference
Dibasic and tribasic phosphate salts	*Colletotrichum lagenarium*	Cucumber	Influences the activity of apoplastic enzymes, such as polygalacturonases, thereby releasing elicitor-active oligogalacturonides from plant cell walls.	[78,79]
			Preceded by a rapid generation of superoxide and hydrogen peroxide.	
	Blumeria graminis f.sp. *hordei*	Barley	Reduces powdery mildew infection by 89%.	[80]
SA Derivatives	TMV	Tomato Tobacco	Establishes plant immunity by an accumulation of PR proteins.	[81]
Isonicotinic acid derivatives	TMV	Tobacco	Decreases the necrotic area on leaves.	[82]
	Colletotrichum lagenarium	Cucumber	Induces chitinase and modifies the physiology of the host.	[83]
Thiadiazole and isothiazole derivative	Powdery mildew, anthracnose, and bacterial leaf spot Alternaria leaf spot, anthracnose, bacterial shot hole	Wheat Pumpkin Cucumber Chinese cabbage Strawberry Peach	Promotes the expression of defense-related genes and SA catabolism. Induces plant defense responses.	[84] [85]
β-Aminobutyric acid	*Alternaria brassicicola*, *Plectosphaerella cucumerina*	Arabidopsis	Promotes callose accumulation by an ABA-dependent defense pathway.	[86]

As the world's population expands, the demand for food production increases. Therefore, agriculture must be able to meet the nutritional needs of people throughout the world, making the protection of crops from plant pests and pathogens paramount. Therefore, new ways appear to be needed to stimulate the defense genes in plants to suppress pathogen attacks. The application of abiotic inducer agents derived from natural factors represents an environmentally friendly way to trigger the induction of resistance in the field. The plant's defense system is highly triggerable; therefore, the existence of an external abiotic inducer factor that has no adverse effect on the environment can play a major role in activating the plant defense system and suppressing pathogens.

Environmentally friendly polymer compounds, especially ALG, are compatible compounds that stimulate plant defense mechanisms. The use of these abiotic materials avoids the known toxic effects of synthetic chemical pesticide agents on humans and other nontarget organisms. These compounds are able to induce plant resistance against pathogens and increase the expression of SA-dependent defense pathways. In what follows, we discuss the advantages of polysaccharides and the mechanisms of ALG in the induction of resistance against plant pathogens.

3.1. *Polysaccharides as Plant Defense Inducers*

The plant's defense system is fundamental to its ability to resist pathogens and is, therefore, an effective target for research on disease management. Plants recognize pathogens using PAMPs with structures or chemical patterns similar to their pathogens [31,87]. Therefore, not surprisingly, oligosaccharides that share structures similar to the components of pathogen cell walls or other structures can also serve as PAMPs to activate the plant immune system [87,88].

The promotion of eco-friendly alternatives is necessary to reduce the environmental effects of present-day chemicals used in agriculture [89]. In recent decades, there have been many reports regarding the induction of defense resistance by the application of plant extracts and essential oils, microbial (bacteria, fungi, and microalgae) extracts, seaweed extracts, and polysaccharides. Polysaccharides with high structural complexity and biological activity have become ideal and environmentally friendly biological resources for inducing resistance against plant pathogens [89–93]. The effects of polysaccharides obtained from microalgae and cyanobacteria on the biochemical and metabolomic markers linked to defense pathways in tomato plants were evaluated by Rachidi et al. [89]. The polysaccharides extracted from *Phaeodactylum triocnutum*, *Desmodesmus* sp., and *Porphyridium* sp. improved the activities of phenylalanine ammonia-lyase, chitinase, β-1,3 glucanase, and peroxidase enzymes in tomato leaves [89]. Further, GC–MS metabolomics analysis revealed that polysaccharides induced the modification of metabolite profiles, such as fatty acids, alkanes, and phytosterol, in tomato leaves [89].

Pettongkhao et al. [94] reported that sulfated polysaccharide from *Acanthophora spicifera*, a red alga, induced defense responses against *Phytophthora palmivora* in a rubber tree (*Hevea brasiliensis*). Their results showed that the extracted crude polysaccharide induced SA and scopoletin accumulation and SA-responsive gene expression but suppressed JA-responsive gene expression [94]. An elicitor from the green algae *Ulva* spp. caused the protection of *Medicago truncatula* against infection by *Colletotrichum trifolii* [95]. A broad range of defense-related transcripts upregulated notable genes involved in the biosynthesis of phytoalexins, PR proteins, and cell wall proteins [95].

One polysaccharide, tramesan, obtained from *Trametes versicolor*, caused an increase in the JA level and the early expression of plant defense genes against Septoria Leaf Blotch complex disease in wheat [96]. The use of biopolymers as elicitors for controlling plant diseases is gaining momentum worldwide due to the eco-friendly and nontoxic nature of polysaccharides. These materials have the added advantage of being sufficiently resistant to degradation by hydrolytic enzymes and by exposure to acidic environments [97,98].

Oligosaccharides are low molecular weight carbohydrates that arise from the degradation of polysaccharides [15]. These compounds have biological activity in many living organisms [99]. In plants, they regulate specific processes, such as cell morphogenesis and the pH-dependent development of flowers or callus, and in general, they modulate plant growth. The use of oligosaccharides can increase soil fertility and activate plant defense against both biotic and abiotic stresses [15].

3.2. *Alginate and Induction of Resistance against Plant Pathogens*

Algal polysaccharides are among the most abundant organic molecules in nature and have great diversity, as well as the potential to induce resistance in plants [27,100,101]. ALG is extracted from the cell walls of brown macroalgae (e.g., *Macrocystis pyrifera*, *Laminaria hyperborean*, *Ascophyllum nodosum*), and several bacteria (*Azotobacter vinelandii*, *Pseudomonas* spp.) contain ALG at up to 40% of their dry weight [102].

ALG is a linear biopolysaccharide copolymer consisting of 1,4-linked β-D-mannuronate (M) and α-L-guluronate (G), which can be arranged in heteropolymeric and homopolymeric blocks (Figure 2) [27,103]. Due to their hydrophilic properties, ALG hydrogels can absorb large amounts of water or biological fluids without losing their structure. ALG is a nontoxic and environmentally friendly polysaccharide that can be used as a delivery vehicle in various applications due to its unique physicochemical properties [102]. Alginic acid is insoluble in water or organic solvents, but its monovalent alginate salts are soluble in water and organic solvents and form stable solutions in water [102].

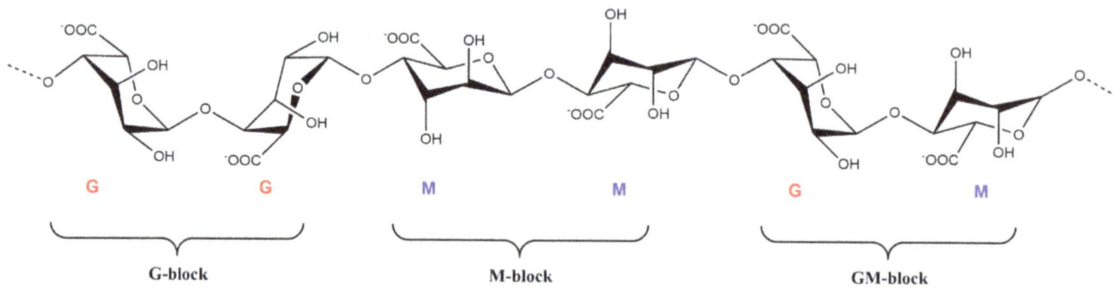

Figure 2. Chemical structure of alginate.

The linear ALG polymer, at physiological temperature and pH, and in the presence of some chemical initiators, can be converted to a three-dimensional polymer by a process called free-radical polymerization [102,104,105]. During this polymerization, some chemicals can be easily combined into the forming hydrogel to generate a liquid–solid phase under physiological conditions [106,107]. ALG is widely used in this way in medicine to encapsulate various drugs for delivery to target organs and tissues. The formation of hydrogels allows the use of ALG as a carrier of proteins, DNA, and live cells while maintaining their biological activity [108]. ALG is also able to stimulate the growth and development of plants and induce resistance to biotic and abiotic stresses [109]. Phenolic compounds (as secondary metabolites) can cross-link with ALG to strengthen plant cell walls against pathogen attack [110]. Figure 3 shows the biological activity of ALG in plants against different stresses and environmental factors.

Figure 3. Biological activity of alginate in plants against different stresses.

ALG has received much attention due to its environmental compatibility and nontoxic properties as an elicitor in the control of plant diseases [27]. In one study, ALG was investigated as a factor in the induction of resistance against *Alternaria solani*, the causal agent of tomato blight disease [27]. Tomato leaves were treated with different concentrations of ALG (0.2, 0.4, and 0.6%) two days before infection with the pathogen. ALG effectively controlled the growth of *A. solani* in the treated tomato plants and significantly enhanced the expression levels of superoxide dismutase (SOD) in response to infection. Staining of infected tomato leaves with Uvitex-2B and observation by fluorescence microscopy showed significant reductions in pathogen colonization following ALG treatment. ALG at a concentration of 0.4% was very effective in controlling fungal hyphal growth. The level of defense enzymes, including SOD, GPX, and CAT, was enhanced in the treated tomato plants [27]. Identification of the induced resistance mechanisms in tomato by ALG against blight disease was further explored by examining the expression changes in defense marker genes, including β-1,3-glucanase (*PR2*), chitinase (*PR4*), nonexpressor of pathogenesis-related protein 1 (*NPR1*; related to SA signaling pathways), 1-aminocyclopropane-1-carboxylate oxidase (*ACO1*; related to ET signaling pathways), and lipoxygenase D (*LoxD*; related to JA signaling pathways). The expression levels of *PR2*, *NPR1*, *LoxD*, and *ACO1* were significantly upregulated in leaves treated with *A. solani* and 0.4–0.6% ALG [27]. *PR4* expression was upregulated in pathogen-infected leaves when compared with uninfected control leaves and 0.4% or 0.6% ALG-pretreated leaves infected with pathogen [27].

The major cell wall components of many phytopathogenic fungi are chitin and glucan. Therefore, plant β-1,3-glucanases and chitinases play antifungal roles by hydrolyzing the fungal cell wall. Further, β-1,3-glucanases and chitinases exhibit indirect effects via the formation of oligosaccharide elicitors, which further induce the expression of other PR proteins [111]. The ALG-induced defense responses, therefore, arise by activation of antioxidant enzymes and PR proteins against *A. solani*, to inhibit disease development in tomato seedlings [27].

Much interest is now expressed in the use of protein elicitors enclosed in a complex with biopolymers, such as ALG, to protect them against adverse external factors, facilitate their interaction with plant cell receptors, and invoke disease resistance [112,113]. Peptidylprolyl isomerases (PPIases) play roles in the folding of synthesized proteins, immune system responses, transcriptional regulation, cell cycle control, and nuclear events [114]. In one study, the FKBP-type PPIase from *Pseudomonas fluorescens*, which has significant eliciting activity regarding a wide range of plant pathogens, was encapsulated in ALG microparticles [26]. Synergistic interaction between ALG and other compounds was promoted by constructing microparticles consisting of 70% ALG, 20% bovine serum albumin (BSA), and 10% PPIase and evaluating three different plant–pathogen models (tobacco–TMV, tobacco–*A. longipes*, and wheat–*Stagonospora nodorum*). In the wheat–*S. nodorum* model system, a significant eliciting activity of the ALG–albumin complex was observed, and the activity of encapsulated PPIase increased, compared with the free PPIase. The ALG–BSA complex had an eliciting activity that suppressed the development of *A. longipes* on tobacco plants. The PPIase ALG biopolymer complex served as an antipathogenic compound and an inducer of resistance against pathogens in a wide range of plants while also helping to promote plant growth [26]. In the TMV–tobacco model system, no significant differences were observed between PPIase and ALG–BSA–PPIase, and in these treatments, the average amount of necroses per leaf decreased 32–35 times. compared with the control. No eliciting activity was revealed in the case of ALG–BSA [26].

The role of AOS in the induction of resistance against *Pseudomonas syringae* pv. *tomato* DC3000 was evaluated in Arabidopsis by Zhang et al. [25]. Arabidopsis were pretreated by spraying with different concentrations of AOS (25, 50, 100, and 200 mg/L) three days before inoculation with *P. syringae* pv. *tomato* DC3000. The disease index, bacterial growth, production of ROS, and qualitative and quantitative detection of NO and SA were then evaluated. The qRT–PCR analysis revealed an increase in induced immunity against this disease in Arabidopsis. The expression of the *avrPtoB* gene, which represents the pathogenic mechanism of this bacterium, was significantly reduced in leaves treated with AOS, compared with the control leaves. AOS also prevented the growth of bacteria on the leaves. At 25 mg/L, AOS induced both NO and ROS production against the pathogen in Arabidopsis. ROS and NO are the primary signals that initiate defense reactions against plant pathogens [115–119]. After pretreatment with AOS, the SA pathway was activated and significantly enhanced *PR1* expression [25].

Zhang et al. [4] also investigated the activity of AOS and its potential application for the protection of rice plants against *Magnaporthe grisea*. Germinating rice seeds were detached from 5–7-day-old seedlings when the sprouts were 1–2 cm in length and then were treated with AOS. The AOS activity on germinating rice was assayed by determining the accumulation of phytoalexin in seed tissues as a marker of plant disease resistance. The activities of PAL, CAT, and POD were determined in the treated leaves of rice with AOS. An enhancement in PAL activity was detected in the rice leaves treated with AOS. PAL activity is considered to represent a direct response of the host plant to suppress a pathogen attack and is associated with disease resistance. This enzyme was induced by the application of exogenous elicitors, such as abiotic inducer agents [4,120]. CAT, POD, and PAL have a synergistic role in plant disease protection. The production of four kinds of phytoalexin—oryzalexin A, oryzalexin C, phytocassane A/D, and phytocassane B/C—was elicited in rice-seed tissues by AOS. The accumulation of oryzalexin C could be considered a more sensitive marker for assaying elicitor activity [4].

In another study, ALG isolated from the brown seaweed *Bifurcaria bifurcata* and AOS were evaluated for their ability to stimulate the natural defenses of tomato seedlings [121]. PAL activity and polyphenol levels were measured in leaves treated with ALG [121]. PAL activity increased 12 h after treatment. Polysaccharides extracted from *B. bifurcata* and the oligosaccharide derivatives of those polysaccharides significantly induced phenylpropanoid metabolism in tomato seedlings. ALG and its oligosaccharide derivatives should, therefore, be considered potential bioresources for plant protection against phytopathogens in the context of eco-sustainable green technology [121].

Other studies have confirmed that an ALG–lentinan–aminooligosaccharide hydrogel induces strong plant resistance against TMV and increases the release of calcium ions to promote the growth of *Nicotiana benthamiana* [122]. Table 3 shows other studies on the role of ALG in the induction of resistance against plant diseases.

Table 3. Other examples of alginates that induce resistance against plant pathogens.

ALG Concentration	Pathogen	Plant	Mechanism	Reference
5 g/L	Tobacco mosaic virus (TMV)	Tobacco (on leaves)	The antiviral activity of ALG on infectivity of TMV on blocking the decapsulation process of TMV protein on the cell membrane surface.	[123]
50 g/L	*Botrytis cinerea*	Kiwifruit (on fruit)	Reduction in the incidence of gray mold and diameter of lesions of kiwifruit during storage; enhancing the activity of polyphenol oxidase, l-phenylalanine ammonia-lyase (PAL), and β-1,3-glucanase related to pathogen defense.	[124]
1 g/L	*Fusarium oxysporum* f.sp. *albedinis*	Date Palm (on roots)	The stimulation of PAL activity in roost; the increased transcriptional level; stimulates expression of the genes involved in phenolic metabolism and burst oxidation.	[125]
2 g/L	*Verticillium dahliae*	Olive (on twigs of 10 cm in length with 16 leaves)	Increase in the enzymatic activity of PAL in the stem; inhibitory rates on mycelial growth of the fungus in vitro.	[126]
0.3 g/L	*Erwinia carotovora* *Xanthomonas campestris*	soybean cotyledon	The accumulation of phytoalexin and inducing PAL in soybean cotyledon.	[127]
5 g/L AOS combined with *Meyerozyma guilliermondii*	*Penicillium expansum*	Pears (on Fruits)	Increase in the activities of superoxide dismutase (SOD), catalase (CAT), polyphenol oxidase (PPO), peroxidase. (POD), phenylalanine ammonia-lyase (PAL), chitinase (CHI), total phenol content, and flavonoid content in pears; reduce spore germination rate and inhibit the germ tube elongation of *P. expansum*.	[128]

Based on the studies mentioned above, ALG and AOS are effective elicitors for inducing resistance in plants against various pathogens including fungi, bacteria, and viruses. Both the SA and JA–ET pathways are triggered by these elicitors, and there is evidence of ABA-dependent pathway activation by AOS [15,129]. Therefore, AOS can induce resistance to abiotic stress, such as drought, salinity, and heavy metals, by triggering the ABA signaling pathway in plants [129–131].

Figure 4 shows a scheme for seed treatment with ALG, pathogen attack, and the defense pathways that are activated.

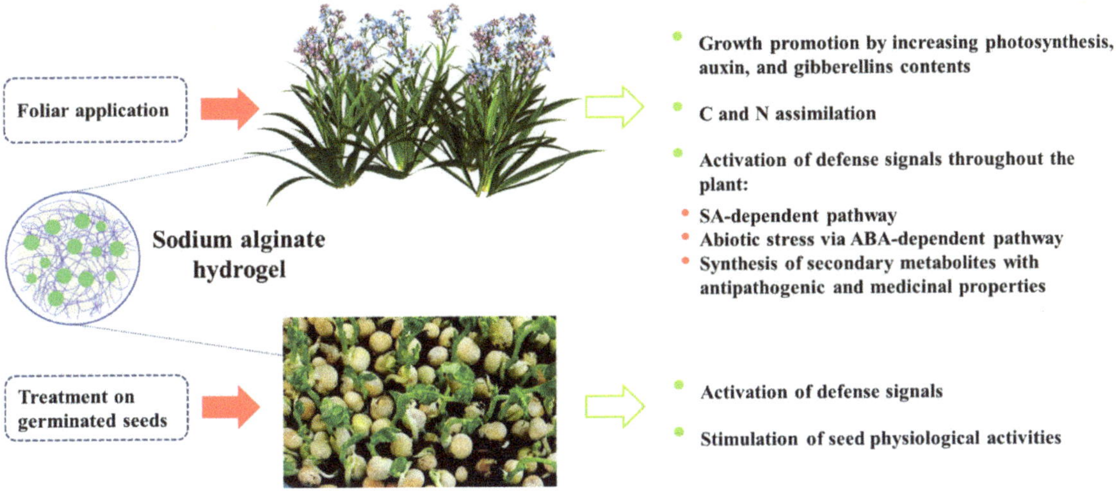

Figure 4. The mechanism of induction of plant disease resistance by alginate.

4. Conclusions

Resistance to plant diseases is a very important issue that should be given great attention. Some plant genotypes and cultivars have a natural resistance to plant pests and diseases. Some have a protective wax-like layer on their surface that prevents damage from pathogens. Others respond to the presence of factors that stimulate the plant's immune system as an effective way to promote resistance to disease. However, the introduction of resistant cultivars and gene transfer to nonresistant cultivars is an extensive plant breeding process. Further, the geographic compatibility of the introduced resistant cultivars must be considered. Biological control agents, such as beneficial bacteria and nonpathogenic strains, have led to the successful control of many pathogens in the greenhouse and laboratory. However, these agents may fail under field conditions due to complex interactions between the environment, pathogens, plants, and biological factors (e.g., PGPR). Therefore, abiotic inducer compounds that are environmentally friendly and can trigger plant resistance under adverse conditions are very important candidates for research on plant disease resistance.

ALG is a natural polymer that, due to its potential properties, has been considered a viable choice for the induction of plant resistance against pathogens. This polymeric compound plays a role by stimulating plant defense signals and activating defense genes. Treatment of plants with this compound leads to the activation of SA and JA pathways that protect against pathogen attacks. Plant defense responses, such as the synthesis of phenolic compounds, lignin, PPO, PAL, and PR proteins, are significantly increased in plants treated with ALG, and these responses induce disease resistance. Extensive applications of ALG in the field confirm its effects on the activation of SAR and ISR against a wide range of pathogens. However, induced resistance is a host response and can be influenced in practice by factors such as plant genotype, crop nutrition, frequency, and the method of elicitor application under field conditions.

Author Contributions: Conceptualization, R.S.R. and Y.A.S.; writing—original draft preparation, R.S.R., M.G.V. and M.E.-Z.; writing—review and editing, Y.A.S. All authors have read and agreed to the published version of the manuscript.

Funding: This research received no external funding.

Institutional Review Board Statement: Not applicable.

Informed Consent Statement: Not applicable.

Data Availability Statement: Not applicable.

Acknowledgments: The authors would like to acknowledge Anastasia Valova (Institute of Macromolecular Compounds of the Russian Academy of Sciences) for assistance with the graphic design of the figures.

Conflicts of Interest: The authors declare no conflict of interest.

References

1. Al-Ani, L.K.T.; Furtado, E.L. The effect of incompatible plant pathogens on the host plant. In *Molecular Aspects of Plant Beneficial Microbes in Agriculture*; Academic Press: Cambridge, MA, USA, 2020; pp. 47–57. [CrossRef]
2. Surico, G. The concepts of plant pathogenicity, virulence/avirulence and effector proteins by a teacher of plant pathology. *Phytopathol. Mediterr.* **2013**, *52*, 399–417. [CrossRef]
3. de León, I.P.; Montesano, M. Activation of defense mechanisms against pathogens in mosses and flowering plants. *Int. J. Mol. Sci.* **2013**, *14*, 3178–3200. [CrossRef] [PubMed]
4. Zhang, S.; Tang, W.; Jiang, L.; Hou, Y.; Yang, F.; Chen, W.; Li, X. Elicitor activity of algino-oligosaccharide and its potential application in protection of rice plant (*Oryza sativa* L.) against *Magnaporthe grisea*. *Biotechnol. Biotechnol. Equip.* **2015**, *29*, 646–652. [CrossRef]
5. Guo, J.; Du, G.; Chen, J.; Chen, X.; Li, X. Oligosaccharides act as elicitors to protect plant against crop disease based on knowledge of plant defence response mechanism. In *Sugar Beet Crops: Growth, Fertilization and Yield*; Nova Science Publishers: New York, NY, USA, 2009.
6. Conrath, U. Priming of induced plant defense responses. In *Advances in Botanical Research*; Academic Press: Cambridge, MA, USA, 2009; Volume 51, pp. 361–395.
7. Aranega-Bou, P.; de la O Leyva, M.; Finiti, I.; García-Agustín, P.; González-Bosch, C. Priming of plant resistance by natural compounds. Hexanoic acid as a model. *Front. Plant Sci.* **2014**, *5*, 488. [CrossRef]
8. Gozzo, F.; Faoro, F. Systemic Acquired Resistance (50 Years after Discovery): Moving from the Lab to the Field. *J. Agric. Food Chem.* **2013**, *61*, 12473–12491. [CrossRef]
9. Cohen, Y.R. β-Aminobutyric Acid-Induced Resistance against Plant Pathogens. *Plant Dis.* **2002**, *86*, 448–457. [CrossRef]
10. Iriti, M.; Castorina, G.; Vitalini, S.; Mignani, I.; Soave, C.; Fico, G.; Faoro, F. Chitosan-induced ethylene-independent resistance does not reduce crop yield in bean. *Biol. Control* **2010**, *54*, 241–247. [CrossRef]
11. Zhou, J.; Sun, A.; Xing, D. Modulation of cellular redox status by thiamine-activated NADPH oxidase confers Arabidopsis resistance to *Sclerotinia sclerotiorum*. *J. Exp. Bot.* **2013**, *64*, 3261–3272. [CrossRef]
12. Song, G.C.; Choi, H.K.; Ryu, C.-M. The folate precursor para-aminobenzoic acid elicits induced resistance against *Cucumber mosaic virus* and *Xanthomonas axonopodis*. *Ann. Bot.* **2013**, *111*, 925–934. [CrossRef]
13. El Hadrami, A.; Adam, L.R.; El Hadrami, I.; Daayf, F. Chitosan in Plant Protection. *Mar. Drugs* **2010**, *8*, 968–987. [CrossRef]
14. Rasul, S.; Dubreuil-Maurizi, C.; Lamotte, O.; Koen, E.; Poinssot, B.; Alcaraz, G.; Wendehenne, D.; Jeandroz, S. Nitric oxide production mediates oligogalacturonide-triggered immunity and resistance to *Botrytis cinerea* in *Arabidopsis thaliana*. *Plant Cell Environ.* **2012**, *35*, 1483–1499. [CrossRef] [PubMed]
15. Moenne, A.; González, A. Chitosan-, alginate-carrageenan-derived oligosaccharides stimulate defense against biotic and abiotic stresses, and growth in plants: A historical perspective. *Carbohydr. Res.* **2021**, *503*, 108298. [CrossRef] [PubMed]
16. Li, J.; Zhu, L.; Lu, G.; Zhan, X.-B.; Lin, C.-C.; Zheng, Z.-Y. Curdlan β-1,3-Glucooligosaccharides Induce the Defense Responses against *Phytophthora infestans* Infection of Potato (*Solanum tuberosum* L. cv. McCain G1) Leaf Cells. *PLoS ONE* **2014**, *9*, e97197. [CrossRef] [PubMed]
17. Shibuya, N.; Minami, E. Oligosaccharide signalling for defence responses in plant. *Physiol. Mol. Plant Pathol.* **2001**, *59*, 223–233. [CrossRef]
18. Vera, J.; Castro, J.; Gonzalez, A.; Moenne, A. Seaweed Polysaccharides and Derived Oligosaccharides Stimulate Defense Responses and Protection Against Pathogens in Plants. *Mar. Drugs* **2011**, *9*, 2514–2525. [CrossRef] [PubMed]
19. Stengel, D.B.; Connan, S. Marine algae: A source of biomass for biotechnological applications. In *Natural Products from Marine Algae*; Humana Press: New York, NY, USA, 2015; pp. 1–37.
20. van Aubel, G.; Buonatesta, R.; Van Cutsem, P. COS-OGA: A novel oligosaccharidic elicitor that protects grapes and cucumbers against powdery mildew. *Crop Prot.* **2014**, *65*, 129–137. [CrossRef]
21. van Aubel, G.; Cambier, P.; Dieu, M.; Van Cutsem, P. Plant immunity induced by COS-OGA elicitor is a cumulative process that involves salicylic acid. *Plant Sci.* **2016**, *247*, 60–70. [CrossRef]
22. Guarnizo, N.; Oliveros, D.; Murillo-Arango, W.; Bermúdez-Cardona, M.B. Oligosaccharides: Defense Inducers, Their Recognition in Plants, Commercial Uses and Perspectives. *Molecules* **2020**, *25*, 5972. [CrossRef]

23. Hu, X.; Jiang, X.; Hwang, H.; Liu, S.; Guan, H. Promotive effects of alginate-derived oligosaccharide on maize seed germination. *J. Appl. Phycol.* **2004**, *16*, 73–76. [CrossRef]
24. Xu, X.; Iwamoto, Y.; Kitamura, Y.; Oda, T.; Muramatsu, T. Root Growth-promoting Activity of Unsaturated Oligomeric Uronates from Alginate on Carrot and Rice Plants. *Biosci. Biotechnol. Biochem.* **2003**, *67*, 2022–2025. [CrossRef]
25. Zhang, C.; Howlader, P.; Liu, T.; Sun, X.; Jia, X.; Zhao, X.; Shen, P.; Qin, Y.; Wang, W.; Yin, H. Alginate Oligosaccharide (AOS) induced resistance to Pst DC3000 via salicylic acid-mediated signaling pathway in *Arabidopsis thaliana*. *Carbohydr. Polym.* **2019**, *225*, 115221. [CrossRef] [PubMed]
26. Popletaeva, S.B.; Statsyuk, N.V.; Voinova, T.M.; Arslanova, L.R.; Zernov, A.L.; Bonartsev, A.P.; Bonartseva, G.A.; Dzhavakhiya, V.G. Evaluation of eliciting activity of peptidil prolyl cys/trans isomerase from *Pseudonomas fluorescens* encapsulated in sodium alginate regarding plant resistance to viral and fungal pahogens. *AIMS Microbiol.* **2018**, *4*, 192–208. [CrossRef] [PubMed]
27. Dey, P.; Ramanujam, R.; Venkatesan, G.; Nagarathnam, R. Sodium alginate potentiates antioxidant defense and PR proteins against early blight disease caused by *Alternaria solani* in *Solanum lycopersicum* Linn. *PLoS ONE* **2019**, *14*, e0223216. [CrossRef] [PubMed]
28. Akimoto, C.; Aoyagi, H.; Dicosmo, F.; Tanaka, H. Synergistic effect of active oxygen species and alginate on chitinase production by *Wasabia japonica* cells and its application. *J. Biosci. Bioeng.* **2000**, *89*, 131–137. [CrossRef]
29. Ausubel, F.M. Are innate immune signaling pathways in plants and animals conserved? *Nat. Immunol.* **2005**, *6*, 973–979. [CrossRef] [PubMed]
30. Bittel, P.; Robatzek, S. Microbe-associated molecular patterns (MAMPs) probe plant immunity. *Curr. Opin. Plant Biol.* **2007**, *10*, 335–341. [CrossRef]
31. Boller, T.; Felix, G. A Renaissance of Elicitors: Perception of Microbe-Associated Molecular Patterns and Danger Signals by Pattern-Recognition Receptors. *Annu. Rev. Plant Biol.* **2009**, *60*, 379–406. [CrossRef]
32. Jones, J.D.G.; Dangl, J.L. The plant immune system. *Nature* **2006**, *444*, 323–329. [CrossRef]
33. Abramovitch, R.; Janjusevic, R.; Stebbins, E.; Martin, G.B. Type III effector AvrPtoB requires intrinsic E3 ubiquitin ligase activity to suppress plant cell death and immunity. *Proc. Natl. Acad. Sci. USA* **2006**, *103*, 2851–2856. [CrossRef]
34. Cui, H.; Wang, Y.; Xue, L.; Chu, J.; Yan, C.; Fu, J.; Chen, M.; Innes, R.W.; Zhou, J.-M. *Pseudomonas syringae* Effector Protein AvrB Perturbs Arabidopsis Hormone Signaling by Activating MAP Kinase 4. *Cell Host Microbe* **2010**, *7*, 164–175. [CrossRef]
35. Grant, S.R.; Fisher, E.J.; Chang, J.H.; Mole, B.M.; Dangl, J.L. Subterfuge and Manipulation: Type III Effector Proteins of Phytopathogenic Bacteria. *Annu. Rev. Microbiol.* **2006**, *60*, 425–449. [CrossRef] [PubMed]
36. Zhou, J.-M.; Chai, J. Plant pathogenic bacterial type III effectors subdue host responses. *Curr. Opin. Microbiol.* **2008**, *11*, 179–185. [CrossRef] [PubMed]
37. Denoux, C.; Galletti, R.; Mammarella, N.; Gopalan, S.; Werck, D.; De Lorenzo, G.; Ferrari, S.; Ausubel, F.M.; Dewdney, J. Activation of Defense Response Pathways by OGs and Flg22 Elicitors in Arabidopsis Seedlings. *Mol. Plant* **2008**, *1*, 423–445. [CrossRef]
38. Lotze, M.T.; Zeh, H.J.; Rubartelli, A.; Sparvero, L.J.; Amoscato, A.; Washburn, N.R.; Devera, M.E.; Liang, X.; Tor, M.; Billiar, T. The grateful dead: Damage-associated molecular pattern molecules and reduction/oxidation regulate immunity. *Immunol. Rev.* **2007**, *220*, 60–81. [CrossRef] [PubMed]
39. Thomma, B.P.H.J.; Nürnberger, T.; Joosten, M.H.A.J. Of PAMPs and Effectors: The Blurred PTI-ETI Dichotomy. *Plant Cell* **2011**, *23*, 4–15. [CrossRef] [PubMed]
40. Nicaise, V.; Roux, M.; Zipfel, C. Recent Advances in PAMP-Triggered Immunity against Bacteria: Pattern Recognition Receptors Watch over and Raise the Alarm. *Plant Physiol.* **2009**, *150*, 1638–1647. [CrossRef] [PubMed]
41. van Loon, L.; Rep, M.; Pieterse, C. Significance of Inducible Defense-related Proteins in Infected Plants. *Annu. Rev. Phytopathol.* **2006**, *44*, 135–162. [CrossRef] [PubMed]
42. Bent, A.F.; Mackey, D. Elicitors, Effectors, and R Genes: The New Paradigm and a Lifetime Supply of Questions. *Annu. Rev. Phytopathol.* **2007**, *45*, 399–436. [CrossRef]
43. Zipfel, C. Early molecular events in PAMP-triggered immunity. *Curr. Opin. Plant Biol.* **2009**, *12*, 414–420. [CrossRef]
44. Baysal, Ö.; Gürsoy, Y.Z.; Ornek, H.; Duru, A. Enhanced tomato resistance to bacterial canker by application of turtle oil. *J. Gen. Plant Pathol.* **2005**, *71*, 204–210. [CrossRef]
45. Choudhary, D.K.; Prakash, A.; Johri, B.N. Induced systemic resistance (ISR) in plants: Mechanism of action. *Indian J. Microbiol.* **2007**, *47*, 289–297. [CrossRef] [PubMed]
46. Durrant, W.E.; Dong, X. Systemic acquired resistance. *Annu. Rev. Phytopathol.* **2004**, *42*, 185–209. [CrossRef] [PubMed]
47. Ryals, J.A.; Neuenschwander, U.H.; Willits, M.G.; Molina, A.; Steiner, H.-Y.; Hunt, M.D. Systemic acquired resistance. *Plant Cell* **1996**, *8*, 1809. [CrossRef] [PubMed]
48. Sticher, L.; Mauch-Mani, B.; Métraux, J.P. Systemic acquired resistance. *Annu. Rev. Phytopathol.* **1997**, *35*, 235–270. [CrossRef]
49. Zhang, Y.; Xu, S.; Ding, P.; Wang, D.; Cheng, Y.T.; He, J.; Gao, M.; Xu, F.; Li, Y.; Zhu, Z.; et al. Control of salicylic acid synthesis and systemic acquired resistance by two members of a plant-specific family of transcription factors. *Proc. Natl. Acad. Sci. USA* **2010**, *107*, 18220–18225. [CrossRef]
50. David, L.; Kang, J.N.; Dufresne, D.; Zhu, D.; Chen, S.X. Multi-Omics Revealed Molecular Mechanisms Underlying Guard Cell Systemic Acquired Resistance. *Int. J. Mol. Sci.* **2021**, *22*, 191. [CrossRef]
51. Wang, C.; El-Shetehy, M.; Shine, M.B.; Yu, K.; Navarre, D.; Wendehenne, D.; Kachroo, A.; Kachroo, P. Free radicals mediate systemic acquired resistance. *Cell Rep.* **2014**, *7*, 348–355. [CrossRef]

52. Conrath, U.; Pieterse, C.M.J.; Mauch-Mani, B. Priming in plant–pathogen interactions. *Trends Plant Sci.* **2002**, *7*, 210–216. [CrossRef]
53. Bakker, P.A.H.M.; Pieterse, C.M.J.; van Loon, L.C. Induced Systemic Resistance by Fluorescent *Pseudomonas* spp. *Phytopathology* **2007**, *97*, 239–243. [CrossRef]
54. De Vleesschauwer, D.; Höfte, M. Rhizobacteria-induced systemic resistance. In *Advances in Botanical Research*; Academic Press: Cambridge, MA, USA, 2009; Volume 51, pp. 223–281.
55. Van Loon, L. Systemic induced resistance. In *Mechanisms of Resistance to Plant Diseases*; Springer: Berlin/Heidelberg, Germany, 2000; pp. 521–574.
56. Heil, M. Induced systemic resistance (ISR) against pathogens–a promising field for ecological research. *Perspect. Plant Ecol. Evol. Syst.* **2001**, *4*, 65–79. [CrossRef]
57. Romera, F.J.; García, M.J.; Lucena, C.; Martinez-Medina, A.; Aparicio, M.A.; Ramos, J.; Alcántara, E.; Angulo, M.; Pérez-Vicente, R. Induced Systemic Resistance (ISR) and Fe Deficiency Responses in Dicot Plants. *Front. Plant Sci.* **2019**, *10*, 287. [CrossRef] [PubMed]
58. Pieterse, C.M.; Zamioudis, C.; Berendsen, R.L.; Weller, D.M.; Van Wees, S.C.; Bakker, P.A. Induced Systemic Resistance by Beneficial Microbes. *Annu. Rev. Phytopathol.* **2014**, *52*, 347–375. [CrossRef]
59. Dong, X. NPR1, all things considered. *Curr. Opin. Plant Biol.* **2004**, *7*, 547–552. [CrossRef] [PubMed]
60. Mauch-Mani, B.; Baccelli, I.; Luna, E.; Flors, V. Defense Priming: An Adaptive Part of Induced Resistance. *Annu. Rev. Plant Biol.* **2017**, *68*, 485–512. [CrossRef] [PubMed]
61. Verhagen, B.W.M.; Trotel-Aziz, P.; Couderchet, M.; Höfte, M.; Aziz, A. *Pseudomonas* spp.-induced systemic resistance to *Botrytis cinerea* is associated with induction and priming of defence responses in grapevine. *J. Exp. Bot.* **2009**, *61*, 249–260. [CrossRef] [PubMed]
62. Takishita, Y.; Charron, J.-B.; Smith, D.L. Biocontrol Rhizobacterium *Pseudomonas* sp. 23S Induces Systemic Resistance in Tomato (*Solanum lycopersicum* L.) Against Bacterial Canker *Clavibacter michiganensis* subsp. *michiganensis*. *Front. Microbiol.* **2018**, *9*, 2119. [CrossRef] [PubMed]
63. Siddiqui, I.A.; Shaukat, S.S. *Pseudomonas aeruginosa*-mediated induction of systemic resistance in tomato against root-knot nematode. *Plant Pathol. J.* **2005**, *4*, 21–25.
64. Chen, C.; Bélanger, R.R.; Benhamou, N.; Paulitz, T.C. Induced systemic resistance (ISR) by *Pseudomonas* spp. impairs pre-and post-infection development of *Pythium aphanidermatum* on cucumber roots. *Eur. J. Plant Pathol.* **1998**, *104*, 877–886. [CrossRef]
65. Xing, Z.; Wu, X.; Zhao, J.; Zhao, X.; Zhu, X.; Wang, Y.; Fan, H.; Chen, L.; Liu, X.; Duan, Y. Isolation and identification of induced systemic resistance determinants from *Bacillus simplex* Sneb545 against *Heterodera glycines*. *Sci. Rep.* **2020**, *10*, 11586. [CrossRef]
66. Akram, W.; Anjum, T.; Ali, B. Searching ISR determinant/s from *Bacillus subtilis* IAGS174 against *Fusarium* wilt of tomato. *BioControl* **2014**, *60*, 271–280. [CrossRef]
67. Nie, P.; Li, X.; Wang, S.; Guo, J.; Zhao, H.; Niu, D. Induced Systemic Resistance against *Botrytis cinerea* by *Bacillus cereus* AR156 through a JA/ET- and NPR1-Dependent Signaling Pathway and Activates PAMP-Triggered Immunity in Arabidopsis. *Front. Plant Sci.* **2017**, *8*, 238. [CrossRef] [PubMed]
68. Martínez-Medina, A.; Fernández, I.; Sánchez-Guzmán, M.J.; Jung, S.C.; Pascual, J.A.; Pozo, M.J. Deciphering the hormonal signalling network behind the systemic resistance induced by *Trichoderma harzianum* in tomato. *Front. Plant Sci.* **2013**, *4*, 206. [CrossRef] [PubMed]
69. Tucci, M.; Ruocco, M.; De Masi, L.; De Palma, M.; Lorito, M. The beneficial effect of *Trichoderma* spp. on tomato is modulated by the plant genotype. *Mol. Plant Pathol.* **2010**, *12*, 341–354. [CrossRef] [PubMed]
70. Alkooranee, J.T.; Aledan, T.R.; Ali, A.K.; Lu, G.; Zhang, X.; Wu, J.; Fu, C.; Li, M. Detecting the Hormonal Pathways in Oilseed Rape behind Induced Systemic Resistance by *Trichoderma harzianum* TH12 to *Sclerotinia sclerotiorum*. *PLoS ONE* **2017**, *12*, e0168850. [CrossRef]
71. Pozo de la Hoz, J.; Rivero, J.; Azcón-Aguilar, C.; Urrestarazu, M.; Pozo, M.J. Mycorrhiza-induced resistance against foliar pathogens is uncoupled of nutritional effects under different light Intensities. *J. Fungi* **2021**, *7*, 402. [CrossRef]
72. Mustafa, G.; Khong, N.G.; Tisserant, B.; Randoux, B.; Fontaine, J.; Magnin-Robert, M.; Reignault, P.; Sahraoui, A.L.-H. Defence mechanisms associated with mycorrhiza-induced resistance in wheat against powdery mildew. *Funct. Plant Biol.* **2017**, *44*, 443–454. [CrossRef]
73. Kauss, H.; Theisinger-Hinkel, E.; Mindermann, R.; Conrath, U. Dichloroisonicotinic and salicylic acid, inducers of systemic acquired resistance, enhance fungal elicitor responses in parsley cells. *Plant J.* **1992**, *2*, 655–660. [CrossRef]
74. Walters, D.; Walsh, D.; Newton, A.; Lyon, G. Induced Resistance for Plant Disease Control: Maximizing the Efficacy of Resistance Elicitors. *Phytopathology* **2005**, *95*, 1368–1373. [CrossRef]
75. Oostendorp, M.; Kunz, W.; Dietrich, B.; Staub, T. Induced Disease Resistance in Plants by Chemicals. *Eur. J. Plant Pathol.* **2001**, *107*, 19–28. [CrossRef]
76. Worrall, D.; Holroyd, G.H.; Moore, J.P.; Glowacz, M.; Croft, P.; Taylor, J.E.; Paul, N.D.; Roberts, M.R. Treating seeds with activators of plant defence generates long-lasting priming of resistance to pests and pathogens. *New Phytol.* **2011**, *193*, 770–778. [CrossRef]
77. Sillero, J.; Rojas-Molina, M.; Avila, C.; Rubiales, D. Induction of systemic acquired resistance against rust, ascochyta blight and broomrape in faba bean by exogenous application of salicylic acid and benzothiadiazole. *Crop Prot.* **2012**, *34*, 65–69. [CrossRef]
78. Orober, M.; Siegrist, J.; Buchenauer, H. Mechanisms of Phosphate-induced Disease Resistance in Cucumber. *Eur. J. Plant Pathol.* **2002**, *108*, 345–353. [CrossRef]

79. Gottstein, H.D.; Kuc, J. Induction of systemic resistance to anthracnose in cucumber by phosphates. *Phytopathology* **1989**, *79*, 176–179. [CrossRef]
80. Mitchell, A.F.; Walters, D.R. Potassium phosphate induces systemic protection in barley to powdery mildew infection. *Pest Manag. Sci.* **2004**, *60*, 126–134. [CrossRef] [PubMed]
81. Zhou, M.; Wang, W. Recent Advances in Synthetic Chemical Inducers of Plant Immunity. *Front. Plant Sci.* **2018**, *9*, 1613. [CrossRef]
82. Czerwoniec, P.; Lewandowski, P.; Smiglak, M. Derivatives of Isonicotinic Acid as New Efficient Systemic Acquired Resistance (SAR) Inducers. *ChemistrySelect* **2020**, *5*, 10759–10764. [CrossRef]
83. Métraux, J.; Ahlgoy, P.; Staub, T.; Speich, J.; Steinemann, A.; Ryals, J.; Ward, E. Induced systemic resistance in cucumber in response to 2, 6-dichloro-isonicotinic acid and pathogens. In *Advances in Molecular Genetics of Plant-Microbe Interactions*; Springer: Berlin/Heidelberg, Germany, 1991; Volume 1, pp. 432–439.
84. Ogawa, M.; Kadowaki, A.; Yamada, T.; Kadooka, O. Applied development of a novel fungicide isotianil (STOUT). *Sumitomo Kagaku* **2011**, *1*, 1–15.
85. Krämer, W.; Schirmer, U. *Modern Crop Protection Compounds, 3 Volume Set*; John Wiley & Sons Ltd.: Hoboken, NJ, USA, 2007.
86. Ton, J.; Mauch-Mani, B. β-aminobutyric acid induced resistance against necrotrophic pathogens is based on ABA-dependent priming for callose. *Plant J.* **2004**, *38*, 119–130. [CrossRef]
87. Zhang, C.; Wang, W.; Zhao, X.; Wang, H.; Yin, H. Preparation of alginate oligosaccharides and their biological activities in plants: A review. *Carbohydr. Res.* **2020**, *494*, 108056. [CrossRef]
88. Macho, A.P.; Zipfel, C. Plant PRRs and the Activation of Innate Immune Signaling. *Mol. Cell* **2014**, *54*, 263–272. [CrossRef]
89. Rachidi, F.; Benhima, R.; Kasmi, Y.; Sbabou, L.; El Arroussi, H. Evaluation of microalgae polysaccharides as biostimulants of tomato plant defense using metabolomics and biochemical approaches. *Sci. Rep.* **2021**, *11*, 930. [CrossRef] [PubMed]
90. Yakhin, O.I.; Lubyanov, A.A.; Yakhin, I.A.; Brown, P.H. Biostimulants in Plant Science: A Global Perspective. *Front. Plant Sci.* **2017**, *7*, 2049. [CrossRef] [PubMed]
91. Wiesel, L.; Newton, A.C.; Elliott, I.; Booty, D.; Gilroy, E.M.; Birch, P.R.J.; Hein, I. Molecular effects of resistance elicitors from biological origin and their potential for crop protection. *Front. Plant Sci.* **2014**, *5*, 655. [CrossRef] [PubMed]
92. Kocira, S.; Szparaga, A.; Kocira, A.; Czerwińska, E.; Wójtowicz, A.; Bronowicka-Mielniczuk, U.; Koszel, M.; Findura, P. Modeling Biometric Traits, Yield and Nutritional and Antioxidant Properties of Seeds of Three Soybean Cultivars Through the Application of Biostimulant Containing Seaweed and Amino Acids. *Front. Plant Sci.* **2018**, *9*, 388. [CrossRef] [PubMed]
93. Stadnik, M.J.; Freitas, M.B.d. Algal polysaccharides as source of plant resistance inducers. *Trop. Plant Pathol.* **2014**, *39*, 111–118. [CrossRef]
94. Pettongkhao, S.; Bilanglod, A.; Khompatara, K.; Churngchow, N. Sulphated Polysaccharide from *Acanthophora spicifera* Induced *Hevea brasiliensis* Defense Responses Against *Phytophthora palmivora* Infection. *Plants* **2019**, *8*, 73. [CrossRef]
95. Cluzet, S.; Torregrosa, C.; Jacquet, C.; Lafitte, C.; Fournier, J.; Mercier, L.; Salamagne, S.; Briand, X.; EsquerrÉ-TugayÉ, M.T.; Dumas, B. Gene expression profiling and protection of *Medicago truncatula* against a fungal infection in response to an elicitor from green algae *Ulva* spp. *Plant Cell Environ.* **2004**, *27*, 917–928. [CrossRef]
96. Scala, V.; Pietricola, C.; Farina, V.; Beccaccioli, M.; Zjalic, S.; Quaranta, F.; Fornara, M.; Zaccaria, M.; Momeni, B.; Reverberi, M.; et al. Tramesan Elicits Durum Wheat Defense against the Septoria Disease Complex. *Biomolecules* **2020**, *10*, 608. [CrossRef]
97. Riseh, R.S.; Skorik, Y.A.; Thakur, V.K.; Pour, M.M.; Tamanadar, E.; Noghabi, S.S. Encapsulation of Plant Biocontrol Bacteria with Alginate as a Main Polymer Material. *Int. J. Mol. Sci.* **2021**, *22*, 11165. [CrossRef]
98. Riseh, R.S.; Ebrahimi-Zarandi, M.; Vazvani, M.G.; Skorik, Y.A. Reducing Drought Stress in Plants by Encapsulating Plant Growth-Promoting Bacteria with Polysaccharides. *Int. J. Mol. Sci.* **2021**, *22*, 12979. [CrossRef]
99. de Moura, F.A.; Macagnan, F.T.; da Silva, L.P. Oligosaccharide production by hydrolysis of polysaccharides: A review. *Int. J. Food Sci. Technol.* **2015**, *50*, 275–281. [CrossRef]
100. Klarzynski, O.; Plesse, B.; Joubert, J.-M.; Yvin, J.-C.; Kopp, M.; Kloareg, B.; Fritig, B. Linear β-1,3 Glucans Are Elicitors of Defense Responses in Tobacco. *Plant Physiol.* **2000**, *124*, 1027–1038. [CrossRef] [PubMed]
101. Mani, S.D.; Nagarathnam, R. Sulfated polysaccharide from *Kappaphycus alvarezii* (Doty) Doty ex P.C. Silva primes defense responses against anthracnose disease of *Capsicum annuum* Linn. *Algal Res.* **2018**, *32*, 121–130. [CrossRef]
102. Abasalizadeh, F.; Moghaddam, S.V.; Alizadeh, E.; Akbari, E.; Kashani, E.; Fazljou, S.M.B.; Torbati, M.; Akbarzadeh, A. Alginate-based hydrogels as drug delivery vehicles in cancer treatment and their applications in wound dressing and 3D bioprinting. *J. Biol. Eng.* **2020**, *14*, 8, Correction in **2020**, *14*, 17. [CrossRef] [PubMed]
103. Venegas, M.; Matsuhiro, B.; Edding, M.E. Alginate Composition of *Lessonia trabeculata* (Phaeophyta: Laminariales) Growing in Exposed and Sheltered Habitats. *Bot. Mar.* **1993**, *36*, 47–51. [CrossRef]
104. Hu, X.; Gao, C. Photoinitiating polymerization to prepare biocompatible chitosan hydrogels. *J. Appl. Polym. Sci.* **2008**, *110*, 1059–1067. [CrossRef]
105. Ifkovits, J.L.; Burdick, J.A. Review: Photopolymerizable and Degradable Biomaterials for Tissue Engineering Applications. *Tissue Eng.* **2007**, *13*, 2369–2385. [CrossRef]
106. Varghese, S.; Hwang, N.S.; Canver, A.C.; Theprungsirikul, P.; Lin, D.W.; Elisseeff, J. Chondroitin sulfate based niches for chondrogenic differentiation of mesenchymal stem cells. *Matrix Biol.* **2008**, *27*, 12–21. [CrossRef]
107. Garagorri, N.; Fermanian, S.; Thibault, R.; Ambrose, W.M.; Schein, O.D.; Chakravarti, S.; Elisseeff, J. Keratocyte behavior in three-dimensional photopolymerizable poly(ethylene glycol) hydrogels. *Acta Biomater.* **2008**, *4*, 1139–1147. [CrossRef]

108. Goh, C.H.; Heng, P.W.S.; Chan, L.W. Alginates as a useful natural polymer for microencapsulation and therapeutic applications. *Carbohydr. Polym.* **2012**, *88*, 1–12. [CrossRef]
109. González, A.; Castro, J.; Vera, J.; Moenne, A. Seaweed Oligosaccharides Stimulate Plant Growth by Enhancing Carbon and Nitrogen Assimilation, Basal Metabolism, and Cell Division. *J. Plant Growth Regul.* **2012**, *32*, 443–448. [CrossRef]
110. Salgado, L.T.; Tomazetto, R.; Cinelli, L.P.; Farina, M.; Filho, G.M.A. The influence of brown algae alginates on phenolic compounds capability of ultraviolet radiation absorption in vitro. *Braz. J. Oceanogr.* **2007**, *55*, 145–154. [CrossRef]
111. Lawrence, C.B.; Singh, N.P.; Qiu, J.; Gardner, R.G.; Tuzun, S. Constitutive hydrolytic enzymes are associated with polygenic resistance of tomato to *Alternaria solani* and may function as an elicitor release mechanism. *Physiol. Mol. Plant Pathol.* **2000**, *57*, 211–220. [CrossRef]
112. Keefe, A.J.; Jiang, S. Poly(zwitterionic)protein conjugates offer increased stability without sacrificing binding affinity or bioactivity. *Nat. Chem.* **2011**, *4*, 59–63. [CrossRef] [PubMed]
113. Roberts, M.; Bentley, M.; Harris, J. Chemistry for peptide and protein PEGylation. *Adv. Drug Deliv. Rev.* **2002**, *54*, 459–476. [CrossRef]
114. Shaw, P.E. Peptidyl-prolyl isomerases: A new twist to transcription. *EMBO Rep.* **2002**, *3*, 521–526. [CrossRef] [PubMed]
115. Rizhsky, L.; Liang, H.; Shuman, J.; Shulaev, V.; Davletova, S.; Mittler, R. When Defense Pathways Collide. The Response of Arabidopsis to a Combination of Drought and Heat Stress. *Plant Physiol.* **2004**, *134*, 1683–1696. [CrossRef]
116. Apel, K.; Hirt, H. Reactive oxygen species: Metabolism, oxidative stress, and signal transduction. *Annu. Rev. Plant Biol.* **2004**, *55*, 373–399. [CrossRef]
117. Smirnoff, N.; Arnaud, D. Hydrogen peroxide metabolism and functions in plants. *New Phytol.* **2018**, *221*, 1197–1214. [CrossRef]
118. Arnaud, D.; Lee, S.; Takebayashi, Y.; Choi, D.; Choi, J.; Sakakibara, H.; Hwang, I. Cytokinin-Mediated Regulation of Reactive Oxygen Species Homeostasis Modulates Stomatal Immunity in Arabidopsis. *Plant Cell* **2017**, *29*, 543–559. [CrossRef]
119. Ribeiro, D.M.; Desikan, R.; Bright, J.; Confraria, A.; Harrison, J.; Hancock, J.T.; Barros, R.S.; Neill, S.J.; Wilson, I.D. Differential requirement for NO during ABA-induced stomatal closure in turgid and wilted leaves. *Plant Cell Environ.* **2008**, *32*, 46–57. [CrossRef] [PubMed]
120. Ramanathan, A.; Vidhasekaran, P.; Samiyappan, R. Induction of defense mechanisms in greengram leaves and suspension-cultured cells by *Macrophomina phaseolina* and its elicitors. *J. Plant Dis. Prot.* **2000**, *107*, 245–257.
121. Aitouguinane, M.; Bouissil, S.; Mouhoub, A.; Rchid, H.; Fendri, I.; Abdelkafi, S.; El-Hadj, M.D.O.; Boual, Z.; Dubessay, P.; Gardarin, C.; et al. Induction of Natural Defenses in Tomato Seedlings by Using Alginate and Oligoalginates Derivatives Extracted from Moroccan Brown Algae. *Mar. Drugs* **2020**, *18*, 521. [CrossRef] [PubMed]
122. Xiang, S.; Lv, X.; He, L.; Shi, H.; Liao, S.; Liu, C.; Huang, Q.; Li, X.; He, X.; Chen, H.; et al. Dual-Action Pesticide Carrier That Continuously Induces Plant Resistance, Enhances Plant Anti-Tobacco Mosaic Virus Activity, and Promotes Plant Growth. *J. Agric. Food Chem.* **2019**, *67*, 10000–10009. [CrossRef] [PubMed]
123. Sano, Y. Antiviral activity of alginate against infection by tobacco mosaic virus. *Carbohydr. Polym.* **1999**, *38*, 183–186. [CrossRef]
124. Zhuo, R.; Li, B.; Tian, S. Alginate oligosaccharide improves resistance to postharvest decay and quality in kiwifruit (*Actinidia deliciosa* cv. Bruno). *Hortic. Plant J.* **2021**, *8*, 44–52. [CrossRef]
125. Bouissil, S.; Guérin, C.; Roche, J.; Dubessay, P.; El Alaoui-Talibi, Z.; Pierre, G.; Michaud, P.; Mouzeyar, S.; Delattre, C.; El Modafar, C. Induction of defense gene expression and the resistance of date palm to *Fusarium oxysporum* f.sp. *albedinis* in response to alginate extracted from *Bifurcaria bifurcata*. *Mar. Drugs* **2022**, *20*, 88. [CrossRef]
126. Ben Salah, I.; Aghrouss, S.; Douira, A.; Aissam, S.; El Alaoui-Talibi, Z.; Filali-Maltouf, A.; El Modafar, C. Seaweed polysaccharides as bio-elicitors of natural defenses in olive trees against verticillium wilt of olive. *J. Plant Interact.* **2018**, *13*, 248–255. [CrossRef]
127. An, Q.D.; Zhang, G.L.; Wu, H.T.; Zhang, Z.C.; Zheng, G.S.; Luan, L.; Murata, Y.; Li, X. Alginate-deriving oligosaccharide production by alginase from newly isolated *Flavobacterium* sp. LXA and its potential application in protection against pathogens. *J. Appl. Microbiol.* **2009**, *106*, 161–170. [CrossRef]
128. Han, J.; Zhao, L.; Zhu, H.; Dhanasekaran, S.; Zhang, X.; Zhang, H. Study on the effect of alginate oligosaccharide combined with *Meyerozyma guilliermondii* against *Penicillium expansum* in pears and the possible mechanisms involved. *Physiol. Mol. Plant Pathol.* **2021**, *115*, 101654. [CrossRef]
129. Liu, H.; Zhang, Y.-H.; Yin, H.; Wang, W.-X.; Zhao, X.-M.; Du, Y.-G. Alginate oligosaccharides enhanced *Triticum aestivum* L. tolerance to drought stress. *Plant Physiol. Biochem.* **2013**, *62*, 33–40. [CrossRef] [PubMed]
130. Tang, J.; Zhou, Q.; Chu, H.; Nagata, S. Characterization of Alginase and Elicitor-Active Oligosaccharides from *Gracilibacillus* A7 in Alleviating Salt Stress for *Brassica campestris* L. *J. Agric. Food Chem.* **2011**, *59*, 7896–7901. [CrossRef] [PubMed]
131. Ma, L.J.; Li, X.M.; Bu, N.; Li, N. An alginate-derived oligosaccharide enhanced wheat tolerance to cadmium stress. *Plant Growth Regul.* **2010**, *62*, 71–76. [CrossRef]

MDPI
St. Alban-Anlage 66
4052 Basel
Switzerland
Tel. +41 61 683 77 34
Fax +41 61 302 89 18
www.mdpi.com

Polymers Editorial Office
E-mail: polymers@mdpi.com
www.mdpi.com/journal/polymers

www.ingramcontent.com/pod-product-compliance
Lightning Source LLC
Chambersburg PA
CBHW040806250926
51850CB00007BA/1091
9783036560472